"If someone points out to you that your pet theory of the universe is in disagreement with Maxwell's equations – then so much the worse for Maxwell's equations. If it is found to be contradicted by observation – well these experimentalists do bungle things sometimes. But if your theory is found to be against the second law of thermodynamics I can give you no hope; there is nothing for it but to collapse in deepest humiliation."

Sir Arthur Stanley Eddington, Gifford Lectures (1927),
The Nature of the Physical World (1928)

"The deepest understanding of thermodynamics comes, of course, from understanding the actual machinery underneath."

Richard Feynman, The Feynman Lectures on Physics 39-1

"The Sands of Time were eroded by the River of Constant Change …"

Genesis, *Firth of Fifth*

Introduction to Entropy

The concept of entropy arises in diverse branches of science, including physics, where it plays a crucial role. However, the nature of entropy as a unifying concept is not widely discussed—it is dealt with in a piecemeal manner within different contexts. The interpretation of the concept is also subtly different in each case. This book will draw these diverse threads together and present entropy as one of the crucial physical concepts. It will cover a range of different applications of entropy, from the classical theory of thermodynamics, the statistical approach, entropy in quantum theory, information theory and finally, its manifestation in black hole physics. Each will be presented in a manner suitable for undergraduates and interested laypersons with no previous knowledge. The book will take an overview of these areas and see to what extent the concept of entropy is being treated in the same way in each, and how it differs.

Key Features:

- Provides an accessible introduction to the exciting topic of entropy, setting out its manifestations in classical thermodynamics, statistical mechanics, and information theory.
- Covers applications in black holes, quantum theory, and Big Bang cosmology.

Jonathan Allday taught physics at a range of schools in the UK. After attending the Liverpool Blue Coat School, he took his first degree in Natural Sciences at Cambridge, then gained a PhD in particle physics in 1989 at Liverpool University. Shortly after this, he started work on *Quarks Leptons and the Big Bang*, now published by Taylor & Francis and available in its third edition, which was intended as a rigorous but accessible introduction to these topics. Since then, he has also written *Apollo in Perspective*, *Quantum Reality* (now in its second edition), and *Space-time*, co-authored a successful textbook and contributed to an encyclopaedia for young scientists. He has also written on aspects of the history and philosophy of science. Outside of physics, Jonathan has a keen interest in cricket and Formula 1.

Simon Hands was also educated at the Blue Coat School and Cambridge University, followed by a PhD in theoretical particle physics at the University of Edinburgh. After research positions at the universities of Oxford, Illinois, Glasgow, and then CERN, he taught and researched theoretical and computational physics at Swansea University for almost 30 years. He moved to the University of Liverpool in 2021, where he helps manage the *DiRAC* high-performance computing facility for astronomy, nuclear and particle physics theory. An elected Fellow of the Learned Society of Wales, Simon enjoys singing and cycling.

Introduction to Entropy

The Way of the World

Jonathan Allday and Simon Hands

CRC Press
Taylor & Francis Group
Boca Raton London New York

CRC Press is an imprint of the
Taylor & Francis Group, an **informa** business

Designed cover image: "Maxwell's Demon contemplates information loss as material from a black hole's accretion disk falls inside the event horizon. The black hole in question is near-extreme with $a_K c^2/GM$=0.999" Oliver James et al 2015 Class. Quantum Grav. 32 065001. DOI 10.1088/0264-9381/32/6/065001 and © Shutterstock_ 1719946366

First edition published 2025
by CRC Press
2385 NW Executive Center Drive, Suite 320, Boca Raton FL 33431

and by CRC Press
4 Park Square, Milton Park, Abingdon, Oxon, OX14 4RN

CRC Press is an imprint of Taylor & Francis Group, LLC

Library of Congress Cataloging-in-Publication Data
Names: Allday, Jonathan, author. | Hands, Simon, author.
Title: Introduction to entropy : the way of the world / Jonathan Allday and Simon Hands.
Description: First edition. | Boca Raton, FL : CRC Press, 2025. |
Includes bibliographical references and index. |
Summary: "The concept of entropy arises in diverse branches of science, including physics where it plays a crucial role. However, the nature of entropy as a unifying concept is not widely discussed - it is dealt with in a piecemeal manner within different contexts. The interpretation of the concept is also subtly different in each case. This book will draw these diverse threads together and present entropy as one of the crucial physical concepts. It will cover a range of different applications of entropy, from the classical theory of thermodynamics, the statistical approach, entropy in quantum theory, information theory and finally its manifestation in black hole physics. Each will be presented in a manner suitable for both undergraduates and interested laypersons with no previous knowledge. The book will take an overview of these areas and see to what extent the concept of entropy is being treated in the same way in each, and also how it differs"– Provided by publisher.
Identifiers: LCCN 2024011551 | ISBN 9780367638689 (hbk) |
ISBN 9780367638665 (pbk) | ISBN 9781003121053 (ebk)
Subjects: LCSH: Entropy.
Classification: LCC QC318.E57 A45 2025 | DDC 536/.73–dc23/eng/20240629
LC record available at https://lccn.loc.gov/2024011551

ISBN: 9780367638689 (hbk)
ISBN: 9780367638665 (pbk)
ISBN: 9781003121053 (ebk)

DOI: 10.1201/9781003121053

Typeset in Times
by Newgen Publishing UK

Dedication

To all the teachers at the Liverpool Blue Coat School when we were inmates during the 1970s, but most especially:

Nick Cowan – the only person who could conceivably have come close to making us chemists.

Dave Sleight & Keith Caulkin, both of the Physics Department; dedicated and inspirational teachers.

Contents

About the Authors

Jonathan Allday taught physics at a range of schools in the UK. After attending the Liverpool Blue Coat School, he took his first degree in Natural Sciences at Cambridge and then gained a PhD in particle physics in 1989 at Liverpool University.

Shortly after this, he started work on *Quarks Leptons and the Big Bang*, now published by Taylor & Francis and available in its third edition, which was intended as a rigorous but accessible introduction to these topics. Since then, he has also written *Apollo in Perspective*, *Quantum Reality* (now in its second edition) and *Space-time*. He co-authored a successful textbook and contributed to an encyclopaedia for young scientists. He has also written on aspects of the history and philosophy of science.

Outside of physics, Jonathan has a keen interest in cricket and Formula 1.

Simon Hands was also educated at the Blue Coat School and Cambridge University, followed by a PhD in theoretical particle physics at the University of Edinburgh. After research positions at the Universities of Oxford, Illinois, Glasgow, and then CERN, he taught and researched theoretical and computational physics at Swansea University for almost 30 years, moving to the University of Liverpool in 2021. An elected Fellow of the Learned Society of Wales, Simon enjoys singing and cycling.

Introduction

What Is Entropy?

In the 1940s, when mathematician and communications engineer Claude Shannon was discussing ideas on quantifying the information inherent in a message, his interlocutor, the mathematician John von Neumann, is reputed to have responded: "You should call it entropy for two reasons: in the first place your uncertainty function has been used in statistical mechanics under that name, so it already has a name. In the second place, and *more important,*[1] nobody knows what entropy really is, so in a debate you will always have the advantage."

When scientists get together over coffee and swap amusing anecdotes, this one usually raises a wry smile precisely because it conveys an uncomfortable truth. "Entropy" is widely held to be a fundamental concept describing the world we live in. It forms an important component of the education of scientists and engineers across many fields: physics, chemistry, biology, mechanical, and electronic engineering. It even holds a certain cultural significance, as we shall explore. Yet, we suspect that most beneficiaries of a science education, despite having dutifully worked through the problems in the textbook, remain hazy about what entropy actually *is*. It doesn't help that there appear to be several different variants on the market—indeed, both Shannon *and* von Neumann now have "entropies" associated with their names!

Over the course of roughly 100 research publications in theoretical physics, one of us has used the word just a handful of times, never with any particular feeling of confidence. The other has written several books purporting to help develop an understanding of the physics of quantum theory, general relativity, and cosmology and has always regarded entropy as unfinished business. Interestingly, both authors had unusually parallel physics educations for writing collaborators, and both felt at the end that something important was being hinted at but not pinned down. Attempting to improve on this haziness and clarify at least two people's understanding was a major driver of this book.

The term "entropy" crops up in an impressively diverse range of areas in science:

- *Classical thermodynamics*: where it is regarded as a function of state, giving it a strong claim to be a quantifiable element of reality
- *Statistical thermodynamics*: where entropy is defined as the logarithmic count over microscopic states that a system can inhabit, without changing its macroscopic properties. This is then generalized into a measure over probability distributions
- *Chemistry*: where changes in an associated property called *free energy* determine the direction in which a chemical reaction proceeds
- *Quantum theory*: where the *von Neumann entropy* yields interesting information about the quantum states of systems, including those that have interacted with each other

DOI: 10.1201/9781003121053-1

- *Information theory*: where the *Shannon entropy* is used as a measure of the average information content in a message
- *Black hole physics*: where the identification of entropy with the horizon surface area gives rise to one of the most astonishing theoretical predictions of the modern era: that the gravitational field of a black hole should thermally radiate into the surrounding space

Is there, however, a *unique underlying concept* at work in all these different applications? Perhaps instead, the word "entropy" refers to a set of partially overlapping ideas without an exclusive conceptual framework to tie them together. In other words, is entropy fundamental or a collective noun for somewhat disparate ideas?

There is another way of viewing the same issue. In classical thermodynamics, the theory of heat and work, entropy naturally arises as a state function (to be defined later) intimately related to temperature and, in a key sense, more fundamental than heat.[2] Understanding entropy transfer proves crucial in a range of contexts, for example in calculating the efficiency of an engine. However, in contrast to other state functions, e.g., internal energy, pressure, spontaneous magnetization, etc., entropy has a more abstract nature. We can calculate with it, but we are left with the impression that we do not really know what we are playing with.

Arguably (and we have argued with each other about this), if physics had never discovered atomic theory, we would plausibly still have a concept of entropy. With the development of statistical thermodynamics, however, which serves to relate the macroscopic properties of systems to the business going on at the underlying microscopic (atomic and molecular) level, entropy gained a more visualizable nature. Boltzmann defined entropy as the logarithmic count of the number of ways in which atoms, for example, could be arranged in a system, given the values of that system's externally measured physical properties, or its state, if you will. With this concept as a template, other workers (notably Gibbs) developed contexts in which this notion could be applied and generalized, ultimately to gravity, but at the same time possibly deviating from the original.

During their parallel physics educations, both authors found classical thermodynamics rather cold,[3] characterized by austere beauty and mathematical elegance, but at the same time rather abstract and difficult to grasp in its entirety. Later, as we learned statistical thermodynamics, we gained a much greater appreciation of what was going on "under the hood" and how the subject could be applied. Entropy suddenly seemed more real. As a consequence, in our more anarchical moods, we think it would be far better to teach statistical thermodynamics before classical thermodynamics!

Interestingly, it is unclear what the state of play might have been had quantum theory not developed under the pressure of anomalous experimental results. The Boltzmann definition of entropy, proposed in 1877 well before the 1920s advent of quantum theory, would still stand, but an important context would have been lost. Without quantum theory, what appears to be a simple count of how atoms might be arranged is technically more difficult in practice. However, does this alter our understanding of entropy itself?

One of the jobs we have set ourselves is to look for an underlying idea and to try to call out situations where it is perhaps being abused. From the outset, it seems that clear language helps to untangle situations where "entropy" has been "badge engineered" to fit another context without necessarily following the physical content. It's also helpful to consider a diversity of applications of entropy across different scientific areas.

STRUCTURE

Each of us approached this book with slightly different perspectives: one is more focussed on the application of concepts to "learn by doing", the other more interested in heading directly towards a fundamental understanding. It's likely that the reader will sense that difference, which we hope will be a strength of the book rather than a path to a bumpy ride.

We have divided the book, like Roman *Gaul*, into three parts.

Part I

Our initial perspective is heavily skewed by the notion that thermodynamics is the branch of physics most intimately related to direct human experience.[4] We begin from this base and undertake a development of classical thermodynamics, covering similar ground to that taught in an undergraduate physics degree at a UK university, but with a focus on building towards an understanding of entropy. With a couple of necessary background chapters to get to that point, the idea is then developed by a light-brush approach to classical statistical thermodynamics.

The conclusion of this section is that entropy is a function of state and therefore just as real as other similar functions, such as pressure and temperature. However, without some statistical underpinnings, we have a much less intuitive grasp of what entropy is than, say, temperature or volume.

Part II

Here, we introduce quantum concepts and use them to revisit and extend statistical thermodynamics. This part is noticeably more technical, but in the service of developing an understanding of the entropic options on the table and how they are applied. Within the chapters, we will also pause to consider the conceptual issues at play. Principally, we will see how entropy is related to our limited ability to control the preparation of a system. The final chapter considers von Neumann entropy, which is a direct outcrop of quantum theory without a classical analogue.

Part III

These chapters are more discursive in nature, though still sequential. We consider applications of entropy beyond the workhorse ideal gas to living systems; information theory; the history, evolution, and potential future of the universe at the largest scales; and finally, to black holes. The purpose here is both to interest and inform the reader and to gauge to what extent the conceptual basis for entropy from Part II can be extended in practice.

We end with a discussion on this topic between the authors.

Finally, there are two Appendices:

MA is a gentle introduction/reminder of the main mathematical techniques needed to follow the arguments in the book. If you have studied calculus before then hopefully you won't find too many surprises and need only use it as an *aide-memoire* or when it's referred to from the main text. If you've never encountered a logarithm before, though, we strongly advise you to start here.

PA summarizes some important technical arguments needed for the developments in Part II.

The overall structure of the book is summarized in the diagram on the following page, which shows linkages between Chapters within each Part as discreet vertical arrows and links between Chapters in different Parts with bolder arrows in lighter shades.

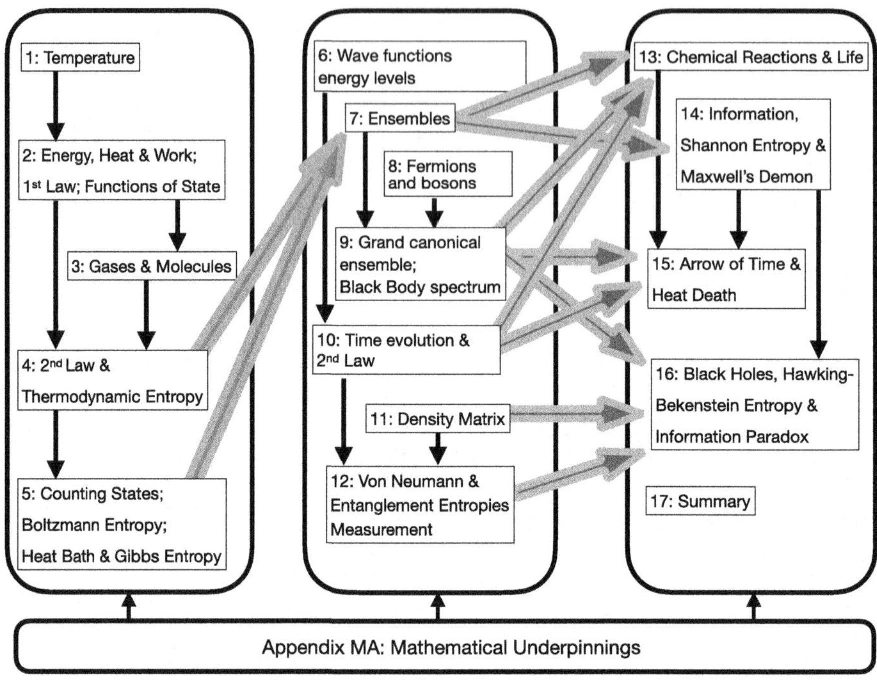

ACKNOWLEDGEMENTS

While individual authors are broadly responsible for each section, we each read, edited, and approved all aspects. Inevitably, some material is reiterated across different sections of the book, but we hope (fondly perhaps) that hearing things twice but with differing styles and emphases will aid understanding. We earnestly hope that there are no contradictions. The book has been over-long in the making, due in part to changing personal circumstances during the Covid pandemic. We are indebted to a variety of people who helped make things better by their enthusiasm, advice, patience, and sympathetic reading of draft chapters. Most notably:

Our respective spouses, Sally and Carolyn, who were all too keen on us getting out of the way to write.

Dr Danny Kielty, Senior Editorial Assistant, Physics CRC Press Taylor & Francis Group, for his gentle and continual prompting and sympathetic ear.

Prof Stephen Fairhurst, School of Physics and Astronomy, Cardiff University, for LIGO advice.

Prof Ian Ford, Dept of Physics and Astronomy, University College London, for writing an inspirational text and supporting our efforts in a variety of ways, including being Our Man in Vienna.

Dr Sarah Harris and Prof Michelle Peckham of the Astbury Centre for Structural Molecular Biology at the University of Leeds; Sarah for sharing her enthusiasm for Maxwell's Demon, and Michelle for advice on all things chemomechanical.

Prof Roger Jones of the Physics Department, Lancaster University, for some cold truths.

Our friends artist Chris Daunt and writer Karlijn Stoffels for their interest and advice.

Special thanks also go to Scott Hayek, Bruce Newhall, John Sweeney, Neal Brower (all of the Johns Hopkins University Applied Physics Laboratory, retired), and John R. Moore (retired), who continue to embody our ideal readers and somehow remain keen to read draft material that we produce.

Of course, the authors remain solely responsible for any errors or omissions that the book retains.

Simon Hands, Liverpool Blue Coat School 1980.
Jonathan Allday, Liverpool Blue Coat School 1979.

NOTES

1 The italics are ours.
2 There is an unresolved difference between the authors, intimately related to our different professions. One of us is happy to use the term "heat" while the other feels that it gives rise to mistaken impressions. If "heat" is a form of energy, then why is "work" excluded from the canon? Is heat a process or a thing? We have left this unresolved partly as the point is not that crucial, but also as the word is used ambiguously in so many other papers and publications.
3 Rather ironically for a theory of heat…
4 With the possible exception of mechanics…

Part I

1 Zero

This opening chapter discusses what it means to describe something as being "hot" or "cold" and why it matters to us. It also introduces the science of thermodynamics. Hotness or coldness is quantified by specifying a temperature, and we learn that two bodies at the same temperature can exist in a special state called thermal equilibrium. Practical means for measuring temperature are reviewed, along with some technical issues associated with comparing results obtained using different thermometers. It is apparent we are no closer to understanding what is meant by "heat".

1.1 HOW HOT?

What do we mean when we ask, "How hot is it?"? It's a commonplace question one might ask before tasting a spoonful of soup, stepping into a bath, or choosing what to wear on a bike ride. The answer is frequently important to us, perhaps determining the successful outcome of a recipe, diagnosing the severity of a fever, or maybe avoiding a painful blister resulting from a burn. Being aware of how hot it is, and making decisions and judgements based on what we learn, is one of the earliest skills we learn when making our way in the world, so we all have a working understanding of what the question means from an early age.

Of course, we often go further than this and quantify the answer. Baking a cake requires an oven heated to 160°C; tomorrow, it will be just 4°C outside, so wrap up warm[1]; the surface of the sun is almost 6000°C... These numbers all measure *temperature*, something we all know and intuitively understand: the larger the number, the hotter it feels. Occasionally, the number is negative; for instance, a supermarket freezer cabinet might be at −18°C. At school, we learn that the funny symbols following the number stand for *degrees Celsius*, so that the question "how hot is it?" can also be phrased as "how many degrees?".[2] Again, we quickly learn to adapt our plans and behaviour according to these temperature numbers, but rarely stop to enquire exactly what the number means or where it comes from. The dictionary definition, "degree or intensity of heat present", doesn't seem to help much, except in reiterating the connection with heat, the quantity possessed by a hot body. There is a distinct feeling of a circularity in the argument, which often arises when struggling to find language to describe very basic things.

The branch of science which deals with such issues is called *thermodynamics*. It is one of the oldest parts of physics, and to this day remains one of its cornerstones. Thermodynamics is deeply rooted in human experience, as hinted in the examples of the preceding paragraphs. Yet, subtle and abstract reasoning is often needed to construct and interpret the arguments it uses. Historically, it has enjoyed a "classical" period, where thermodynamic reasoning was developed very often in close tandem with state-of-the-art technology, namely steam power driving the Industrial Revolution (where the physical qualities of interest are the pressure and volume of a gas, water vapour, confined within a

DOI: 10.1201/9781003121053-3

cylinder by a moving piston connected to a crankshaft driving a wheel), succeeded by a "statistical" age motivated by fundamental and revolutionary advances in atomic and quantum theories of matter, where the homogeneous fluids of the classical world were replaced by vast numbers[3] of identical bullet-like molecular entities seething in rapid random motion. Telling the story of this evolution and uncovering the link between these two viewpoints is one of the principal aims of this book.

1.2 EQUILIBRIUM AND THE ZEROTH LAW

For now let's return to the question of temperature. The way forward focuses on how we compare temperatures of different bodies or systems.[4] If someone hands us two identical-looking bread rolls, one fresh from the oven and the other from yesterday's batch, we can tell which is which immediately by touching them, identifying the fresh roll as having the higher (warmer) temperature. The important concept here is the physical contact. For obvious reasons, it's not always practicable to rank the temperature of objects by touch, but we can place them in physical contact with each other. Unless we take precautions, such as putting the objects in a thermos flask or putting a layer of polystyrene foam between them, then if the two objects are initially at different temperatures, the situation is not stable and will change with time. We describe the objects as being in *thermal contact*, so that heat can transfer between them. Sometimes the resulting change is obvious; e.g. an ice cube placed on a hotplate will readily lose its shape and rigidity to quickly become a small puddle. In other cases, the change is more subtle; if we carefully wrap our bread rolls together in aluminium foil and seal them in a foam box so they are thermally isolated from their surroundings,[5] then when we return sometime later we find their temperatures as gauged by touch are indistinguishable. We might even convince ourselves that the fresh roll is slightly cooler than it was when first wrapped, and yesterday's slightly warmer. By this point, no further change is occurring, and some kind of stability, i.e. reluctance to further change, has set in, and we say the two bodies are in *thermal equilibrium*. The property of thermal equilibrium can also be attributed to isolated systems once all their component parts, or subsystems, have achieved equilibrium among themselves. One of the most important aspects of systems in equilibrium is that their properties don't change with time; that is, there is no bulk or *macroscopic* property, such as pressure, density (or even more nuanced properties, such as colour, opacity, compressibility, or rigidity) that would show evolution if monitored closely. There is something special about the equilibrium state, and the study and description of its properties play a central role in thermodynamics.

As the touch test confirms, two objects in thermal equilibrium have the same temperature. However, this statement concerns two particular objects. At this stage, it's logically conceivable that two bread rolls could be at the same temperature, as determined by the onset of equilibrium, but that one, placed in contact with an ice cube, might cause it to melt more quickly than the other. That this does *not* happen is an important empirical truth about the world, encapsulated in the *Zeroth Law of Thermodynamics*[6]:

If two systems are each in thermal equilibrium with a third system, then they are in thermal equilibrium with each other.

Three or more systems in equilibrium all share the same temperature—it is this law of experience that makes temperature such a universal and powerful concept, underlying our innate and intuitive understanding. The temperature difference between two bodies tells us which is "hotter" and, as we shall see, determines in which direction heat will flow if the bodies are in thermal contact.[7]

1.3 THERMOMETRY

As an immediate application, consider *thermometers*. A thermometer is a device with a physical property that depends on temperature in a predictable and unique way. We can place a thermometer in thermal contact with a system until equilibrium is established, and then measure that property.

If we repeat the process with a second, remote object, which might be out of thermal contact with the first, and obtain the same reading, we can be confident the two objects are at the same temperature. The best-known example, still found in many kitchens and medicine cabinets, is a "mercury-in-a-glass" thermometer, consisting of a thin-walled glass bulb (permitting diathermal contact with the system under study) containing a liquid (frequently the metal mercury, which is liquid for temperatures above −38.83°C) connected to a thin gradated tube, sealed at the far end. As temperature rises, the mercury expands; that is, it changes its volume, forcing more and more liquid to move from the bulb into the tube, where the increase in volume can be read off by how far the liquid thread reaches along its length. There are other practical thermometers based on different properties: the electrical resistance of a piece of wire, the differential expansion of two different metals fastened together to form a thin flexible strip, the voltage developed in an electrical circuit formed from wires of two different metals in contact, or even a remote-sensing device known as a *pyrometer* which relates temperature to the properties of the infrared radiation emitted by a body. In each case, the key is that the response to a temperature change is unique and readily measured.

In order to convert a thermometer reading to a numerical value, fit to appear in a recipe book or tomorrow's weather forecast, we need to *calibrate* it by recording its reading at at least two temperature *fixed points* agreed by everyone, which ideally should be readily capable of being reproduced in a laboratory, or indeed the workshop of the factory where the thermometer is made. A convenient choice is the melting and boiling points of water (ideally as pure as we can find) at 1 atmosphere pressure.[8] The *centigrade* scale assigns zero degrees to the temperature at which ice melts into liquid water and 100 degrees to the temperature at which liquid water boils to form steam. We can mark the extent (call it x_0) of the mercury thread on the scale of a glass thermometer placed in a beaker containing ice and water in equilibrium, another (x_{100}) when the thermometer bulb is held in the jet of steam escaping the spout of a boiling kettle. If the mercury thread extends to a position x when held in thermal contact with a sample object, then the object's centigrade temperature θ is given by:

$$\theta(x) = \frac{x - x_0}{x_{100} - x_0}. \tag{1.1}$$

The name follows because a scale with 100 equally spaced gradations between x_0 and x_{100} can be engraved on the tube.[9] Notice that if x lies below (i.e. closer to the bulb than) the zero mark x_0, then θ is a negative number, consistent with what we know about supermarket freezer cabinets. Similarly, x lying beyond x_{100} corresponds to a temperature over 100°C. There is no problem in principle with extending the temperature scale beyond the range between the fixed points.[10] The Centigrade scale is still commonly used in English-speaking countries, but as the same term is sometimes use to measure (small) angles, since 1948 it's been standard in scientific work to refer to it as the *Celsius* scale.[11] In either usage we write temperatures the same way, e.g. 30°C.

Let's review the progress so far. We've identified temperature as a useful measure of how hot things are and set out practical means for its measurement. The Zeroth Law implies that temperature is an essential characteristic of systems in thermal equilibrium, whose properties don't evolve in time, and enables two different systems to be compared in a useful way. Pause for thought, however, and you might notice at least two important issues hanging over us.

Firstly, how universal is the Celsius temperature scale, as measured by the above-mentioned procedure and encapsulated in Equation (1.1)? We chose to base our measurement on the length of a mercury thread. If we used another working substance such as ethanol, would we still observe the same readings? One could argue the answer must be yes, by definition, at the two fixed points, but what about points in between? For mercury and ethanol thermometers to give the same readings for all temperatures, it has to be the case that mercury and ethanol respond to temperature changes through thermal expansion in precisely the same way[12]; this might be true up to a point, but it seems to require a knowledge of the behaviour of materials we haven't up to now specified or worried about. And even if that were true, what about other thermometers based on electrical or radiative properties of materials? Until we know better, it seems prudent to specify our temperature scale in relation to the particular thermometer used to make measurements. In later chapters, we'll see that thermodynamics provides a way out of this problem and that it is indeed possible to rigorously define temperature so that everyone must agree on the reading.

That said, practical thermometry can still present problems. The hottest temperatures routinely produced in laboratories today occur in particle accelerators, such as the Relativistic Heavy Ion Collider (RHIC) at Brookhaven National Laboratory (USA) or the Large Hadron Collider (LHC) at CERN in Geneva (Switzerland). These facilities use an intricate combination of electric and magnetic fields to accelerate charged particles, in both directions around a near-circular beam pipe many kilometres in circumference, to very very close to the speed of light.[13] High temperatures are reached when the particles in question are the nuclei of heavy elements, such as lead (Pb) or gold (Au), which have been completely ionised, i.e. stripped of their attendant electrons, and thus very highly charged. When two such heavy ions collide head-on, most of their kinetic energy is converted (via collisions among the O(1000) constituent quarks occurring within a tiny instant) into thermal energy, raising the temperature in a region roughly 10^{-14} m in diameter to $1–5 \times 10^{12}$ C. Clearly a mercury-in-a-glass thermometer is of no use here! Rather, we infer these conditions by measuring the energies of the many thousands of particles emerging in the aftermath of the collision[14] and using knowledge of the thermal properties of the hot medium calculated using *Quantum Chromodynamics*, the fundamental theory of strongly interacting matter.[15]

The second issue is so big as to be almost invisible. The essential property we have attributed to temperature is that temperature differences direct the transfer of heat. Heat is the property of a body, which makes it "hot", yet we haven't got any closer to saying what this means. That work starts in the next chapter.

NOTES

1 At the time of writing, there has been much concern in the UK about the first recorded outside temperature exceeding 40°C.

2 Older readers, or those living in the USA or Liberia, may prefer to answer the question using *degrees Fahrenheit*, wrapping up warm when temperatures drop to 40°F. To convert from °C to °F, multiply by 9, divide by 5, then add 32.

3 The number of atoms or molecules involved in a tangible macroscopic system is unimaginably, but not unquantifiably huge. Typical values are conveniently expressed in terms of *Avogadro's number* $N_A \simeq 6.022 \times 10^{23}$, i.e. roughly 6 followed by 23 zeros.

4 In thermodynamics physical things are called "bodies", "systems", "substances", or "objects" almost interchangeably—we'll make no attempt to be consistent.

5 The technical adjective describing a wrapping or "wall" which inhibits the transfer of heat is *adiabatic*, in contrast to *diathermal* walls, such as the metal body of a saucepan, which permit heat transfer.

6 Why "zeroth"? The law was first articulated in the 1930s, long after Laws 1 and 2 to be discussed in what follows. It's placed at the head of the list because logically it underpins subsequent developments.

7 We will spare the reader the full rigour of the argument underpinning this deduction, which can be found in *The Elements of Classical Thermodynamics* by A.B. Pippard (Cambridge University Press 1957).

8 The temperature at which the melting and boiling of most substances take place depends on the applied pressure, so it's important to make this stipulation.

9 Yes, we know there are really 99.

10 In practice, the freezing and vaporization temperatures of mercury provide constraints; a polar explorer will prefer to use a thermometer containing ethanol dyed red for easy visibility, which provides useful readings down to −70°C.

11 Named after astronomer Anders Celsius, who introduced the scale in 1742.

12 Suppose that we denote the temperature measured by an ethanol thermometer (or indeed any other thermometer) by Θ; the assumption that all thermometers and fixed-point choices measure the same temperature using (1.1) is equivalent to the statement that Θ is a simple linear function of θ, i.e. $\Theta = A\theta + B$ where the numerical constants A, B can be determined by careful measurement. In general, the thermal properties of materials are more complicated than that.

13 At such *relativistic* speeds, it is much more useful to state the particle's energy in units of GeV. 1 GeV = 10^9 eV = 1.6×10^{-10} J. For a nucleus containing a total number of protons plus neutrons equal to A, RHIC accelerates it to an energy of about 100AGeV while LHC reaches over 2500AGeV.

14 Often referred to as the *Little Bang*.

15 The calculations require high-performance computer resources; a state-of-the-art calculation might involve over 10^{22} individual floating-point operations, bizarrely close to Avogadro's number.

2 One

The notion of "energy" has increasingly entered everyday language since the latter half of the 20th century. Energy, the capacity to do work, takes many different forms, which are capable of being interconverted but crucially are all measured using the same units. The First Law of Thermodynamics asserts that heat is another form of energy, and we review empirical evidence for this. The work done either by or on a gas can be quantified in terms of its pressure p and volume V, and this is developed to relate the work done by an engine to the area of a cyclic path in a p–V plot. The privileged role of functions of state, whose value only depends on where you are in the plot on not on the path taken to get there, is pointed out.

2.1 ENERGY CRISIS

This book's authors met at secondary school in the 1970s, a relatively turbulent period in the UK's modern history. The bulk of the UK's electricity supply was generated by coal-fired power stations. As a consequence of industrial action by coal miners in the winter of 1972, the nation experienced power outages for both commercial and domestic users, leading to enforced candle-lit evenings and the declaration of a State of Emergency during particularly cold weather. The following year, further disputes resulted in a three-day working week, strong restrictions on non-essential power consumption, and TV broadcasts ending each evening at 10:30. The decade was also punctuated by petrol supply crises as world production peaked. Oil prices rose dramatically, leading to economic stagnation. At one stage, petrol coupons were issued in preparation for possible rationing, but in the event, the UK muddled through, in part because of increasing oil production from the newly opened reserves beneath the North Sea.[1]

This was a period, then, when issues relating to energy supply were at the forefront of people's minds—as never before there was a sense that "energy" was a scarce resource, and that supplies might be far less plentiful in the future. Nowadays our fears about resources becoming exhausted have been to a great extent supplanted by the looming threat of crisis due to the modification of the atmosphere's composition that results from our persistent reliance on fossil fuels (coal, natural gas, and oil), which is building up a blanket of gases trapping heat near the earth's surface and causing significant, in many cases harmful, changes to the climate. Energy supply in years to come has to be "sustainable" or "carbon-neutral", and there is a huge and ongoing investment throughout the world in newer, "cleaner" technologies such as wind, solar, and nuclear power generation.

Our point here is not to review energy policy in detail but simply to point out that people today have a relatively sophisticated understanding of what is meant by "energy"; we all know it is a finite resource needed to fuel all aspects of our activity, that it comes in many forms (fossil fuel, nutrition, electrical energy, nuclear power,...), that some effort is needed for it to be to be harvested or

DOI: 10.1201/9781003121053-4

generated, that these different forms are inter-convertible, that it can be quantified (and monetized), and that some forms of energy are more useful and better for us than others. These concepts would have seemed quite alien and abstract to a well-educated person living in the 18th century. It is a consequence of developments in the science of thermodynamics that "energy" has entered everyday language[2]: this chapter aims to set out the terminology with a little more care.

2.2 DIFFERENT FORMS OF ENERGY

Let's begin by reviewing some different forms of energy and how to quantify them. Much of this material is familiar material in high-school physics curricula. The simplest to specify is *kinetic energy*, the energy something has by reason of its motion. If a body of mass m travels with speed v, then its kinetic energy is

$$\text{kinetic energy} = \frac{1}{2}mv^2. \tag{2.1}$$

So, the faster something moves, the more energy it has. A cricket ball[3] of mass 0.163 kg propelled by a fast bowler at 40 ms^{-1} has a kinetic energy of 130.4 J—here, the letter J stands for *Joule*,[4] a widely used unit of energy.[5] Now, in order to deliver the cricket ball at such a fearsome pace, the bowler has to do some work, by imparting a force F to the ball. In mechanics, work is done when the point at which the force is applied (the bowler's hand) moves through some distance d

$$\text{work done} = Fd. \tag{2.2}$$

Let's make some simplifying assumptions: the bowler's hand describes an arc of a circle of radius 0.7 m, and moves through an angle of 180 = π radians during the delivery. The average force imparted is therefore $130.4/(\pi\times 0.7) \simeq 59.3\,\text{N}$; the N stands for *Newtons*, the SI unit of force. This is roughly the force required to lift three 2 kg bags of sugar.[6] In order to reach this conclusion we equated the energy of the ball with the work done on it—in most cases this is done almost without thought. We conflate the state of the system, as described by its *energy*, with the *work done* on the system to get it into that state. Both quantities are measured with the same units, i.e. Joules. As we'll see in what follows, once we start thinking about heat this naive identification of energy with work no longer holds.

A body can possess *potential energy* because of its position or state, classic examples being a coiled spring under tension or a body teetering at the edge of a cliff. In the second case the potential arises because the body is subject to a downward force due to the Earth's gravitational field, which close to the Earth's surface imparts a uniform acceleration of $g \simeq 9.8$ ms^{-2} to all freely falling objects. If the cliff height is h, then relative to the beach, a body of mass m has

$$\text{potential energy} = mgh. \tag{2.3}$$

If the body is pushed over the edge it moves rapidly downwards, picking up speed as it falls. During this time potential energy is converted to kinetic energy as a consequence of the work done on a body in motion by the gravitational force. We can calculate the speed it has attained by the time it reaches the sand by equating kinetic and potential energies (2.1, 2.3) to find $v = \sqrt{2gh}$.[7]

Conversion of gravitational potential to kinetic energy can, of course, contribute to our discussion of energy generation; think of a watermill, an early manifestation of hydroelectric power. If the mill is modelled as a collection of N buckets arranged evenly around the circumference of the mill wheel, each holding a mass m of water, then every complete rotation delivers an amount of work equal to $Nmgd$, where d is the wheel's diameter.

The next important example is electrical energy. The force on a wire of length ℓ carrying an electric current I and lying perpendicular to a magnetic field B is given by $BI\ell$.[8] The resulting force is at right angles to both the wire and the magnetic field. Work is done when the wire moves; in an electric motor, an arrangement of wires in a coil mounted on a rotating spindle known as an armature connected to a current source with a clever switching device known as a commutator is able to convert this force into rotation. The motor can then be used to do mechanical work on other objects, perhaps by increasing their gravitational potential energy if the motor is part of a crane or a lift. Now, as a consequence of Faraday's law of electromagnetic induction, a coil rotating in an external magnetic field experiences an opposing electromotive force, so a voltage[9] V must be applied across the circuit to maintain a steady current. If the motor is run for time t, then

$$\text{electromotive work} = IVt. \tag{2.4}$$

It is frequently more helpful to specify the rate at which work is being done, otherwise known as *power*. Power is measured in *Watts* (W), such that anything working at a rate of 1 Js^{-1} delivers 1 W of power. The power delivered by an electrical circuit[10] is thus IV. We are used to comparing electrical devices and assessing their energy demands on the basis of their power rating or *wattage*. This also gives rise to other (non-SI) units for energy/work—the energy needed to sustain a power of 1000 W over a period of 60 minutes is called a *kilowatt hour* (kWh). $1\ \text{kWh} = 3.6 \times 10^6\ \text{J} = 3.6\ \text{MJ}$. Most of us are introduced to this way of quantifying energy when we first carefully read our electricity bill.

The last form worth discussing at this stage is the *chemical energy*, which becomes available due to a rearrangement of the component atoms forming a molecule, during a chemical reaction, which has its origin in the electrical forces holding the molecule together. The chemical in this context is often known as a *fuel*, and the best-known examples are essential to power generation: firewood, coal, natural gas, and oil. Usually in this context the reaction is known as "burning". In each case the amount of energy released when say, a fuel hydrocarbon with generic composition chemical C_nH_{2n+2} combines with oxygen O_2 in the air to form the waste products carbon dioxide (CO_2) and water (H_2O) is known—in fact, we can get a rough-and-ready idea of the quantities involved by inspecting the packaging on another hydrocarbon commonly found in our fridge, namely butter. The nutritional value is listed as 306 kJ per 10 g serving, which translates to roughly 7 kWh per litre[11] of fuel, not too dissimilar to the value for petrol. Your gas bill uses the same kWh units as the one for electricity. Of course, this highlights the fact that our food, too, is a fuel, and people concerned about what they weigh need an encyclopaedic knowledge of the calorific values of their favourite treats.[12] A qualitatively similar energy release occurs when strongly interacting particles known as *nucleons* (protons and neutrons to you and me) rearrange the nuclear system they are bound within, either by the fission of a single large *mother* nucleus into two smaller *daughters*, or through fusion of two small nuclei into a single larger product. When a nucleus of uranium (with 235 component nucleons) fissions to form a nucleus of rubidium (93) and caesium (141), together with a spare neutron left over, then 214 MeV $\simeq 3.43 \times 10^{-11}$ J is released, principally into the kinetic energy of the daughter fragments. This may not sound a lot until you recall that 1 kg of uranium contains over 10^{22} such nuclides; persuading just a small fraction of them to fission in a controlled way is the idea driving nuclear power generation. On a per-event basis, the energy released in nuclear fission dwarfs the energy released in chemical reactions by a roughly a factor of a million. In fact, the energy release manifests itself as a measurable difference between the mass of the mother and the combined masses of daughters and product neutrons—the conversion factor in this case being the celebrated relation of energy–mass equivalence $E = mc^2$ from the theory of relativity.[13]

2.3 THE FIRST LAW

Probably much of this material is already familiar, at least in part. You may have noticed we went to some pains not to mention heat at any point, which seems a little perverse when considering the burning of fuel. The reason is to set the stage for the next big idea in thermodynamics—the equivalence of work and heat. The statement was first made by Joule in 1843 (his discussion was restricted to mechanical work) and then generalized by Rudolf Clausius in 1850. There are several differing wordings, and several other scientists can also claim a partial stake in the formulation of what is now known as the *First Law of Thermodynamics*:

Energy is conserved, if heat is taken into account.

In other words, heat is another form of energy, and is measured in the same units. We quantify heating devices such as kettles and electric radiators in the same way we quantify the performance of an electric motor driving a crane or an engine propelling a ship. Implicit in the wording of the First Law is the idea that heat and work can be interconverted; the electric kettle exemplifies the conversion of electrical work done by a current flowing through a heating element into heat to raise the temperature of the water. Already, you're probably thinking up further interesting and useful examples of such interconvertibility, perhaps this time involving heat to work. But the First Law says so much more—it says that "energy is conserved", implying that it can be neither created nor destroyed, and that the amount of energy in the world is the same as it ever has been, and will be for always. Why, then, do we always seem to be talking about an "energy crisis"?

In physics, a conserved quantity has a particularly exalted status. The statement of conservation often hints at some underlying symmetry or way of viewing the problem that opens up a new perspective. If energy is conserved, it implies that any system which is sufficiently isolated from its surroundings, so that no work is being done on it, with adiabatic walls so that no heat is transferring in or out, has a constant energy, usually denoted U and referred to as *internal energy*. For now let's leave aside the thorny issue of how to measure (or calculate) U: it suffices to think about how it can change. According to the First Law, any changes to U are of the form[14]

$$\Delta U = \Delta W + \Delta Q \tag{2.5}$$

Here, ΔW is the work performed on the system by some exterior agency and ΔQ the heat entering the system. Note that $\Delta W, \Delta Q$ can be either positive or negative valued (e.g., heat can flow either in or out of the system); of course, each term in equation (2.5) is measured in a common unit, e.g. J. If the system is contained within adiabatic walls, so no heat can enter or leave, then $\Delta Q = 0$ and $\Delta U = \Delta W$, i.e. the change in internal energy equals the work done on the system. It's also possible to think about changes where no work is done, but the walls are now diathermal so that $\Delta U = \Delta Q$.

In order to flesh out these rather abstract considerations, let's consider Joule's own experimental work. He placed some water in a container known as a *calorimeter*, which is adiabatically sealed (so that we know $\Delta Q = 0$), accompanied by an arrangement of paddle wheels on a rotating spindle designed to stir the liquid thoroughly once connected to a driving device. The original driver was a weight on a line falling over a pulley turning the spindle, but we could equally replace it with a motor driven by an electric current—in either case it is possible to quantify the amount of mechanical work ΔW applied to turn the spindle. At the end of the experiment, once the water motion has dissipated so that we can be confident thermodynamic equilibrium is reached, the temperature θ is measured and found to have increased over its value at the start of the experiment.[15] The same temperature change results independently of the nature of driving mechanism used, provided the amount of work supplied ΔW is always the same. Such a temperature change could equally well be induced by heating the calorimeter over a flame or indeed by an electric heating element in which current passing through a wire with resistance R imparts heat through an effect known as *Ohmic heating*.[16] This time, we know that $\Delta W = 0$, $\Delta Q > 0$. In either case, if due care is taken, we are left with the same mass of water, occupying the same volume, with the only measurable difference at the end being an increase in temperature θ. There is no reason to suppose that the internal energy depends in any way on whether the change has been effected by application of work ΔW

or heat ΔQ, implying that both possibilities must be taken into account when considering changes in internal energy, as stated by the First Law. The same apparatus can thus also be used to measure ΔQ accurately, calibrating it in relation to the observed change in θ.

There are many variations on Joule's experiment demonstrating the equivalence of work and heat. We remember a school laboratory demonstration of the heating effect on lead shot enclosed in a long tube, which is repeatedly inverted so that the shot falls the length of the tube to crash into the far end. If the tube has length ℓ and there are N repetitions, the applied work $\Delta W = Nmg\ell$. Another famous experiment predating Joule's work was due to Count Rumford in 1798, who demonstrated that the heat produced when boring a cannon using a blunt tool could be used to boil a specified amount of water in a measured time. One of the most romantic was conducted by Joule himself on a honeymoon trip in the French Alps in 1847, where he attempted to measure the heating of water falling some 270 m over the famous Cascade de Sallanches. The experiment was unsuccessful—the breakup of water into spray at the base made accurate thermometer readings impossible.[17]

In order to develop our understanding, however, it helps to focus on a particularly simple system—a gas[18] contained within a cylinder with a moveable piston at one end so that the volume of the enclosure can be varied by moving the piston. The very terminology employed here—cylinder, piston—is a reminder that the prevalent technology when these ideas were first developed was motive power through steam engines. What properties of a gas do we need to know to begin? First, it's a fluid, that is it has no rigidity and changes shape to occupy any vessel it happens to be contained in. Next, unlike a liquid, if not contained, a gas has a tendency to disperse and drift away; the walls of the container therefore need to supply an inwards force to keep it confined. Within the bulk (i.e. away from the walls) the gas transmits and communicates this force equally in all directions, at right angles to the surface of any solid object that happens to be in the way—including the walls of the container. The word used for this phenomenon is *pressure*, namely the force transmitted per unit area of the surface. Pressure is measured in units of N m^{-2}, although a frequently used alternative (and SI unit in its own right) is the *Pascal*: 1 Pa = 1 N m^{-2}. As we shall see in the next chapter, the state of a gas can be specified in terms of two quantities: its temperature θ and pressure p.

2.4 THE $p - V$ PLOT

Right now each of us is experiencing a pressure of roughly 10^5 Pa exerted by the air surrounding us as a consequence of the weight of the Earth's atmosphere above us pressing down. This sounds unexpectedly large! The reason we don't get pushed sideways, or onto the floor, is that pressure in a fluid acts equally in all directions, even upwards, so the forces acting on our bodies balance and cancel out. We don't get crushed because there is air (or blood) in many internal cavities in our bodies under near-atmospheric pressure, which again provide an opposing outward force enabling us to keep in shape. As a corollary, astronauts working in conditions where atmospheric pressure is effectively zero need (besides a supply of breathable oxygen) to wear special suits maintaining an inwards-acting pressure over their body surface to stay healthy. Outside this extreme example, on the whole we're remarkably insensitive to atmospheric pressure. By contrast, if air is sucked out of a vessel[19], be it a cardboard carton, plastic bottle or steel drum, atmospheric pressure can manifest itself dramatically, leaving the formerly rigid container in a crumpled mess. Now, the pressure in a fluid confined by gravity, such as the atmosphere, depends on height. When we climb to a greater altitude there is less atmosphere above us, so the air pressure is less than at ground level. If you're a mountaineer or an airline pilot these effects matter. On the scale of the cylinders and containers found in a laboratory or engine shed, however, the influence of gravity is negligible, and in what follows we'll assume the pressure is constant throughout the vessel.

Now return to the cylinder with the moveable piston, as shown in the diagram of Figure 2.1. The part of the piston in contact with the gas is a circular disk fitting snugly in the cylinder walls so that no gas can leak around the edges (this is an idealization, but we'll go with it for now…). Suppose

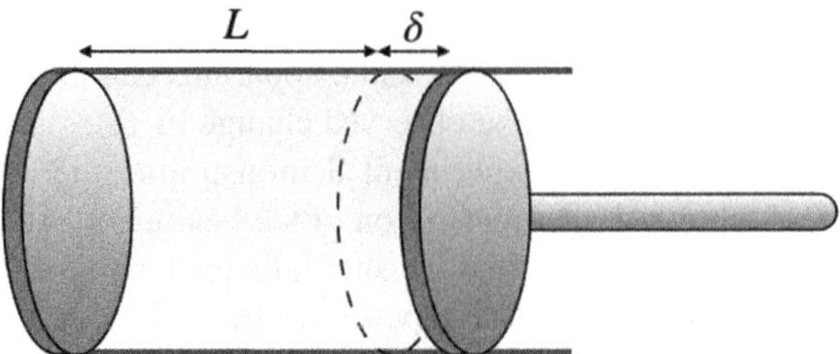

FIGURE 2.1 A gas contained within a cylinder by a moveable piston.

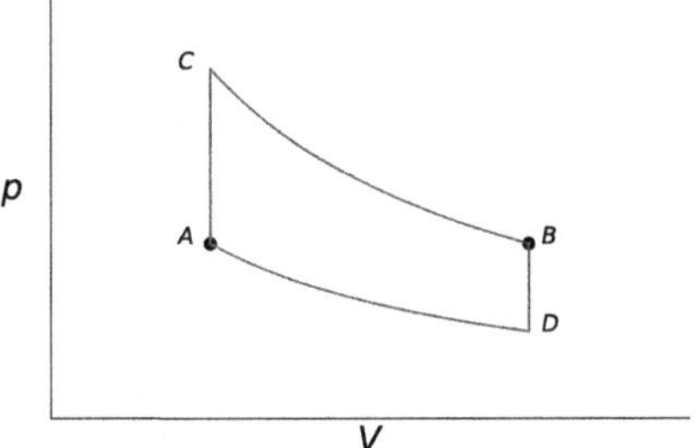

FIGURE 2.2 Thermodynamic cycle for a gas in the *p–V* plane.

the area of this disk is *A*, then the force exerted on the piston by a gas under pressure *p* is equal to *pA*. This force acts at right angles to the face of the piston, and if things are to stay as they are must be balanced by an equal and opposite external applied force, perhaps by putting a weight on top of the piston or turning a screw mounted on a bracket. If the forces don't balance the piston will start to move outwards, increasing the volume in which the gas is enclosed. The full story of what happens when a gas increases its volume is surprisingly complicated, depending on how the piston's motion is controlled. An obvious question, to be answered soon, is how the change in volume affects the pressure. For this reason, we'll restrict our attention for now to very tiny piston movements, corresponding to very tiny changes in the volume, and assume that the pressure will be roughly constant as a consequence. The first thing to do is to relate the piston movement to the change in volume. Suppose the piston moves a distance δ ($\delta \ll L$, where L is the cylinder's length). The volume ΔV swept out by the piston face is a short stubby cylinder of volume $A\delta$, the base of which is indicated by a dashed line in Figure 2.1. The work done by the gas in moving the piston is the force it exerts multiplied by the distance it moves, namely $pA\delta$. We conclude the work done *by* the gas $= p\Delta V$. Now the ΔW appearing in the First Law equation (2.5) is the work done *on* the gas, and so is the negative of the quantity. Hence

$$\Delta W = -p\Delta V \Rightarrow \Delta U = \Delta Q - p\Delta V. \tag{2.6}$$

This is the statement of the First Law for whenever the volume change ΔV is sufficiently small that any resulting change in *p* can be ignored. Although we have derived (2.6) for the particular case of a straight-sided cylinder, the result can be shown to hold for a vessel of any shape.

Because it's possible to express the performed work ΔW in terms of the pressure *p* and changes in the volume ΔV, it makes sense to consider the behaviour of the gas in terms of these two quantities. We can do this using a *p–V* plot in which the horizontal axis extends along a range of *V* values, and the vertical axis *p*. Each point in the plane marks a particular pair of *p* and *V* values. As we'll confirm later, gases are such simple systems that this combination serves to completely specify the state of a fixed amount of gas in equilibrium[20]. Each point in the plane therefore

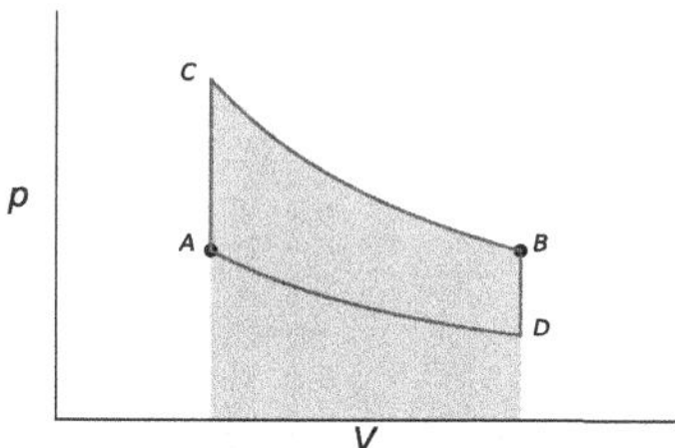

FIGURE 2.3 The shaded area represents the total work ΔW_{CB} performed by the gas on the *CB* leg.

corresponds to a particular value of the internal energy $U = U(p,V)$. The difference between U values at two different points A,B can be labelled ΔU_{AB}. The plot in Figure 2.2 shows points *A* and *B*, and also, two different paths we could take to go from one state to the other. One path starts by a vertical line up to *C*, followed by a longer, curving path finishing at *B*. The second path starts with a curved section finishing at a point *D* aligned below *B*, so that the finishing segment is a vertical straight line.

First let's consider the two straight segments *AC* and *DB*. Every point on each of these lines corresponds to the same *V* value; in other words $\Delta V_{AC} = \Delta V_{DB} = 0$. Using the First Law in the form (2.6) we immediately conclude

$$\Delta U_{AC} = \Delta Q_{AC};\ \Delta U_{DB} = \Delta Q_{DB}; \tag{2.7}$$

in other words, along these paths changes to the gas's state are entirely due to adding heat. As we'll see in the next chapter, the change in *U* following an upwards displacement in the plane is positive, so $\Delta Q_{AC} > 0, \Delta Q_{DB} > 0$: in both cases heat is being added to the system. The discussion along the two curved segments *CB* and *AD*, is more complicated because now the volume is changing, so the work terms $\Delta W_{CB}, \Delta W_{AD} \neq 0$. In fact we can calculate ΔW via the sum over lots and lots of contributions of the form[21] $-\sum_i p_i dV$, where *i* labels a rectangle of (tiny) base width *dV* and height p_i. In the limit that *dV* is made arbitrarily small and the number of *i* values contributing to the sum correspondingly large, this expression becomes precisely the area of the *p*–*V* plane lying between the curve and the *V* axis, as shown in Figure 2.3, giving an appealing geometric interpretation of the work done. For both *CB* and *AD*, the resulting answer for ΔW is negative, meaning that on both of these sections the gas is performing work on its surroundings, literally by pushing the piston outwards. Without a detailed knowledge of the internal energy function $U(p,V)$, it's impossible to be more precise at this stage. Similarly, it's impossible at this point to make a precise calculation of the contributions $\Delta Q_{CB}, \Delta Q_{AD}$ along these paths, but both turn out to be positive, i.e., heat is entering the gas along each segment.

Let's now think about what happens if we now make the gas traverse the path *ACB*, followed by the path *BDA*, i.e. the lower path *in reverse*. Since we returned to the starting point, the gas finishes in the same state in which it started—we have performed a *cycle*. In principle we could repeat the cycle as many times as we wish, each time taking the gas through the same sequence of states specified by the coordinates in the *p*–*V* plane. However, along the return path we need to take into account a sign flip for the heat and work contributions constrained by $\Delta U_{BA} = -\Delta U_{AB}$ and the First Law:

$$\Delta Q_{BD} = -\Delta Q_{DB};\ \Delta Q_{DA} = -\Delta Q_{AD};\ \Delta W_{DA} = -\Delta W_{AD}. \tag{2.8}$$

This means that while heat is entering the gas along *AC*, it is now leaving the gas during the passage *BD*. Although we can't yet calculate it, this is also true for the curved paths; heat enters the system

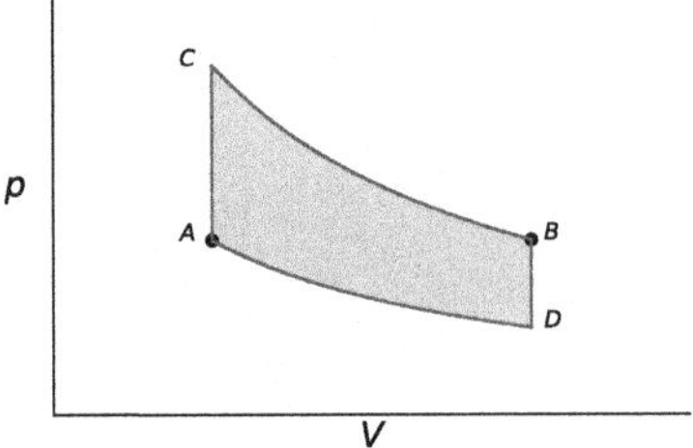

FIGURE 2.4 The shaded area represents the total work performed during one cycle.

along CB and departs along DA. During the cycle, heat "flows" through the system, in a way which we still have to quantify. We can, however, be more specific when discussing the work contributions. The (negative) work done on the gas during CB is the same, but now the work done along DA is *positive* due to the sign flip (2.8). The total work done *on* the gas is now the negative of the area enclosed within the path defining the cycle; more straightforwardly, the work done *by* the gas during the cycle equals the enclosed area in the p–V plane. This is represented in Figure 2.4. We are describing a system in which heat, possibly originating from burning a fuel such as coal, petrol, or fissioning nuclei, flows through a system, resulting in a periodic force communicated through a piston moving within a cylinder, performing work on the outside world.

Almost by accident, we have described an engine,[22] that is a device which converts heat into mechanical work, a process compatible with the First Law. Thermodynamic cycles of the form shown in Figure 2.4 can be drawn for steam engines, internal combustion engines, and the turbines operating within nuclear power plants—indeed, any device where the *working substance* is a gas. In later chapters, we'll return to the cycle to tackle the vexed question of engine efficiency: just how much work can be extracted from the heat arising from consuming a given amount of fuel? As we'll see, as well as being of enduring importance in our everyday lives, this question is of profound importance in the science of thermodynamics.

2.5 FUNCTIONS OF STATE

It's now time for an important technical point. The final state of the gas at B is the same regardless of whether we choose to get there along a path ACB, path ADB, or indeed any other path drawn connecting the endpoints A,B. The internal energy U of the gas only depends on the location of the point in the p–V plane. Any aspect of a thermodynamic system that doesn't depend on the detailed history of how it was set up but merely on a fixed number of measurable quantities has a special status; it's known as a *function of state*. Functions of state typically depend on just a small number of parameters of state (in this case p and V) and completely specify the equilibrium properties of the system. It's a very different story, however, for heat Q and work W: changes to these quantities depend on the path taken to get from A to B, and in general $\Delta Q_{ACB} \neq \Delta Q_{ADB}, \Delta W_{ACB} \neq \Delta W_{ADB}$. We can't study a sample of gas and deduce how much work and how much heat has been supplied to achieve its current state. The First Law (2.5) guarantees, however, that

$$\Delta Q_{ACB} + \Delta W_{ACB} = \Delta Q_{ADB} + \Delta W_{ADB} = \Delta U_{AB}. \tag{2.9}$$

As path-dependent quantities, heat and work are therefore not functions of state, so there is an important sense in which the terms in equation (2.5) don't all share equivalent status.[23] This makes physical sense, but it also makes subsequent mathematical analysis awkward. There is, however, a special set of circumstances that assists calculations: when changes made to the system are sufficiently slow and steady that throughout the process the system remains in an equilibrium state. We already made an assumption of this sort when deriving (2.6), namely that under very small motions of the piston the pressure in the cylinder can be taken to be constant. In the example of (2.6), it meant that ΔW could be expressed in terms of equilibrium properties and small changes therein, i.e., the parameters of state p and V. Since ΔU is also expressed in terms of a function of state, it follows from the First Law that in this case the change in heat ΔQ must also be in principle calculable as a function of p and V. Changes of this form are called *quasistatic*, and in such cases it is conventional to rewrite the First Law as a relation between infinitesimal changes[24]:

$$dU = dQ - pdV. \tag{2.10}$$

In our gas-in-a-cylinder example, quasistatic conditions require that the pressure applied by an external force acting through the piston must always match the gas pressure within the cylinder. This could be upset if the piston is jerked suddenly, producing a local compression or rarefaction within the gas, which would then need time to settle down, probably with the production of extra heat not accounted for in (2.10). Similarly, any friction between the piston and the cylinder walls will cause a mismatch between external and internal pressure, invalidating the second term on the right-hand side of (2.10), and also producing excess heat. The quasistatic condition is something of an idealization, but fortunately for well-machined cylinders and pistons, and real gases, particularly at low pressures, we can get pretty close in practice.

To summarise, in this chapter, we have discussed a familiar concept, energy, in perhaps rather unfamiliar, formal terms. The key idea is the equivalence of work (mechanical, electrical, or whatever) and heat, expressed in the First Law. Heat has played centre stage in this chapter, just as temperature did in the previous, but there are still big questions to answer. Using a calorimeter and working carefully, we now know how to quantify changes ΔQ in heat, but as yet have said nothing about how to relate this to temperature readings θ made using a thermometer. And then there is still the question of just how efficiently we can make engines run.

From a modern perspective, it's a little hard to appreciate how revolutionary this unification of two apparently unrelated everyday quantities heat and work in an overarching abstraction called "energy" must have seemed at the time. Arguably, this is just as great (and fertile) a conceptual leap as those associated with the unification of electricity and magnetism comprising electromagnetic theory in the late 19th century, the unification of space with time, and matter with energy in the Relativity theory developed in the early 20th century, and the unification of weak and electromagnetic phenomena underlying the Standard Model of particle physics developed in the late 20th century. All three examples are rightly celebrated as towering achievements in the development of modern physics. Why then don't we regard the First Law of Thermodynamics in the same light? We suggest there are two factors responsible: firstly, as we've tried to illustrate, the First Law is firmly rooted in facets of our everyday experience, and there's a sense in which it feels like just a formal language developed to tell us stuff we already "know in our bones". At this point (let's get it out of the way sooner rather than later), recall the famous nostrum[25]: "*Science owes more to the steam engine than the steam engine owes to science*". The development of steam-powered technology largely predates the formal science of thermodynamics, and we can't point to any major technological breakthrough that has resulted as a direct consequence, as we can, say, for electromagnetism (radio transmission) or relativity (nuclear power generation).

More importantly, the picture we have outlined so far is a *macroscopic* one, in which current electricity is the flow of a mysterious substance called charge, a solid a space-filling object resistant to changes in shape and frequently opaque, and liquids and gases continuous media of varying density and compressibility capable of filling arbitrarily shaped vessels and imparting pressure, evenly, wherever they go. This picture is about to be supplanted in the next chapter by another profound shift: the idea that all things are made of discrete entities called atoms.

NOTES

1 On the bright side, there was Monty Python…
2 The original term, introduced by Leibniz in the 17th century, was *vis viva*, i.e. living force. Leibniz is perhaps a little unfortunate not to have a unit of measurement named after him, but rather like his famous contemporary and rival "Fig" Newton, his name is immortalised via the tasty Choco Leibniz biscuit.
3 Check the laws of the game! Baseball fans should instead use 0.145 kg.
4 Throughout the book, unless stated otherwise, we will use the International System (SI) units for physical quantities. The rule is that you can translate an abstract equation like (2.1) into a numerical prediction provided all quantities involved are specified in SI units (in this case, kg, m, s, and J). SI units are particularly helpful in discussing energy, since as a result of its interconvertible nature "energy" crops up in many many different equations.
5 Named in honour of the physicist and thermodynamics pioneer James Prescott Joule (1818–1889).
6 No-one said fast bowling was easy…
7 The remarkable absence of the body's mass m in this relation is the basis for Galileo's legendary experiment dropping two unequal objects from the Leaning Tower of Pisa and observing them hit the ground at the same time.
8 SI units for electric current are Amperes (A) and for magnetic field Tesla (T).
9 Yes, you guessed right, the SI unit is Volt (V).
10 Note for simplicity, we assume direct current (DC) throughout—many important real-world applications use alternating (AC) circuits, introducing some annoying factors of two into the equations.
11 This instance of "guerrilla physics" is lifted from David MacKay's light-hearted but serious-minded book *Sustainable Energy—Without the Hot Air* (UIT Cambridge, 2009). Essential reading for our time.
12 The "(large) Calorie" traditionally used in diet books is, confusingly, 1000 times larger than the definition of calorie used in physics texts. The conversion factor is 1 Cal = 1000 cal = 4184 J.
13 The speed of light c in SI units is roughly 3×10^8 ms^{-1}.
14 The notation Δ means "the change in", i.e. $\Delta U = U_{after} - U_{before}$, etc.
15 Joule came from a brewing family, and it is thought his experience in this art helped him with the very precise temperature measurements required by the experiment.
16 The celebrated Ohm's Law $V = IR$ implies that heating power due to electric current flow is given by I^2R. This is another way of using electric current to engineer $\Delta U \neq 0$, conceptually completely different to using a motor to turn the paddle. Oh, and by the way, the SI unit of electrical resistance is Ohm (Ω).
17 The recollections of Joule's bride Amelia have unfortunately not survived for posterity.
18 Thermodynamics courses, particularly foundational ones, spend a lot of time covering gases, because they're simple physical systems very accurately described by mathematical models, as we shall see. This has been a source of dismay to generations of students who naively imagined they'd chosen physics to study quarks, lasers and the Big Bang…. Our advice: stay patient.
19 The technical term is "evacuate", i.e. "form a vacuum within", the word's original sense.
20 In this case this means no circulating currents in the gas, or temperature differences between different portions of the cylinder.
21 Readers familiar with integral calculus (cf. 0.14) will recognize $\Delta W = -\int_{V_A}^{V_B} pdV$. The negative sign is the same as that in Equation (2.6).
22 A two-stroke engine, in fact, because during the cycle the piston moves twice, one inwards, once out.
23 Some physics texts use special notation to distinguish the Δ operators in (2.5) acting on Q, W from that acting on U; (at least one of) the authors find this confusing.
24 The notation d meaning "infinitesimal change" is taken from differential calculus.
25 Due to biochemist Lawrence Joseph Henderson (1878–1942).

3 Ideal

Gases are great vehicles for developing thermodynamic ideas: here, we review what is known about them experimentally, and show that their behaviour in terms of p, V and temperature (now written as T) is described by a simple equation of state. The same result is obtained using a model in which the gas is viewed as a collection of point-like molecules continually colliding with the walls of the container and with each other; it proves possible to express the properties of all ideal gases with a universal equation of state. The range of molecular speeds requires a statistical approach in order to fully describe the gas. After briefly considering how closely these idealized properties describe real gases, we show that an ideal gas's internal energy only depends on temperature. Finally, we introduce the notions of heat capacity, adiabatic and isothermal changes, and an engine based on the so-called Carnot cycle, for which we are able to calculate the work performed during one cycle.

3.1 EMPIRICAL PROPERTIES OF AN IDEAL GAS

So far, our treatment of thermodynamics has used the behaviour of gases to illustrate the ideas. In this chapter, we will flesh out this picture by describing gases in a bit more detail, from two perspectives. First, we use experimentally observed gas properties to write an *Equation of State* linking the pressure p and volume V of a gas to its temperature. The rather abstract considerations of, say, the work done in a p–V cycle presented in the previous chapter will then be able to be worked out in detail. Next, we will derive the same results starting from a microscopic model of the gas in terms of molecules in which for the first time we will use statistical language to discuss what's going on—relating the macroscopic to the microscopic is something we will learn to become familiar with.

Boyle's Law[1] states that the pressure of a gas kept in a container at constant temperature is inversely proportional to its volume, that is $p \propto V^{-1}$, so that as V is reduced, e.g. by pushing on a piston, p increases and more force must be applied to keep going. The gas in the cylinder therefore "pushes back", just like a spring, and indeed this principle is put to work in shock absorbers for vehicles travelling over rough ground, particularly when low weight is desired, e.g. on a mountain bike. The equation encapsulating this behaviour is

$$pV = \text{constant}. \tag{3.1}$$

For gas at constant temperature the line in the p–V plane, known as an *isotherm*, is therefore a rectangular hyperbola, i.e. of the form $p \propto 1/V$, which has the limiting behaviour that $p \to \infty$ as $V \to 0$ and $p \to 0$ as $V \to \infty$. In fact, the curves CB and AD in Figure 3.1, showing the same cycle discussed in Chapter 2 are each of precisely this form, so they are both isotherms. The relations $p_C V_C = p_B V_B; p_A V_A = p_D V_D$ immediately follow.

DOI: 10.1201/9781003121053-5

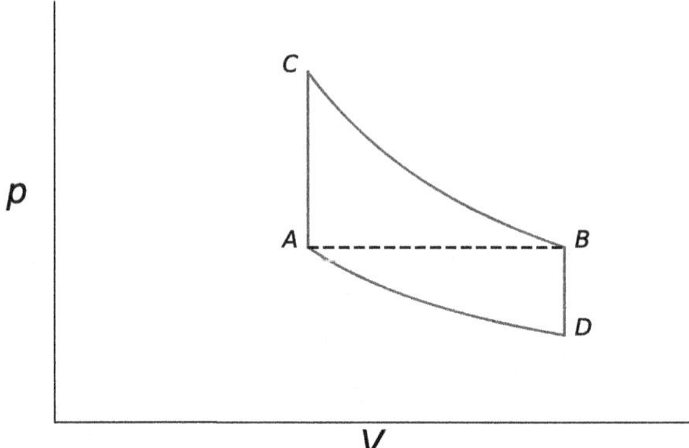

FIGURE 3.1 The same thermodynamic cycle discussed in Chapter 2, but now with the origin at $(V = 0, p = 0)$. The dotted line shows an isobar.

Charles's Law[2] states that for a gas held at constant pressure, the volume V is directly proportional to T, a linear function of temperature θ defined by

$$T = \theta - \theta_0, \tag{3.2}$$

where θ is the Celsius temperature and the constant $\theta_0 = -273.15^\circ$ C has the same value for all gases. Mathematically, $V \propto T$ or

$$\frac{V}{T} = \text{constant}. \tag{3.3}$$

Equation (3.3) implies that gases expand as their temperature rises, so a fixed mass of gas held at constant pressure occupies a larger volume at higher temperatures and hence will be less dense and more buoyant. We recognize this fact of experience in the statement, "warm air rises, cold air sinks", and it is the key element in hot-air ballooning. On the $p-V$ plot a constant pressure process corresponds to a horizontal line known as an *isobar.* In the cycle shown in Figure 3.1, there are no segments corresponding to isobars; note, however, that we have drawn the plot so it is possible to draw an isobar connecting points A and B. Since V increases as we move to the right, we deduce $T_B > T_A$ and therefore that the temperature all the way along the CB isotherm is greater than along AD.

At this stage, relation (3.2) between T and θ can be regarded as simply a means to relate temperature scales given by different thermometers; indeed, we will see later that the "perfect gas thermometer" gives particularly precise and uniform readings. Temperatures specified using T are given units called Kelvins,[3] and are written, e.g., 300 K (read out aloud as "three hundred Kelvin", without any mention of "degrees") corresponding to 26.85°C, i.e. the outside temperature on a warm summer's day. A moment's thought, however, raises awkward questions about what happens as $T \to 0$; to make mathematical sense (3.3) predicts V must become infinitesimally small in the same limit. Also, by analogy with the Celsius scale, what happens when T is a negative number? Eq. (3.3) then predicts V has to also be negative, which makes no physical sense. For now, we can evade these problems by remembering that all gases liquefy or solidify long before such low temperatures are reached,[4] becoming far less compressible as a consequence. It is therefore unreasonable to expect Charles's Law to apply as $T \to 0$, but in a sense, this is just brushing things under the carpet.

Boyle's Law and Charles's Law can be combined in a more general and powerful formula known as the *equation of state*:

$$\frac{pV}{T} = \text{constant.} \tag{3.4}$$

Boyle's Law (3.1) is a special case with the constant on the right-hand side set equal to the constant in (3.4)$\times T$. Similarly, Charles's Law (3.3) has right-hand side equal to the constant in (3.4)$\div p$. The interesting question is what determines the value of this constant, and whether and how it takes different values for different samples of gas.

3.2 MOLECULES AND THE KINETIC THEORY

This is the natural place to introduce the microscopic description of a gas, often known as *the kinetic theory of gases*. A gas is modelled as a collection of N molecules distributed throughout the container, each having a tiny mass and tiny spatial extent. For now, we'll assume that all the molecules are the same kind, but extending the model to describe mixtures of different gases (such as air) is not too hard. The model takes into account the molecule's mass but neglects the size by considering the molecules to be point-like—a mathematical idealization. A typical molecular size is 10^{-9} m or less, so that 10 million of them laid end-to-end would barely extend across your thumbnail. We'll see presently to what extent the point-like assumption is justified. The molecules are in rapid random motion, flying in straight lines in every direction until they collide, either with the walls of the container or with each other. Between collisions, the molecules have negligible interaction with the walls or each other, so they fly freely. When the molecules rebound from the wall, a tiny outward force is imparted, and the average force over many collisions gives rise to pressure.

To make the picture quantitative, let's first discuss the mass. The molecule's mass m is the sum of the masses of its component atoms and is conventionally measured in atomic mass units (u):[5] 1 u = 1.660539×10^{-27} kg. This value is approximately the same as the mass of a proton or neutron. Atomic masses are therefore close to being integer multiples of 1 u, and for simplicity, we will assume this is true in the following, which is usually a good approximation.[6] For instance, the molecular masses of some well-known constituents of air are given in Table 3.1.

Note that the noble gases (argon, neon) consist of molecules formed from just a single atom and are *monatomic*. In contrast, other elemental gases (nitrogen and oxygen) are bound states of two identical atoms and are, hence, *diatomic* and some compounds (carbon dioxide and water) are even *polyatomic*. For now, we'll assume that gas comprises just one kind of molecule.

Now consider a molecule flying with speed v approaching the wall of the container. Immediately before the collision, the molecule has kinetic energy $\frac{1}{2}mv^2$ and momentum mv. Let's assume that the collision with the wall is perfectly elastic so that none of the particle's kinetic energy is lost and that the collision is head-on, i.e., the molecule's trajectory is at right angles to the wall. Because the wall is so much more massive than the molecule, following the collision, the molecule rebounds

TABLE 3.1
Molecular Masses of Gases Found in Earth's Atmosphere

Gas	Chemical Formula	Molecular Mass (u)
Nitrogen	N_2	28
Oxygen	O_2	32
Argon	Ar	40
Carbon dioxide	CO_2	44
Water	H_2O	18
Neon	Ne	20

with the same speed but in the opposite direction,[7] i.e., its kinetic energy is still $\frac{1}{2}mv^2$, but its momentum is now $-mv$.

Since momentum is conserved in every collision a total momentum of $mv-(-mv)=2mv$ is transferred to the wall. Still, because the wall is so massive, it can be considered fixed, so this doesn't result in any discernible motion. However, a momentum change in Newtonian mechanics implies a force is involved. Suppose the wall has an area *A*, and another parallel wall of the same area situated a distance *d* in the direction the molecule started from. After the rebound, the molecule will travel to the far wall, reaching it in time d/v; after that, it rebounds again and returns along the same path to reach the first wall a further time d/v later. In other words, the time interval between collisions at the first wall is $2d/v$. Now, the force transmitted to the wall through repeated collisions of this one molecule is equal, by Newton's Second Law, to the rate of change of momentum, i.e., $\frac{2mv}{2d/v}=\frac{mv^2}{d}$. Suppose that *N* molecules all exhibit this back-and-forth motion, each repeatedly striking at different times and points on the wall. The resulting pressure on the wall equals the imparted force divided by its area:

$$p=\frac{Nmv^2}{dA}. \tag{3.5}$$

This is a very naive model. It assumes the molecules are all flying along parallel trajectories with equal speeds. It completely ignores the possibility of collisions in which the molecule approaches the wall at an angle and is reflected from the wall at the same angle but in the opposite sense to its direction of incidence. We can mitigate the first criticism to some extent by assuming three equal populations are flying along mutually perpendicular directions, each containing $N/3$ molecules. Noting that for a cubic container, with three sets of parallel but mutually orthogonal faces, $dA=V$, we arrive at

$$pV=\frac{1}{3}Nmv^2. \tag{3.6}$$

To address the second issue, we can relax the assumption that the molecules all have the same speed by replacing v^2 with $\langle v^2\rangle$, the *mean square speed* of the molecule population, defined by

$$\langle v^2\rangle=\frac{1}{N}\sum_i v_i^2, \tag{3.7}$$

where by v_i we mean the speed of the *i*th molecule, and the summation symbol Σ_i is introduced in (MA.19) of Appendix MA. Finally, the third issue can be dealt with by doing a much more careful job of averaging over angles of incidence and reflection, assuming that the molecular trajectories are distributed evenly over every possible direction.[8] We'll spare you the details,[9] and jump straight to the result, which also holds for arbitrarily shaped containers:

$$pV=\frac{1}{3}Nm\langle v^2\rangle. \tag{3.8}$$

Equation (3.8) is the central result of kinetic theory. We can immediately compare it with the empirical equation of state (3.4):

$$pV=\text{constant}\times T=\frac{1}{3}Nm\langle v^2\rangle. \tag{3.9}$$

Notice that for a given sample of gas, both N and m are constant so that when we vary temperature T, it causes a proportionate change in $\langle v^2 \rangle$; in other words, the temperature of gas depends on the distribution of molecular speeds, and *vice versa*. Before elucidating further, we need to answer the question posed in Equation (3.4) by finding a convenient way of quantifying the amount of gas present. The solution comes from chemistry[10]: we define an "amount of substance" called a *mole*[11] as having a mass which is the molecular weight expressed in grams. For example, a mole of nitrogen has a mass of 0.028 kg, a mole of carbon dioxide has a mass of 0.044 kg, etc. The wording is a little clumsy, but the key idea is that a mole of anything contains the same number of molecules; this number is given by the ratio

$$\frac{\text{mass of mole}}{\text{mass of molecule}} = \frac{0.001 \times \text{molecular mass}}{1.660539 \times 10^{-27} \times \text{molecular mass}} \simeq 6.022 \times 10^{23} = N_A \tag{3.10}$$

and is none other than Avogadro's number, fleetingly introduced in Chapter 1. The number of molecules N in n moles is therefore nN_A, enabling us to recast Equation (3.9):

$$pV = \frac{1}{3} nM \langle v^2 \rangle = nRT. \tag{3.11}$$

Here, M is the mass of one mole, which depends on which gas we're dealing with and R is the *universal gas constant*, which as the name implies applies equally and impartially to all gases. Its value is

$$R = 8.314 \text{ JK}^{-1}\text{mol}^{-1}. \tag{3.12}$$

Note that R must be experimentally determined since it, in effect, translates between units of energy (and work, and heat, etc.) and units of temperature, which we haven't been able to do until now. So this is a major step forward. Its universal nature follows from the simplicity of the kinetic theory; all gases containing the same number of molecules have the same equation of state when written in terms of T (or up to a scaling factor directly proportional to the molecular mass, if written in terms of $\langle v^2 \rangle$).

Let's pause to take stock and use the kinetic theory to work out some typical numbers. If a gas is a collection of molecules whizzing around, then how many are there? To estimate the number of air molecules in 1 m^3 at atmospheric pressure $p \approx 10^5$ Pa and room temperature $T \approx 300\,\text{K}$, first evaluate the number of moles using (3.11)

$$n = \frac{pV}{RT} = \frac{1 \times 10^5}{8.31 \times 300} \approx 40. \tag{3.13}$$

The number of molecules is nN_A, so the number density of molecules is $2.4 \times 10^{25}\ \text{m}^{-3}$. If all were arranged regularly in a cubic lattice, the resulting inter-molecular spacing would be $(nN_A/1m^3)^{-\frac{1}{3}} \approx 3.5 \times 10^{-9}\text{m}$. We can also use (3.11) to estimate the characteristic speed of the molecules, using Table 3.1 to estimate the mean molecular mass for air of 30 u:

$$v_{\text{typical}} \sim \sqrt{\langle v^2 \rangle} = \sqrt{\frac{3RT}{M}} = \sqrt{\frac{3 \times 8.31 \times 300}{0.030}} \approx 500 \text{ ms}^{-1}, \tag{3.14}$$

It is roughly the same as a bullet from a gun. Is this a reasonable number? We can corroborate it by evaluating a comparable quantity, the speed of sound, using the classical equation of state $pV = nRT$ without any reference to molecules. The fact that sound has a finite propagation speed is well known to anyone who's attended a cricket match, where there is a discernible delay between seeing a batter execute a shot and hearing the corresponding crack of leather on willow; this is because the sound has to travel anything up to 50 m to reach the spectator at the boundary. As Boyle's Law demonstrates, air is an elastic medium,[12] and as such can sustain travelling waves consisting of alternating regions where the pressure is either slightly higher or slightly lower than the bulk average. It is a standard exercise to show that the velocity u of this travelling wave is related to how the pressure of one mole of gas responds to small changes in volume:

$$u^2 = -\frac{V}{\rho}\frac{\Delta p}{\Delta V}. \tag{3.15}$$

Here, ρ is the gas density, equal to M/V in the molecular picture. If T is kept constant during the change, we can use (3.11) to deduce $p\Delta V + V\Delta p = 0$ and find $u^2 = RT/M$. The assumption of constant temperature is unrealistic; sound waves move too quickly to restore thermal equilibrium over the scale defined by a wavelength. So, it is better to model the local compressions and rarefactions as *adiabatic* (see below). This leads to a corrected formula:

$$u = \sqrt{\gamma\frac{RT}{M}}, \tag{3.16}$$

where γ, which will be defined more completely in the following, has the numerical value $\gamma \approx 1.4$ for air. Equation (3.16) yields $u \approx 340\ \text{ms}^{-1}$ for air which is of the same order as the typical molecular speed already calculated. It would be hard to understand sound transmission if it had turned out that $u \gg v_{typical}$. Now, we can use the wave formula $u = f\lambda$ to work out a typical sound wavelength λ given the frequency f. With concert tuning, the A above middle C on a piano keyboard has $f = 440$ Hz; this note corresponds to sound waves with wavelength $\lambda \approx 80$ cm.

Before going further, we need to clarify an important point. Unless all the molecules have the same speed, the mean square speed $\langle v^2 \rangle$ defined in (3.7) is *not* the same as the mean speed[13] squared $\langle v \rangle^2$. We can demonstrate this with a trivial example: a population of three molecules with speeds 3 ms^{-1}, 4 ms^{-1}, and 5 ms^{-1}:

$$\langle v \rangle^2 = \left(\frac{3+4+5}{3}\right)^2 = 16\text{m}^2\text{s}^{-2}; \langle v \rangle^2 = \frac{3^2+4^2+5^2}{3} = 16\frac{2}{3}\text{m}^2\text{s}^{-2}. \tag{3.17}$$

In general, $\langle v^2 \rangle > \langle v \rangle^2$. We have to learn to make such fine distinctions when dealing with distributions of quantities which vary across a population. One of the challenges facing any microscopic description of thermodynamic phenomena is the necessarily *statistical* nature of the physical description. Due to the practical impossibility of keeping track of N_A individual particles, we are forced to focus on the characteristics of the population as a whole. Figure 3.2 plots the distribution[14] of molecular speeds in nitrogen gas in thermal equilibrium[15] for three representative temperatures. The peak of each curve is located at the most probable speed, but note there are significant numbers of molecules with speeds both lower and higher than this. For each temperature, the dotted line marks off the mean speed $\langle v \rangle$ on the horizontal axis, and the dashed line is the RMS (root-mean-squared) speed $\sqrt{\langle v^2 \rangle}$, which always lies to the right, i.e., at a higher value as advertised. Additionally, note the magnitude of the speeds involved; e.g., for nitrogen at 300 K, the RMS speed is 517 ms^{-1}, rising to 2110 ms^{-1} at $T = 5000$ K, although there is a significant fraction with speeds more than double this. Indeed, there is no theoretical upper limit to molecular speed in an ideal gas.[16]

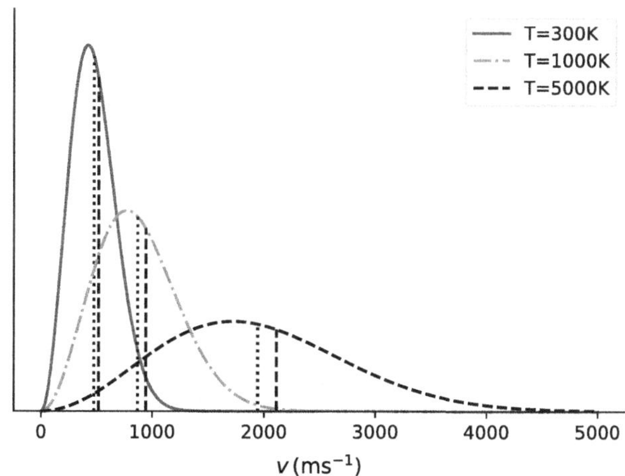

FIGURE 3.2 Distribution of molecular speeds for nitrogen gas at three different temperatures.

3.3 REAL GASES

The model we have developed, encapsulated in the equation of state (3.11), is known as the *ideal gas*. Due to its simplicity, the ideal gas is a fantastic playground for developing our understanding of thermodynamics. Theoretical physics is fun! At some point, though, we should stand back and ask how well the model describes real gases.

In deriving the ideal gas equation of state, we neglected the physical size of molecules and the effects of any interaction between them. In general, this interaction is attractive but falls away steeply $\propto r^{-7}$ where r is the inter-molecular separation.[17] Both effects are taken into account in a corrected form of the equation of state proposed in 1873 by J.D. van der Waals.[18] For one mole of gas:

$$\left(p+\frac{a}{V^2}\right)\left(V-b\right)=RT. \tag{3.18}$$

Because of inter-molecular attraction a particle colliding with the container wall has a slightly smaller speed than it would otherwise, so the force transfer and hence the pressure p are smaller than in an ideal gas. The term proportional to a corrects for this effect—the factor of V^{-2} ensures that this correction quickly becomes less important as the gas density decreases and the inter-molecular separation accordingly grows. Similarly, because each molecule occupies a non-zero volume, the other molecules have a slightly smaller volume in which to move around, and the "excluded volume" b corrects for that. Experimentally determined values for the constants a and b for our familiar atmospheric gases are shown in Table 3.2. Using (3.11), a mole of gas at room temperature ~300 K and atmospheric pressure $\sim 10^5\,\text{Pa}$ occupies a volume ~0.025 m^3. Using the data of Table 3.2 we calculate the typical van der Waals correction to p to be less than 1% and to V about 0.1%. Under everyday conditions the constituents of our atmosphere are pretty good approximations to ideal gases.

Equation (3.18) also shows that ideal gas behaviour will be recovered at constant temperature in the limit $V \to \infty$, or equivalently $p \to 0$. Precision thermometry can be accomplished with a device known as a *perfect gas thermometer*, shown schematically in Figure 3.3: the volume V of gas held in a vessel with diathermal walls is kept fixed by adjusting the height of a mercury-filled U-tube, and pressure p determined by the excess height of the mercury column[19] in the other branch of the tube. Temperature is then given by $T = pV/nR$, where n is the number of moles of gas contained in

TABLE 3.2
Van der Waals Constants for Gases Found in Earth's Atmosphere

Gas	a (m^6 Pa mol^{-2})	b (m^3 mol^{-1} × 10^{-5})
Nitrogen	0.1370	3.87
Oxygen	0.1382	3.19
Argon	0.1355	3.20
Carbon dioxide	0.3640	4.27
Water	0.5536	3.05
Neon	0.0214	1.71

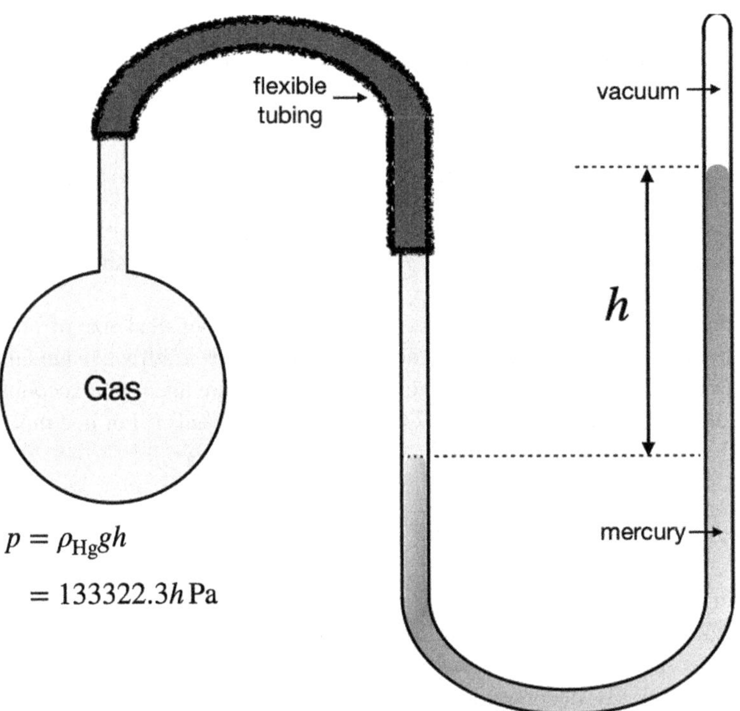

FIGURE 3.3 Constant volume gas thermometer.

the vessel. As n is reduced so that the pressure in the vessel falls, readings made by the thermometer become universal (i.e., independent of the nature of the gas) as the ideal gas limit is recovered. If we had to choose a thermometer to define a universal temperature scale, surely this would be the one? Remember though, that no ideal gases exist as $T \to 0$.

One advantage of treating real gases rather than ideal ones is that we can have a sensible discussion about inter-molecular collisions. It's very hard to do this using hypothetical point-like objects, because there's no notion of how close two molecules have to get to take part in a collision. In the van der Waals gas a molecule is assumed to occupy a spatial volume b/N_A, so that it presents a cross-sectional area to other molecules $\sigma \sim (b/N_A)^{\frac{2}{3}}$. The *mean free path* λ_{mfp} is the length of the straight-line path taken by a molecule between one collision and the next. We can estimate its value using guerrilla physics; the mean free path will be the length of a long thin cylinder of cross-section σ which contains on average just one other molecule, in other words having the volume of a mole of gas divided by N_A. If we take $b = 3.5\times10^{-5}\,\mathrm{m}^3$, we deduce

$$\lambda_{\mathrm{mfp}} \sim \frac{RT}{pb^{\frac{2}{3}}N_A^{\frac{1}{3}}} = 2.8\times10^{-7}\,m. \tag{3.19}$$

So $\lambda_{\mathrm{mfp}} \gg (b/N_A)^{\frac{1}{3}}$, the molecular size, but $\lambda_{\mathrm{mfp}} \ll \lambda$, the wavelength of sound deduced following (3.16). The first inequality supports a basic assumption of kinetic theory that collisions are relatively infrequent; the second justifies in retrospect our modelling of sound propagation as an adiabatic phenomenon, since there is little prospect of molecular collisions restoring thermal equilibrium over scales of an acoustic wavelength.

3.4 INTERNAL ENERGY OF A GAS

One of the major payoffs of kinetic theory is that it permits us to calculate the internal energy of the ideal gas from first principles. In the simplest case where the molecules are monatomic, the energy of the gas is entirely due to the kinetic energy of the molecules, so

$$U = \frac{1}{2}m\sum_i v_i^2 = \frac{1}{2}Nm\langle v^2\rangle, \tag{3.20}$$

where in the second step we have again replaced the sum over the molecule population with the mean square speed. Using (3.11) we get

$$U_{\mathrm{monatomic}} = \frac{3}{2}pV = \frac{3}{2}nRT. \tag{3.21}$$

This result is particularly simple—the internal energy of a monatomic ideal gas depends *only* on its temperature. For molecules containing more than one atom the situation is more complicated because a significant fraction of U is now held in the form of rotational kinetic energy due to the molecules' ability to spin in three dimensions. Teasing out the full story requires quantum mechanics, so for now we'll simply give the result[20]:

$$U_{\mathrm{diatomic}} = \frac{5}{2}nRT; U_{\mathrm{polyatomic}} = 3nRT. \tag{3.22}$$

Now, for any body it is useful to know how much heat is needed in order to raise its temperature a given amount. The *heat capacity*, measured in units J K^{-1}, is the amount of heat needed to raise its temperature by 1 Kelvin. Different substances have differing heat capacities, and we are sensitive to this variation among the many things we encounter. Water feels cooler than air even when both are at the same temperature, because water has much the higher heat capacity per unit volume. The heat capacity of one litre of water is about 4 J K^{-1}; that of dry air is about 1.2 J K^{-1}. When you put your hand into a bowl of cold water heat will continue to ebb out of you for much longer than it would if you replaced the bowl with a similar-sized vessel containing air. You'll continue to feel cold until equilibrium between your hand and its surroundings is reached: the higher the heat capacity of the surroundings, the more heat needs to be transferred to achieve this.

Because knowledge of a gas's volume alone, as a consequence of (3.11), doesn't in itself tell us how many molecules are present, it makes more sense to discuss the *molar heat capacity*, i.e. the amount of heat needed to raise one mole of gas through 1 Kelvin. Moreover, it matters whether the heating is performed while keeping the volume of gas fixed, which requires the gas to be kept in a sealed vessel, or the pressure fixed, more often what's needed in practical situations such as those encountered in meteorology. We therefore distinguish between the molar heat capacity at constant volume C_V and the molar heat capacity at constant pressure C_p.

Let's start by calculating C_V for one mole of an ideal monatomic gas. If heating is performed at constant volume, then $\Delta W = -p\Delta V = 0$, and the First Law tells us $\Delta U = \Delta Q$. So,[21]

$$C_V = \frac{dQ}{dT} = \frac{dU}{dT} = \frac{3}{2}R, \qquad (3.23)$$

where in the final step we used (3.21). Fortunately, the universal gas constant R defined in (3.12) has the correct units for this to make sense! The molar heat capacity of an ideal gas is therefore a constant independent of temperature, but this need not be the case in general.

Now let's think about C_p. If we raise T through 1 K, then the change in internal energy ΔU is the same irrespective of whether we hold V or p constant, because for an ideal gas U only depends on T. What's different this time, however, is that work is being done, as the gas moves along the isobar AB shown in Figure 3.1:

$$\Delta W = -p\Delta V = -R\Delta T, \qquad (3.24)$$

where in the first step we used the First Law and in the second the equation of state (3.11) with p held fixed. Hence[22]

$$C_p = \left.\frac{\partial Q}{\partial T}\right|_p = \left.\frac{\partial(U-W)}{\partial T}\right|_p = \frac{3}{2}R + R = \frac{5}{2}R. \qquad (3.25)$$

Because all ideal gases, including diatomic and polyatomic ones, obey the same equation of state (3.11) the result $C_p = C_V + R$ holds in general.

3.5 ADIABATIC VERSUS ISOTHERMAL CHANGE AND THE CARNOT CYCLE

The next piece of thermodynamic theory to develop is the idea of changes of state under specified constraints. For a gas, we have already seen a convenient arena in which to discuss changes is the p–V plane. Our ideal gas model has now supplied the theoretical ammunition to work out the thermodynamic consequence of moving around within this plane. One simple kind of trajectory to specify are the isotherms along which T remains constant: we've already seen that CB and AD in Figure 3.1 are of this form, obeying the relation $pV = \text{constant} = RT$, which we encountered when discussing Boyle's Law. Another important category of trajectory is *adiabatic*, along which $\Delta Q = 0$ (or $dQ = 0$ for quasistatic change). To work out the condition for adiabatic change requires a bit more calculus technoflash that we might like, but here goes. From the First Law with $dQ = 0$ and one mole of gas we deduce

$$0 = dU + pdV = C_V dT + pdV, \tag{3.26}$$

where in the second step we used the definition $C_V = \frac{dU}{dT}$ from (3.23). Next rewrite the equation of state (3.11) in differential form:

$$d(pV) = d(nRT)$$

$$\Rightarrow pdV + Vdp = RdT \Rightarrow dT = \frac{p}{R}dV + \frac{V}{R}dp, \tag{3.27}$$

where we have set $n = 1$, and substitute the resulting dT into (3.26):

$$0 = \left(C_V \frac{p}{R} + p\right)dV + C_V \frac{V}{R}dp = C_p \frac{p}{R}dV + C_V \frac{V}{R}dp. \tag{3.28}$$

To get this far we used the relation between C_p and C_V for an ideal gas. Finally use the results (3.23, 3.25) and rearrange to find

$$0 = \gamma\frac{dV}{V} + \frac{dp}{p} \tag{3.29}$$

where the ratio of specific heats $\gamma \equiv \frac{C_p}{C_V}$. Equation (3.29) is a differential equation of standard form, with general solution

$$\gamma \ln V = \text{constant} - \ln p, \tag{3.30}$$

where "ln" denotes the natural logarithm introduced in the Appendix MA, and we have used relation (MA.15b). Exponentiating both sides, we find the general condition specifying an adiabatic change[23,24]

$$pV^\gamma = \text{constant}. \tag{3.31}$$

In this context γ is called the *adiabatic index*. For ideal gases following from (3.21,22) there are three cases:

$$\gamma = \begin{cases} \frac{5}{3} & \text{monatomic;} \\ \frac{7}{5} & \text{diatomic;} \\ \frac{4}{3} & \text{polyatomic.} \end{cases} \tag{3.32}$$

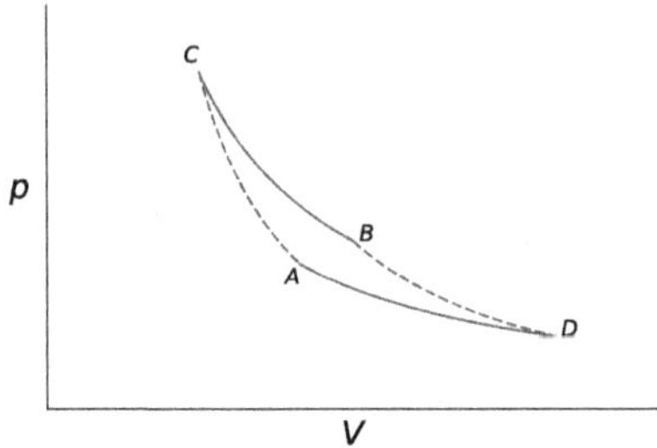

FIGURE 3.4 The Carnot cycle for a monatomic ideal gas with $T_B = 1.4\ T_A$.

We close by describing another thermodynamic cycle, depicted in Figure 3.4. Just as in Figure 3.1, the gas is taken around a closed path *ACBD* in the *p–V* plane, but now we stipulate that while the segments *CB* and *DA* are isotherms (shown as full lines), as in Figure 3.1, the segments *AC* and *BD* correspond to adiabatic changes (shown as dashed lines) obeying (3.31); they too are curved paths. We have drawn the plot to scale with the origin at $(V, p) = (0,0)$, assuming that the isothermal expansion along *CB* is such that $V_B = 2V_C$, that $T_B = 1.4T_A$, and that the adiabatic index $\gamma = \frac{5}{3}$.

Our goal is to calculate the work done for one circuit of the cycle, using the arguments sketched in Chapter 2, supplemented with the results for the ideal gas derived in this chapter. First consider the adiabatic portions *AC*, *BD*. Because $\Delta Q = 0$ along each path, the First Law tells us that

$$\Delta W_{AC} = \Delta U_{AC};\ \Delta W_{BD} = \Delta U_{BD}. \tag{3.33}$$

However, recall that for an ideal gas $U = U(T)$, and that by definition *T* is constant along the isotherm *CB* so that $T_C = T_B$ and also along the isotherm *DA* so that $T_D = T_A$. We conclude that $\Delta U_{AC} = -\Delta U_{BD}$ and therefore

$$\Delta W_{AC} + \Delta W_{BD} = 0; \tag{3.34}$$

that is, the work done ($-\int pdV$) by the gas moving along the two dashed paths cancels out.

It remains to calculate the work done along the two isotherms. For *CB*,

$$\Delta W_{CB} = -\int_{V_C}^{V_B} pdV = -RT_B \int_{V_C}^{V_B} \frac{dV}{V} = -RT_B \ln \frac{V_B}{V_C}, \tag{3.35}$$

where in the second step we used the ideal gas equation of state at constant temperature T_B. Similarly, along *DA* we find

$$\Delta W_{DA} = +RT_A \ln \frac{V_D}{V_A}. \tag{3.36}$$

Next note the following relations holding along the adiabatic trajectory *AC* (cf. equation (3.31)):

$$p_C V_C^\gamma = p_A V_A^\gamma \Rightarrow RT_C V_C^{\gamma-1} = RT_A V_A^{\gamma-1} \Rightarrow \frac{V_C}{V_A} = \left(\frac{T_A}{T_C}\right)^{\frac{1}{\gamma-1}}, \tag{3.37}$$

Once again, we have used the ideal gas equation of state in an intermediate step. An analogous relation can be derived for *BD*. Since the isothermal segments yield $T_A/T_C \equiv T_D/T_B$, we conclude

$$\frac{V_C}{V_A} = \frac{V_B}{V_D} \Rightarrow \frac{V_D}{V_A} = \frac{V_B}{V_C}. \tag{3.38}$$

Finally, recall that the work done *by* the gas around the cycle is the negative of the work done *on* the gas, to conclude

$$\text{work done in one cycle} = -\Delta W_{CB} - \Delta W_{DA} = R\left(T_B - T_A\right)\ln\frac{V_B}{V_C}. \tag{3.39}$$

Although the intermediate steps were a little convoluted, the result (3.39) has a simple form. Since we have already established $T_B > T_A$, and $V_B > V_C$ from the way the cycle was set up, we conclude the work done is positive, which is a good start in designing an engine! This kind of closed cycle, obtained by connecting a pair of isotherms at different temperatures with two adiabatic trajectories, will turn out to be of fundamental importance in the next stage of our development. It's known as a *Carnot cycle.*[25]

We've come a long way in this chapter; having developed a mathematically tractable yet realistic model of a gas in terms of entities called molecules, we're now able to do genuine thermodynamic calculations to extract, say, the work done by an engine in terms of the expansion of its piston and its operating temperatures during different stages of the cycle. Armed with this knowledge, we might dream of designing ever more efficient engines, perhaps exploiting alternative working substances, to maximize the yield of our scarce fuel resources. We've also made first contact with a description of thermodynamic behaviour emerging from a microscopic picture of a gas in terms of constituent molecules and learned that in order to work this way, it's necessary to develop statistical techniques to describe the properties of a population consisting of huge numbers of individual constituents. However, we still have much to learn; there are further Laws of Thermodynamics to take on board. The next chapter introduces Number Two.

NOTES

1 Originally stated by Robert Boyle (1627–1691), although his assistant Robert Hooke (1635–1703) helped with the experiments. Hooke's subsequent achievements included the law of elasticity governing the behaviour of amongst other things a coiled spring.

2 Found by Jacques Charles (1746–1823).

3 Named after William Thomson (1824–1907), professor of Natural Philosophy at the University of Glasgow and ennobled in 1892 as the 1st Baron Kelvin. The River Kelvin flows close to the university site; the Kelvin is thus the only SI unit to be named after a river.

4 The most difficult gas to liquefy is helium, the second-lightest element, first obtained in liquid form in 1908, requiring a temperature of 4.15 K at atmospheric pressure. As the name implies, helium was first discovered in the Sun's atmosphere via its spectral line during a solar eclipse in 1868.

5 Also known as the Dalton (Da) after the atomic theory pioneer John Dalton (1766–1844). A chemist, physicist, and meteorologist, he is also well-known for studying colour blindness.

6 We'll also ignore the complication that the gas molecules may contain different *isotopes* so that for a gas such as chlorine, there are significant fractions of three molecules with different masses: $^{35}Cl_2$, $^{37}Cl_2$, or $^{35}Cl^{37}Cl$. In kinetic theory these are simply treated as different species.

7 a good visualization of this kind of interaction occurs when a snooker ball rebounds off a cushion.

8 We've also ignored inter-molecular collisions which is a good idea for now—but note that these play an important role in randomizing the direction of flight and ensuring a uniform distribution of direction.

9 Really, it's just maths, and not that interesting in itself. Had the numerical factor of $\frac{1}{3}$ turned out to be different between Equations (3.6) and (3.8), we still wouldn't have much to complain about; very often, plausible physical reasoning can get most of the way towards an acceptable answer, and we should learn not to be discouraged by the need for over-the-horizon maths.

10 another area of science in which at a microscopic level the key concepts resolve into discrete events involving integer numbers of atoms or molecules.
11 The mole (mol) is a dimensionless SI unit. Think about it.
12 Think of jelly.
13 We're treating speed as the magnitude of the velocity vector, which is always positive.
14 The area under each curve is the same: formally, the fraction of molecules found with speeds between v and $v+\Delta v$ is the ratio of the area under the curve falling between these limits to the total area.
15 This theoretical curve is known as the Maxwell–Boltzmann distribution after the physicists James Clerk Maxwell (1831–1879) and Ludwig Boltzmann (1844–1906). Maxwell's formulation of equations describing electromagnetic fields and their interactions with charges and currents is to this day considered a towering achievement of physics, but his pioneering work in statistical thermodynamics alone would also ensure him fame among contemporary physicists. Boltzmann will feature in future chapters.
16 Of course, the speed of light is an actual bound: as particle speeds increase the Maxwell–Boltzmann distribution is eventually supplanted by the relativistic Planck black body distribution of particle energies, where all particles have the same speed c.
17 The attraction is electromagnetic in origin. Think of a molecule as a sloshy cloud of electrons surrounding a positively charged nucleus. The force between two electrically neutral molecules arises from the interaction between the clouds, each inducing an electric dipole moment on the other, a naturally weak "tidal" effect.
18 An interesting prediction of the van der Waals equation is that for $p < \frac{a}{27b^2}$, $T < \frac{8a}{27Rb}$ it's possible to find two solutions with different densities, corresponding to vapour and liquid. This separation into physically distinct phases is another realistic property of fluids.
19 The conversion factor is $10^5 \text{ Pa} = 750.062 \text{ mmHg}$.
20 Even this is an over-simplification—we have neglected the possibility of energy being held in the vibrational modes of polyatomic molecules. Again, the resolution lies in quantum theory.
21 For calculus *aficionados*: usually relations such as (3.23) are written using partial derivatives (i.e., $\partial U/\partial T\,|_V$) with a lot of attention paid to which physical quantities are kept fixed under the change (V in this case). Because for the ideal gas $U = U(T)$ is a function of just one variable we can get by without this (for now).
22 Told you! The subscript at the base of the vertical line indicates which quantities are held fixed.
23 While we have derived Equation (3.29) for an ideal gas, the result $\left.\frac{\partial p}{\partial V}\right|_{adiabatic} = \gamma \left.\frac{\partial p}{\partial V}\right|_{isothermal}$ can be proved in general, following just from the First Law. See Chapter 6 of the book by Pippard.
24 For adiabatic sound propagation the correct step following (3.15) is $V^{\gamma}\Delta p + \gamma p V^{\gamma-1}\Delta V = 0$ using similar calculus-style reasoning. See if you can finish off the proof of (3.16).
25 Named after Sadi Carnot (1796–1832), a mechanical engineer serving in the French army.

4 Two

We initially compare the efficiency of two different engines via their p–V cycles and ask whether it is possible to know which cycle generally yields the most efficient engine. The Second Law of Thermodynamics asserts the impossibility of complete conversion of heat into work. Application to the Carnot cycle permits a definition of absolute temperature T based on the cycle's efficiency, independent of the working substance. Reversing the cycle describes a heat pump rather than an engine. The result of integrating Q/T (where Q is reversible heat transfer) along a path is independent of the route taken but simply depends on the start and endpoints, enabling a new state function called entropy S to be defined. When plotted in T–S coordinates, the Carnot cycle is rectangular, which we use to prove it has the maximum efficiency of all heat engines. We describe an irreversible process called Joule expansion in which entropy increases without heat transfer. Finally, it is proved that the entropy of an isolated system can never decrease.

4.1 CULTURE WAR

In 1959, the academic and civil servant C.P. Snow[1] delivered the annual Rede lecture before an audience of his Cambridge peers. In a wide-ranging talk, subsequently published,[2] Snow addressed many issues, principally relating to the quality of scientific and technical education in the UK and other countries, but also drawing attention to what he perceived as a damaging breakdown in communication and mutual understanding between on the one hand scientists and engineers with, on the other, the prevailing literary and intellectual circles of the time—Snow coined the memorable phrase *The Two Cultures* to describe what he saw as opposing camps. As a published author of both scientific research papers and novels perhaps he saw himself as particularly qualified to comment on this. While Snow's thoughts on education merit attention, particularly the over-specialism at an early age still prevalent in the UK, his feelings concerning the underappreciation of scientific "culture" are harder to sympathize with in the present day, when the distinction between science communicator, stand-up comedian and rock star is becoming increasingly blurred. The lecture is principally remembered today for the following quote:

> A good many times I have been present at gatherings of people who, by the standards of the traditional culture, are thought highly educated and who have with considerable gusto been expressing their incredulity at the illiteracy of scientists. Once or twice I have been provoked and have asked the company how many of them could describe the Second Law of Thermodynamics. The response was cold: it was also negative. Yet I was asking something which is about the scientific equivalent of: *"Have you read a work of Shakespeare's?"*

Snow's concerns are deeply felt, and his choice of thermodynamics as the science arena in which to go into battle an interesting one, perhaps influenced by his background in physical chemistry—but

DOI: 10.1201/9781003121053-6

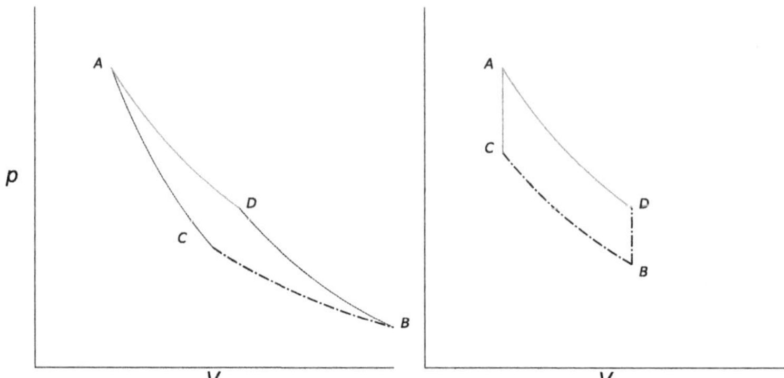

FIGURE 4.1 p–V diagrams for (left) the Carnot cycle, and (right) another cycle delivering the same work. Full lines show when heat is entering the system, and dash-dotted lines when heat is leaving. The dotted lines are adiabatic.

why not use the shinier examples already discussed, such as electromagnetism or relativity?[3] In this chapter we'll try to explain what underlay Snow's choice, and why someone educated in science might consider it such an urgent issue, by introducing the Second Law.

To prepare the ground, we need to return to our discussion of engines. In previous chapters, we've seen how to represent their operation in terms of a working cycle portrayed on a p–V diagram, which enables us to keep track of where work is done. Figure 4.1 shows two different possible cycles.

On the left is a Carnot cycle, characterized by two isotherms, AD and BC, at different temperatures, and two adiabatic trajectories, DB and CA; our analysis of the previous chapter showed that assuming the working substance is an ideal gas the work delivered by one complete cycle is given by (3.39)

$$\oint dW_{\text{Carnot}} = R\left(T_A - T_C\right)\ln\frac{V_D}{V_A}. \tag{4.1}$$

Equation (4.1) introduces a new symbol $\oint$ which is a mathematical shorthand for the sum of contributions around a closed loop.

On the right is an alternative cycle resembling the one first introduced in Figure 2.2. In this case, the two isotherms are connected by vertical trajectories along which volume V remains constant. The two plots have been drawn with the same axis scales, so that in fact the isotherm AD is exactly the same on both; moreover, the temperatures T_A and T_C characterizing the isotherms are also chosen the same. It is not hard to repeat the analysis of the previous chapter to find

$$\oint dW = \Delta W_{AD} + \Delta W_{BC} = RT_A\ln\frac{V_D}{V_A} - RT_C\ln\frac{V_B}{V_C} = R\left(T_A - T_C\right)\ln\frac{V_D}{V_A} = \oint dW_{\text{Carnot}}. \tag{4.2}$$

Geometrically, the area enclosed within the curve is the same for both cycles. If anything the calculation for the second cycle is a little easier, since manifestly no work is done along either CA or DB. Why then have we devoted so much attention to the Carnot cycle?

4.2 ENGINE EFFICIENCY

So far, we've not considered the heat required to run the engine, which remember comes from burning expensive fuel. In the Carnot cycle, no heat enters or leaves the system on either adiabatic leg, i.e. $\Delta Q_{CA} = \Delta Q_{DB} = 0$ During isothermal expansion AD heat enters the system, and during the compression stroke *BC* heat is leaving. For an ideal gas along any isotherm $\Delta Q = -\Delta W$.[4] We immediately conclude the net heat entering the system in one cycle equals the work done by the engine, as it must. However, this is made up of two terms of opposite sign:

$$\Delta Q_{AD} = RT_A \ln \frac{V_D}{V_A}; \ \Delta Q_{BC} = -RT_C \ln \frac{V_D}{V_A}. \tag{4.3}$$

While sharing similar mathematical form, for a real engine, these two contributions have very different physical meaning. The heat entering the system $|\Delta Q_{AD}|$ is obtained from consuming a certain amount of fuel, corresponding to a certain expense (we're only too familiar with this metric—think of "miles-per-gallon"). The heat leaving the system $|\Delta Q_{BC}|$ on the other hand is *not* providing work—it is, literally, "exhaust", resulting in heat (and for an internal combustion engine other waste products) being injected into the surrounding environment at temperature T_C. For someone making a living from operating the engine, therefore, a much more useful way to write the equation is

$$\oint dW_{\text{Carnot}} = |\Delta Q_{AD}| - |\Delta Q_{BC}|, \tag{4.4}$$

where, now since all the quantities involved are positive, we immediately conclude $\Delta Q_{AD} > \oint dW_{\text{Carnot}}$. The Carnot engine does not succeed in converting all the heat needed for its operation into useful work. We can quantify this in terms of the engine's *efficiency*:

$$\eta = \frac{\text{work done per cycle}}{\text{heat supplied per cycle}} = \frac{\oint dW_{\text{Carnot}}}{|\Delta Q_{AD}|} = \frac{R(T_A - T_C)\ln(V_D/V_A)}{RT_A \ln(V_D/V_A)} = \frac{T_A - T_C}{T_A} < 1. \tag{4.5}$$

Remarkably, for the Carnot cycle η only depends on the temperatures T_A, T_C between which the cycle operates. We only approach ideal efficiency $\eta = 1$ as the ratio T_C/T_A approaches zero.

For comparison, let's check the efficiency of the engine in the right-hand panel of Figure 4.1. The calculation of the heat entering and leaving the system along the isothermal legs *AD* and *BC* goes through as before (although the isothermal compression along *BC* begins and ends at different volumes, the ratio V_B/V_C still equals V_D/V_A), and we obtain the same answer for ΔQ_{AD}.

Now, however, we need to take into account heat entering the system along the constant volume leg *CA* (in this context, we don't need that leaving along *DB*, but it's easy to convince yourself it has the same magnitude). If we assume the working substance is a monatomic ideal gas, then

$$\Delta Q_{CA} = C_V \int_{T_C}^{T_A} dT = \frac{3}{2} R (T_A - T_C), \tag{4.6}$$

where we used the result $C_V = \frac{3}{2} R$ (3.23) for the constant volume specific heat. The answer is

$$\eta = \frac{\oint dW}{|\Delta Q_{AD} + \Delta Q_{CA}|} = \frac{T_A - T_C}{T_A + \dfrac{3}{2} \dfrac{(\mathrm{T_A} - \mathrm{T_C})}{\ln(\mathrm{V_D}/\mathrm{V_A})}} < \eta_{\text{Carnot}}. \tag{4.7}$$

Perhaps there *is* something special about the Carnot cycle. It shares with any real engine the idea that the temperatures T_A at the stage of the cycle where heat is supplied and T_C where heat is dumped

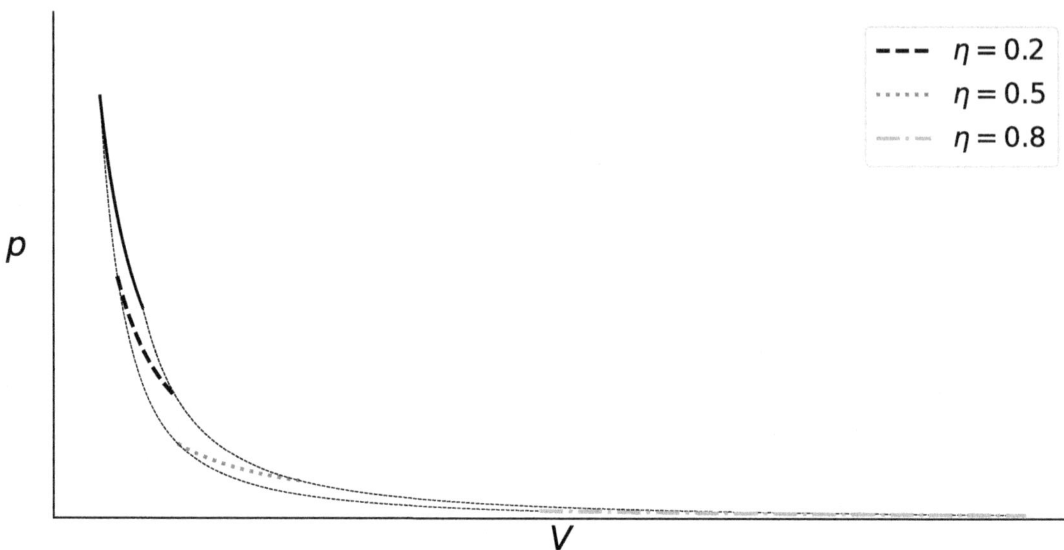

FIGURE 4.2 Carnot cycle for a monatomic ideal gas for three different exhaust temperatures.

must be different; equation (4.5) further implies that to boost fuel efficiency, these temperatures should be as widely separated as is practicable.

Figure 4.2 shows three Carnot cycles of ever-increasing efficiency as the exhaust temperature T_C approaches zero, with the temperature T_A of the isothermal expansion stroke (shown as a thick black line) held fixed. In each case, the volume ratio at the beginning and end of the isothermal stages of the stroke is 2:1. Leave aside, for now, the (enormous) technical difficulties involved in working with low exhaust temperatures. The shape of the adiabatic trajectories is such that as $\eta \to 1$ the volume of the piston during the compression stroke is much, much larger than that during expansion,[5] and in fact becomes infinite as $T_C / T_A \to 0$. There is something singular about the ideal limit $\eta \to 1$.

4.3 THE SECOND LAW

The stage is set for the entrance of the *Second Law of Thermodynamics*. Like the other thermodynamic laws it is empirical, being founded on experiment and experience, and can't be proved from any more basic starting point. There are also several different, not obviously equivalent ways of stating it. It's natural to begin with the version due to Kelvin[6]:

It is impossible to devise an engine which, working in a cycle, shall produce no effect other than the extraction of heat from a reservoir and the performance of an equal amount of mechanical work.

If, following all the buildup, your initial reaction to this is that you'd much prefer a spirited discussion of *The Merry Wives of Windsor,* well, we have some sympathy. The sentence is dry, long-winded, and apparently devoted solely to the problem of heat engine efficiency. Contrast it with the power and economy of the First Law of Chapter 2, proclaiming the existence of a universally conserved quantity in a mere nine words. Why should Snow, or anyone else, expect this to be an attractive topic for the dinner table?

Let's begin to unpick it a little. Kelvin's statement stresses the importance of working in a cycle; that is, the working substance (the ideal gas in our case) shall be periodically returned to the identical state from which it started, ready for the next cycle. Suppose we start the cycle at the point *C*, specified by a particular pair of (p, V) values; after one cycle, the gas is returned to C, and there are no measurements we can perform which give us any information about the points A, D, B visited on the way around and, by extension, the heat which has flowed in and out of the system, and the work which has resulted since the last time we passed through C. This is another way of saying heat and work are not state functions. Kelvin also introduces the idea of a *reservoir*; this is a body in periodic diathermal contact with the engine, with such a large heat capacity that effectively no change in its temperature results whenever a finite amount of heat is exchanged with the engine in either direction. Whenever a system undergoes isothermal change, we conceptualize this in terms of its being in contact with such a reservoir at the same temperature, often also known as a *heatbath*. Finally, let's introduce the idea of *reversible* change, that is, one whose direction can be exactly reversed by an infinitesimal change in external conditions, such as the force applied through the piston or the temperature of the heatbath. This is really just a formalization of the quasistatic processes discussed in Chapter 2, but it will prove to be a crucial component of the arguments to come. All the analyses of engine cycles we have discussed so far have assumed reversibility, implying that the piston and cylinder are very finely machined so that friction can be neglected and that the working substance remains homogeneous throughout the cycle.[7] Maintaining reversibility of the vertical CA and BD legs of the right-hand cycle of Figure 4.1 is tricky—strictly it requires bringing a fixed volume of gas into thermal contact with a succession of heatbaths, each with temperature infinitesimally greater or smaller than the preceding one. We'll brush this difficulty under the carpet for now.

What follows are some arguments of great power and subtlety with far-reaching consequences. We'll mainly use diagrams to guide us, and it should be understood that while using an ideal gas as the working substance enables calculations to be worked through for helpful illustration, we will try as far as possible to reach our conclusions without relying on its specific properties (the most notable being that the internal energy U depends solely on temperature T). For the ideal gas Carnot cycle, we showed above that the ratio of heat flow ΔQ_1 into the system at temperature T_1 to the heat flow ΔQ_2 out of the system at T_2 satisfies

$$-\frac{\Delta Q_1}{\Delta Q_2} = \frac{T_1}{T_2}. \tag{4.8}$$

The minus sign is there to remind us that in thermodynamics heat flow out of a system has a negative sign; notice though, that (4.8) does not need us to specify which is which, or indeed which temperature out of T_1, T_2 is the hotter and which the cooler. We will now show this result holds for *all* Carnot cycles, regardless of the nature of the working substance.

Figure 4.3 shows two Carnot cycles, each operating between temperatures T_1 and T_2 with $T_1 > T_2$. Physically, the cycle is realized by a system of two cylinders, each containing a different working

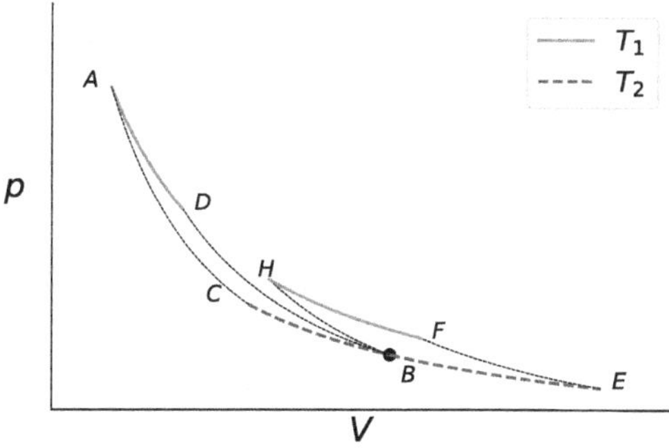

FIGURE 4.3 Two Carnot cycles employing different working substances.

substance, each coupled via a system of pistons and cranks to a device capable of delivering work, and each placed alternately in thermal contact with heatbaths at temperatures T_1 and T_2. The cycles have been drawn with a common point *B* at the lower temperature T_2, but this is not essential to the argument. It does highlight, however, that the working substances are distinct, so the that the adiabatic trajectories *BD* and *BH* emerging from the common point *B* are distinct. In fact, the left-hand cycle *ADBC* corresponds to a polyatomic ideal gas with adiabatic index $\gamma = \frac{4}{3}$ and the right-hand *HFEB* to a monatomic ideal gas with $\gamma = \frac{5}{3}$ (see Equation (3.32)), although the precise values are not essential to the argument. The legs along *AD* and *HF* at temperature T_1 have been adjusted so that the same amount of heat ΔQ_1 is absorbed by each sub-system when isothermal expansion takes place.[8] Now consider a combined reversible cycle *BCADBEFHB*, in which the left-hand cycle is traversed clockwise and the right-hand anti-clockwise. By construction the total heat absorbed at T_1 is $\Delta Q_{AD} + \Delta Q_{FH} = \Delta Q_1 - \Delta Q_1 = 0$. At T_2 the heat absorbed $= \Delta Q_{BC} + \Delta Q_{BE}$. Since the system is in the same state at the end of the cycle as at the start, $\Delta U = 0$ and it follows from the First Law that the work done *by* the system during the complete cycle also equals $\Delta Q_{BC} + \Delta Q_{BE}$. Kelvin's statement of the Second Law, however, asserts the impossibility of this work being a positive quantity, since if it were the combined cycle would have performed work purely as a result of extracting heat from the reservoir at T_2. We deduce an inequality:

$$\Delta Q_{BC} + \Delta Q_{BE} \leq 0. \tag{4.9}$$

Now consider the same cycle performed in reverse, i.e. *BHFEBDACB*. Again no net heat is transferred at T_1, but now the work done in one cycle $\oint dW = \Delta Q_{BE} + \Delta Q_{CB} = -\Delta Q_{BC} - \Delta Q_{BE}$ and we deduce from the Second Law that

$$-\Delta Q_{BC} - \Delta Q_{BE} \leq 0. \tag{4.10}$$

Clearly the only way to satisfy both (4.9) and (4.10) is to insist

$$\Delta Q_{BC} = -\Delta Q_{BE} \equiv \Delta Q_2, \tag{4.11}$$

implying $\Delta Q_1 / \Delta Q_2$ takes a universal value for *all* reversible Carnot cycles, irrespective of the choice of working substance. Equation (4.8) immediately follows if we specify that one of the cycles employs an ideal gas as working substance.

4.4 ABSOLUTE TEMPERATURE

The universal nature of (4.8) has a profound consequence; it permits us to define the *absolute* or *thermodynamic* temperature scale, with temperature defined to be proportional to the heat flux along the isothermal segment, so that in principle we can determine the ratio of the absolute temperatures of two reservoirs by running a Carnot cycle between them. It is denoted symbolically by *T* and measured in Kelvins.[9] Unlike the scales discussed so far, it doesn't rely on the properties of any particular kind of thermometer or working substance. In order to calibrate it we only need one fixed point; it is convenient to define *T* so that the ice-point is 273.15 K.[10] A far more important feature of the thermodynamic temperature is that zero means zero—there is no lower temperature! We conceive of *absolute zero* as the limit $T \to 0$ enabling the Carnot cycle to become a perfect engine with $\eta = 1$. We've already remarked that even if we had access to an ideal gas sample which never liquefied, in order to reach this limit we'd need to work with a cylinder and piston capable of infinite extension! Just as in our discussion of the perfect gas thermometer it is both sensible and useful to consider the approach to the limit even it can never be reached in practice.[11] Thermodynamics will have more to say about absolute zero, as we'll see.

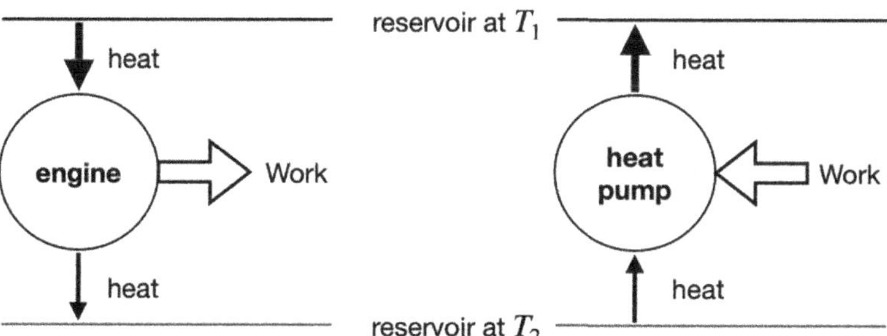

FIGURE 4.4 Two devices operating between reservoirs at temperatures T_1 and $T_2 < T_1$ which satisfy the Second Law.

If you followed the argument concerning the combined cycle of Figure 4.3 closely you'll have noted that on one half of the cycle the system is undergoing a backwards-running Carnot cycle, in which the system extracts heat from a reservoir at the lower temperature T_2, has external work performed on it, and then dumps heat at the higher temperature T_1. There's nothing here contradicting thermodynamics; the very concept of reversibility demands this process is possible. Such backwards-running devices are in fact built and go under the name of heat pumps[12] (where the focus is on getting more heat into a warm place) or refrigerators (where the goal is to remove heat from a cold place). The key point for both is that in order to make them operate, it is necessary to supply external work, e.g. via an electric motor powering the compression stroke.

The distinction between heat engines and heat pumps is displayed schematically in Figure 4.4. Without the work supplied to a heat pump, it would be impossible for heat to transfer from a colder reservoir to a hotter one. It costs money to run a fridge. This is essentially the content of an equivalent statement of the Second Law, due to Clausius[13]:

It is impossible to devise an engine which, working in a cycle, shall produce no effect other than the transfer of heat from a colder to a hotter body

Although engines and cycles are still mentioned, this feels more universal, concisely expressing what everyone knows about heat transfer from immediate human experience along the lines discussed in the opening chapter. It's important to stress, though, that Clausius and Kelvin statements are completely equivalent, linked through the Carnot cycle arguments we've reviewed.

Equation (4.8) relating the heat transfers during the isothermal legs of a Carnot cycle to the temperatures of the two reservoirs can be rewritten in the following form:

$$\frac{\Delta Q_1}{T_1} + \frac{\Delta Q_2}{T_2} = \oint \frac{dQ}{T} = 0, \tag{4.12}$$

where the second expression again employs a mathematical shorthand, which stresses we're summing up the $\frac{\Delta Q}{T}$ contributions around a closed cycle, all the way keeping track of whether heat is flowing into ($\Delta Q > 0$) or out of ($\Delta Q < 0$) the system. It's possible to generalize this result to a system Γ moving through an arbitrary cycle, but in order to do this we again have to introduce an auxiliary system $\hat{\Gamma}$, diathermally coupled to Γ, which does perform a Carnot cycle. Consider the following sequence of operations—follow it closely to see that $\hat{\Gamma}$ is indeed following a Carnot cycle:

1. Γ is in its initial state, $\hat{\Gamma}$ is at a temperature T_0
2. $\hat{\Gamma}$ is brought adiabatically to temperature T

3. Γ moves to the next stage of its cycle (in the process absorbing heat dQ from $\hat{\Gamma}$ (we use dQ rather than ΔQ to stress that conceptually this step can be infinitesimal if needed), while $\hat{\Gamma}$ itself moves along an isothermal at temperature T
4. $\hat{\Gamma}$ is returned adiabatically to T_0 and then compressed or expanded isothermally until it recovers its initial state.

Because $\hat{\Gamma}$ executes a Carnot cycle, in stage 4 it recovers an amount of heat from a reservoir at temperature T_0 equal to $dQ(T_0/T)$, using (4.8). The whole sequence 1-4 is then repeated as many times as necessary for Γ to move through a complete cycle and recover its initial state.

During the course of the whole cycle, the total heat lost by the reservoir $\oint dQ_0 = T_0 \oint \frac{dQ}{T}$, where the integral is taken around the whole of Γ's cycle. This is a generalization of the relation (4.12) derived for the two Carnot cycle setup of Figure 4.3, but now since the steps 1–4 can be repeated many times, the integral represented by the $\oint$ symbol in general corresponds to the sum over many many small, perhaps even infinitesimal, terms of the form $dQ \times T_0/T$. While T_0 is kept constant, T may vary at each step of the cycle, but by completion, Γ has recovered its original state so that its internal energy U is the same as at the beginning. Accordingly, we know from the First Law that the total work performed *on* the combined $\Gamma + \hat{\Gamma}$ cycle is $W = -\oint dQ_0 = -T_0 \oint dQ/T$. Note that strictly T is the temperature of $\hat{\Gamma}$ at the point where heat exchange with Γ occurs; we nowhere specify the temperature of Γ or even make any assumptions about its being in equilibrium. Since T_0 is positive, we can now apply Kelvin's statement of the Second Law to Γ, namely that the work $-W$ done *by* the system cannot be positive, to deduce the *Clausius inequality:*

$$\oint \frac{dQ}{T} \leq 0. \tag{4.13}$$

If and only if Γ's cycle is reversible can we reuse our earlier argument involving running the whole process backwards, to deduce $-\oint \frac{dQ}{T} \leq 0$ and hence $\oint \frac{dQ}{T} = 0$, and conclude the equals sign in (4.13) holds for the special case of reversible cycles.[14]

While we used the compound system $\Gamma + \hat{\Gamma}$ to derive (4.13), it only actually depends on quantities dQ, T related to the cycle performed by Γ, and makes no reference to $\hat{\Gamma}$ or T_0. We deduce (4.13) is universally true for *all* thermodynamic cycles. Let's focus on the case where Γ's cycle is reversible and consider two different points *A* and *B* on it. In order to get from *A* to *B*, one could choose to go along one branch of the cycle that connects them, or one could choose to go along the other branch *in reverse* (see Figure 4.5).

Using the property of the integral around the loop and the saturated form of (4.13), we can write

$$0 = \oint \frac{dQ}{T} = \int_{A\,\text{path1}}^{B} \frac{dQ}{T} + \int_{B\,\text{path2}}^{A} \frac{dQ}{T} \Rightarrow \int_{A\,\text{path1}}^{B} \frac{dQ}{T} = \int_{A\,\text{path2}}^{B} \frac{dQ}{T}. \tag{4.14}$$

In other words, the result of integrating $\frac{Q}{T}$ between any two points in the space of state variables (e.g., p, V) specifying the system does not depend on the path chosen to perform the integral, but just on the endpoints. This implies the existence of another state function, denoted *S*, with the defining property that

$$S_B - S_A = \int_A^B \frac{dQ}{T}, \tag{4.15}$$

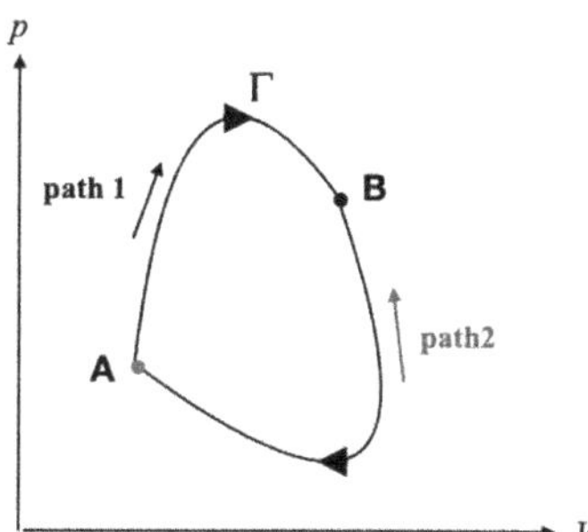

FIGURE 4.5 Two alternative integration paths between A and B based on a reversible cycle Γ. For a function of state, the result just depends on the endpoints A, B and is independent of the path.

the integral being taken over *any* reversible path connecting A to B. Clausius gave the name *entropy* to the new quantity S, from the Greek $\varepsilon\nu\tau\rho o\pi\eta$ meaning "transformation".[15]

4.5 ENTROPY AT LAST

Entropy is a new tool in the thermodynamic toolbox, and to start with it will be helpful to revisit our by now familiar friend the ideal gas and see how it behaves using entropic language. For infinitesimal changes (4.15) can be written

$$dS = \frac{dQ}{T}, \tag{4.16}$$

so that we can re-express the differential form of the First Law (2.10) as[16]

$$dU = TdS - pdV. \tag{4.17}$$

Because this is a differential equation, we can use it to calculate changes in entropy without necessarily knowing the absolute entropy of the system at the start.[17] Very often this is sufficient to solve interesting problems. Since $U = U(T)$, along an isotherm $dU = 0$ and we deduce (using the equation of state $pV = nRT$)

$$S_B - S_A = \Delta S_{AB} = \int_A^B dS = \int_A^B \frac{p}{T} dV = nR\int_A^B \frac{dV}{V} = nR\ln\frac{V_B}{V_A}. \tag{4.18}$$

We stress that although (4.18) is calculated along a reversible trajectory, because entropy is a state function the result obtained holds regardless of how the change from A to B is effected.

First, some basic housekeeping: entropy is proportional to n, the number of moles and hence the amount of gas present. The descriptor for quantities which depend on the amount of substance present is *extensive*, other examples being mass, volume, and internal energy. It's helpful to distinguish these from *intensive* quantities, such as pressure, temperature and density, which take the same value throughout a system in equilibrium regardless of whether we focus attention on just a small subsample, or consider the system as a whole. Secondly, entropy is measured in the same units JK^{-1} as the universal gas constant R. Equation (4.18) says that entropy change for isothermal expansion of an ideal gas is proportional to the logarithm of the ratio of the system volume after and before.

Next, consider a vertical trajectory in the p–V plane, such as CA in the right-hand plot in Figure 4.1. Along such a trajectory $dV = 0$ so $dU = TdS$ and we find

$$\Delta S_{AB} = \int_A^B dS = \int_A^B \frac{dU}{T} = \frac{3}{2}nR\int_A^B \frac{dT}{T} = \frac{3}{2}nR\ln\frac{T_B}{T_A}, \tag{4.19}$$

where we have specialized to a monatomic gas and used equation (3.21) $U = \frac{3}{2}RT$. This time (4.19) says that entropy change is proportional to the logarithm of the ratio of the system temperature after and before. The ratio of any two quantities of the same form is a dimensionless number, so there is no problem in taking the logarithms in (4.18,19).

We can use these results to redraw the two cycles from the start of the chapter in the *T–S* plane, shown in Figure 4.6. As before, we've used the same axis scales on each graph, so that the isothermal trajectory *AD* is identical in both plots. The Carnot cycle shown in the left plot takes a particularly simple form: isotherms with $dT = 0$ are horizontal lines, and adiabatic trajectories with $dQ = 0$ and hence (using (4.16)) $dS = 0$ are vertical lines. The area enclosed within the resulting rectangle is given by

$$\oint TdS = \oint dQ, \tag{4.20}$$

i.e., the net heat entering the system during the cycle. From the First Law (4.17) and the fact that *U* is a function of state this is equal to the work done by the system $\oint dW$ calculated in (3.39) at the end of Chapter 3, as you should be able to check using (4.18). Of course, Equation (4.20) holds for any reversible closed cycle in the *T–S* plane. We can also interpret the plot in terms of entropy flow: entropy enters the system from a high temperature reservoir along *AD* and leaves the system to be transferred to a reservoir at lower temperature along *BC*. These entropy flows are of equal magnitude $R\ln(V_D/V_A)$, and in opposite directions, so that the net entropy change around a closed cycle cancels to zero, as befits a function of state.

Now let's study the right-hand cycle. The vertical segments in the *p–V* plot of Figure 4.1 have become oblique, slightly curved lines determined by Equation (4.19). Equation (4.2) demonstrated that $\oint dW$ evaluated around this cycle is the same as for the Carnot cycle, provided it is performed reversibly. Confident thermodynamic reasoning enables us to assert that therefore $\oint dQ$ also yields the same value as the Carnot cycle.[18] However, there is an important distinction between the two cases. We've already seen that heat enters the system along both the isotherm *AD* and also the constant volume trajectory *CA*. Mathematically, heat entering is expressed as $dQ > 0 \Rightarrow dS > 0$ (since *T* is always positive). Heat therefore enters the system at all points in the *T–S* plane where the cycle

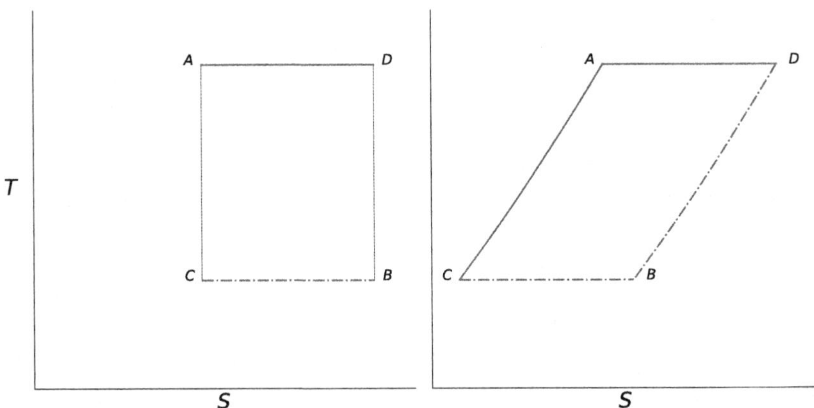

FIGURE 4.6 The same two cycles as Figure 4.1, drawn in the *T–S* plane.

is being traversed from *left to right as drawn.* It is precisely this heat, obtained by burning expensive fuel, that enters the efficiency equation (4.5). To minimize the heat input and therefore maximize efficiency, for a given amount of work done per cycle $-\oint dW = \oint dQ$, corresponding to the area enclosed by the cycle in the T–S plane, we need to choose a cycle which minimises the area lying between its uppermost portion and the $T = 0$ axis, shown for each cycle by the shaded regions in Figure 4.7. The non-Carnot cycle shown on the right requires an excess heat input, represented by the area of the roughly trapezoidal region beneath CA, to deliver the same work as the Carnot cycle shown on the left.

The rectangular shape of the Carnot cycle in these T–S coordinates provides the optimal solution, as we will now prove. In Figure 4.8 the rectangle and ellipse are both cycles working between temperatures T_1 , and T_2, delivering the same work—the rectangle is Carnot and the ellipse represents an arbitrary reversible cycle extending over a larger range along S; since the heat absorbed and the work delivered by either is the same they enclose the same area. The curvature of the ellipse signifies that a non-adiabatic path is taken from T_2 to T_1 and *vice versa*. We've deliberately offset the ellipse slightly to the left to stress that symmetry is not part of the argument. The heat entering the system during the cycle is the area $\int TdS$ enclosed between the upper portion of each cycle and the $T = 0$ axis. For the Carnot cycle this equals the area of the rectangular cycle + the shaded area 3 lying below $T = T_2$; for non-Carnot it is the area of the ellipse + the shaded areas 1 + 2 + 3. Since both cycles have equal area, we see that the excess heat input to the non-Carnot cycle is given by 1 + 2, necessarily greater than zero, and hence the cycle inevitably less efficient. This is a geometrical proof that the Carnot cycle provides the most efficient engine possible for a given pair of working temperatures T_1, T_2.

4.6 JOULE EXPANSION

Because entropy is a function of state, we can use it to discuss changes in thermodynamic systems without needing to worry about the intermediate stages of the transition; in particular, it frees us from having to focus on reversible changes. The classic example illustrating this is the so-called *Joule expansion* of a gas. A gas is kept in a container with adiabatic walls so that it is thermally isolated from the environment, ensuring $\Delta Q = 0$ throughout. The container is divided into two equal halves by a sliding partition, and initially the gas is contained on just one side of the partition, while the other half contains a vacuum. The experimenter then removes the partition, allowing the gas to swiftly expand and fill the whole volume. How should we analyse this? The methods we've developed so far appear at first sight to give contradictory answers. For instance, assuming the shutter mechanism is well-lubricated we can argue that essentially no work is done on the system

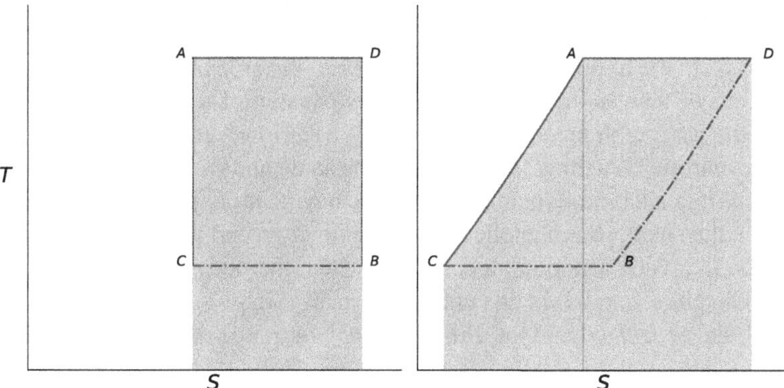

FIGURE 4.7 The same two cycles as Figure 4.6: the shaded regions denote the heat input to power the cycle.

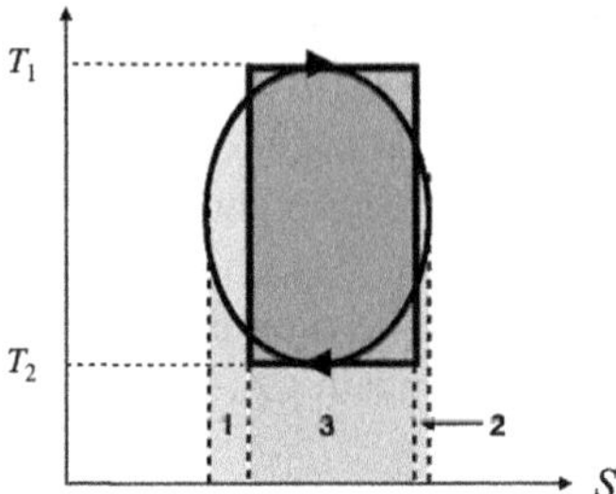

FIGURE 4.8 Graphical proof that the Carnot cycle is the basis of the most efficient heat engine.

when removing the partition, so $\Delta W \approx 0$, but clearly $p\Delta V > 0$. The resolution is to remember the change is not quasistatic. On a timescale of roughly the order of the dimensions of the container divided by the mean molecular speed the gas is reconfiguring itself to occupy the entire available volume; during this time the pressure is not uniform throughout the container and we are not entitled to equate ΔW with $-p\Delta V$. In other words, the gas has not been in an equilibrium state throughout the process, which accordingly is *not* reversible.

For an ideal gas, we can use $\Delta W = 0$, $\Delta Q = 0$ and the First Law to deduce $\Delta U = 0$ and therefore predict that the temperature T is found to be unchanged once equilibrium is restored. We can also use Boyle's Law pV = constant to predict that the pressure of the gas once it has re-equilibrated is half what it was originally since the volume doubles. For real gases, the situation is more interesting, because during the expansion the mean molecular separation increases by a factor $\sqrt[3]{2}$, and work needs to be done against inter-molecular forces. In most cases, real gases cool slightly under Joule expansion, but above a certain *inversion temperature* the energetics of the process is dominated by the short-distance repulsive portion of the inter-molecular potential probed in higher energy collisions, and the opposite is true.

Another thermodynamic variable that changes under Joule expansion is entropy. Because the system is thermally isolated from its surroundings $\Delta Q = 0$, but here we can't equate ΔS with $\Delta Q / T$ since temperature is not meaningful for a system out of equilibrium. Although derived by consideration of a reversible process, Equation (4.18) $S = nR\ln\,(V_B/V_A)$ is more generally valid, and we can use it to work out the increase in entropy $\Delta S = nR\ln 2$, which we expect to be accurate if the gas is sufficiently close to ideal. So from this example, we learn both that: version (4.17) of the First Law, which deals solely in terms of state functions is more generally applicable than the form (2.5) we encountered in Chapter 2; also that entropy can change even *without* heat being exchanged. In this case, as a consequence of an irreversible process, the total entropy of the system increases.

In this chapter, we've learned that heat tends to flow from hotter bodies towards cooler ones (the basic content of the Second Law) and that analysis of the efficiency of thermal cycles is streamlined via the introduction of a new state function called entropy. Phrasing arguments and calculations in terms of S frees us from the requirement to only discuss reversible processes, which are often physically unrealistic, as we saw in the case of thermal expansion of a gas at constant volume (recall the discussion in the paragraph preceding Equation (4.8)). However abstract the derivation, we are forced to conclude entropy "is a thing", measured in the same units as heat capacity and the universal gas constant R, and from a thermodynamic perspective having more justification to be considered an element of reality[19] than heat, which in effect has been downgraded from a *noun* naming an inherent property of a body, to a *verb* denoting an action by which the body changes its state. We've come a long way from Chapters 1 and 2. Is this enough to justify Snow's putting the Second Law on the same cultural pedestal as *Hamlet* and the 18th Sonnet? Not quite yet, but consider the following.

4.7 ENTROPY CAN ONLY INCREASE

Recall the Clausius inequality (4.13), applied this time to an irreversible cycle:

$$\oint \frac{dQ}{T} < 0. \tag{4.21}$$

Next, as we did previously, identify two points A and B on the cycle dividing it into two branches, one of which is deemed to be reversible. We can therefore write

$$\int_{A_{\text{irrev.}}}^{B} \frac{dQ}{T} + \int_{B_{\text{rev.}}}^{A} \frac{dQ}{T} < 0. \tag{4.22}$$

The integral along the reversible branch can be replaced by the entropy change $S_A - S_B = \Delta S_{BA} = -\Delta S_{AB}$, since $dQ = TdS$ on this portion. Relation (4.22) therefore becomes $\int_{A_{\text{irrev.}}}^{B} dQ/T < \Delta S_{AB}$; in general, for any change (including reversible ones, hence the inclusion of an equals sign in the following) we can write[20]

$$\Delta S \geq \int \frac{dQ}{T}. \tag{4.23}$$

Now let the system be thermally isolated from its surroundings, so that $dQ = 0$. The conclusion is

$$\Delta S \geq 0, \tag{4.24}$$

or in words, under any process of change or evolution (in this case from A to B), **the entropy of an isolated system can never diminish.** The argument is extremely general; we have made no stipulations about the size or complexity of the system. The only requirements are that the system can be thermally isolated from outside sources of heat, and there must exist at least one reversible path from A to B so that the entropy change can be defined. At this point, boldness is called for. Let's assume the system consists of the entire universe. The universe's entropy is getting bigger…

When we introduced the First Law we remarked that physicists sit up and take notice whenever a physical quantity, in this case energy, is proclaimed as being conserved. A quantity that can only increase over time, however, is simply jaw-dropping. Clausius, the person who introduced the entropy concept and first noted equation (4.24) summarized the first two Laws rather chillingly in 1865:

> The energy of the universe is constant.
> The entropy of the universe tends to a maximum.[21]

There's no comparable phenomenon in the rest of fundamental science. If we accept entropy as a physical quantity, then its ever-increasing presence in our world implies the world itself must be changing around us: things in the past are not the same as things in the future. How fast is this change occurring and how does it manifest itself? Why is it happening? How will it end? These and related questions will preoccupy us for the rest of the book.

NOTES

1 Charles Percy Snow (1905–1980) was also a novelist and physical chemist.
2 *The Two Cultures and the Scientific Revolution*, Cambridge University Press (1959).
3 The Special and General theories of Relativity entered public consciousness following the dramatic confirmation of Einstein's prediction of the bending of light rays around a massive object by Eddington's observations during a total solar eclipse in 1919. It is likely that Snow might have encountered a more positive response had he chosen this exemplar.
4 from the First Law combined with $U = U(T)$.
5 The relative piston volumes of the isothermal compression and expansion strokes are $(T_A/T_C)^{1/(\gamma-1)}$.
6 Kelvin's formulation of the Second Law dates from 1851.
7 This doesn't necessarily imply things have to move slowly, which would be a bit of a drawback for a real engine. So long as the speed of the piston is much less than the speed of sound, the gas contained in the piston should obey Boyle's Law reversibly.
8 For the ideal gas we can ensure this by choosing the ratios V_D/V_A and V_F/V_H to be equal, but for a general working substance the condition $\Delta Q_{AD} = \Delta Q_{HF}$ must be obtained by careful tuning.
9 We jumped the gun somewhat by introducing this unit and symbol in the previous chapter, but it didn't seem helpful to introduce yet another symbol for temperature. As we've seen, for all practical purposes, the temperature scale defined by the perfect gas thermometer is equivalent.
10 In 1968, the Kelvin was (re)defined to be exactly 1/273.16 of the thermodynamic temperature of the *triple point* of water, an isolated point in the (p,T) plane where ice, liquid water, and water vapour can coexist in equilibrium. This standard can be reproduced by any physicist with a suitably equipped laboratory independent of the local atmospheric pressure on the planet where they're situated.
11 The current laboratory record for sustained low temperature is 5×10^{-6} K, see G.R. Pickett, Cooling metals to the microkelvin regime, then and now. *Physica B: Condensed Matter* **280** (2000) 467.
12 Until recently, heat pumps were quite unfamiliar to most of us, but their use as heating devices powered by sustainably generated electricity promises to be a key component in our collective move towards de-carbonization.
13 Rudolf Clausius (1822–1888) was a physicist and mathematician—his statement of the Second Law dates from 1854.
14 A highbrow way of saying this is that the Clausius inequality is *saturated* for reversible cycles.
15 Perhaps the cognitive chasm separating the Two Cultures was not so wide in 19th-century Germany.
16 known to generations of jaded students as the "tedious" (TdS) equation. Geddit?
17 In other words, we can calculate the difference $S_B - S_A$ without necessarily knowing the value S_A. The absolute entropy of an ideal gas is given by the *Sackur-Tetrode equation*, whose derivation requires quantum mechanics and is postponed until Chapter 7.7.
18 A demonstration by direct calculation is left as an exercise should you wish to test your calculus skills.
19 Bear in mind there would have been a time when energy would also have been regarded as an abstract quantity whose principal purpose was as a bookkeeping device in thermodynamic calculations involving heat and work.
20 Strictly the temperature T appearing in (4.23) is the temperature of the reservoir supplying the system with heat, since because of irreversible processes the temperature of the system may be hard to specify.
21 It sounds even more impressive in the original German: *Die Energie der Welt ist constant. Die Entropie der Welt strebt einem Maximum zu.*

5 Omega

We analyse the gas in terms of a statistical description of its microstates, at first expressed in terms of the positions of its component molecules. Enumerating all possible microstates produces enormous numbers, but taking the logarithm produces a manageable quantity that appears to correspond to an extensive property of the gas. It's demonstrated that statistical fluctuations away from classical expectations typically fall as the inverse square root of the number of particles and become negligible in the thermodynamic limit. Boltzmann's definition of entropy as proportional to the logarithm of the number of microstates is introduced and applied to both the Joule expansion and constant volume thermal expansion of a gas. The formalism is developed to include systems in contact with a heatbath at a constant temperature, and the Boltzmann distribution, Gibbs entropy and Helmholtz Free Energy are thus defined. The Boltzmann distribution is used to derive the distribution of molecular speeds in a gas, first encountered in Chapter 3.

5.1 COUNTING MICROSTATES

The previous chapter saw the introduction of a new state function, entropy S, which proved to be a useful bookkeeping device for understanding and quantifying the efficiency of heat engines, and in addition, provides a means (at least for isolated systems) to distinguish reversible processes, for which S remains unchanged, from irreversible processes where S increases. Strictly, we only outlined how changes ΔS are given by equation (4.15) and have not at this point specified the calculation of absolute entropy. This shouldn't worry us unduly; after all, we also don't at this stage have an unambiguous method in general for calculating absolute values of internal energy U. Nonetheless, however neat the theoretical structure that's been set up, at this point it's hard to appreciate what entropy actually *is*, in any sense. It turns out we will need to develop a picture of entropy in a microscopic, atomistic setting.

The natural place to start is, of course, the ideal gas where we have a model which is at once simple, tractable and remarkably realistic. Consider the Joule expansion process discussed in the previous chapter; one mole of a gas is initially confined within a volume V corresponding to the left-hand side of the container in Figure 5.1, while the equal volume on the right-hand side is evacuated. Kinetic theory tells us the ideal gas consists of N_A molecules, randomly distributed within the volume and moving with a distribution of speeds (shown in Figure 3.2) evenly spread over all possible directions.[1] When the partition separating left and right is removed, the molecules comprising the gas rapidly redistribute themselves so as to occupy the entire volume $2V$. A useful way of describing this process is to focus on which half of the vessel any particular molecule is situated. Let's specify labels "0" if the molecule is in the left-hand side of the vessel, and "1" if it's on the

DOI: 10.1201/9781003121053-7

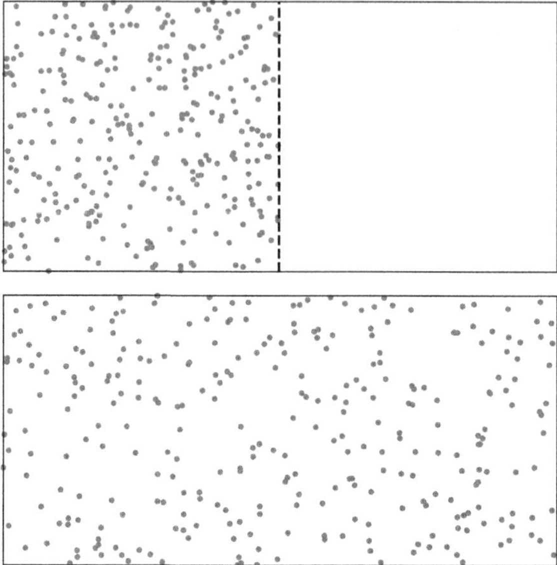

FIGURE 5.1 Typical distributions for a gas of 300 molecules before and after removal of partition.

right. While the partition is still in place, every molecule lies on the left, so we can represent this state of the gas[2] by a string of N_A zeros, of which just a short segment is shown here:

$$....000000000000000000000000000000000000000000000000..... \tag{5.1}$$

Now, remove the partition and allow the molecules to redistribute themselves. Imagine (this is a thought experiment…) we could freeze the motion of the molecules for a while so that we could visit each in turn[3] and record which half of the vessel they occupy. Since some molecules have by now made their way to the right, we expect this time to generate a more complex string:

$$....0100110100111010101011011010010001000101010001011 1..... \tag{5.2}$$

Of course, if we waited some more time and repeated this procedure, we'd generate a completely different string of 0s and 1s.

The strings can also be thought of as composed of binary digits, or *bits*, familiar from computing, in which the position of each digit represents a power of two.[4] In binary notation the numbers from 0, 1, 2,…, 7 read as follows: 000, 001, 010, 011, 100, 101, 110, and 111. There is a one-to-one correspondence between the bit strings representing the gas and numbers, in other words, every state of the gas in this language can be represented by a number between 0, corresponding to (5.1) and $2^{N_A}-1$, corresponding to

$$....11111111111111111111111111111111111111111111111111....., \tag{5.3}$$

i.e., all the molecules on the right-hand side. Let's introduce some terminology. Each bit string corresponds to a distinct *microstate* of the gas, and there are 2^{N_A} possible microstates in total. This number is called the *statistical weight* of the gas following removal of the partition, represented symbolically by an upper-case Greek letter Ω :[5]

$$\Omega = 2^{N_A}. \tag{5.4}$$

Remember that Avogadro's number $N_A \simeq 6.022\times10^{23}$, so that $\Omega \simeq 10^{2\times10^{23}}$. It's customary to refer to such very large numbers as *astronomic*, but really we're unaware of anything in astronomy whose quantification requires something as big as this. One step towards making sense of it is to instead consider its logarithm: using (MA.6c) we find

$$\ln\Omega = \ln 2^{N_A} = N_A \ln 2 \simeq 0.693 N_A. \tag{5.5}$$

Interestingly the logarithm of the statistical weight, while still an enormous number, is proportional to the number of molecules present. If we were to repeat the analysis for a vessel containing n moles of gas we'd find $\ln\Omega = nN_A\ln 2$; $\ln\Omega$, therefore, looks like an extensive property of the gas.

So far all we've done is label the microstates of the gas and counted them; there doesn't seem to be much physics involved. This is input from the assumptions underlying the ideal gas; molecules occupy negligible space, and interact with other molecules only when their separation is infinitesimal. If we focus on the molecules' position for now, this means that the likelihood of finding a molecule at a particular location doesn't depend at all on where its neighbours are; *all distributions of molecules are equally likely.* In particular, once the partition is removed and molecules have had time to explore the full space available, we're equally likely to find any particular molecule on the right as on the left of the vessel. We conclude that were we to "freeze" the system and analyse the state of the gas as described above, then all 2^{N_A} possible outcomes are equally likely. At first sight this is a little surprising: a few seconds following removal of the partition we surely don't expect to find all the molecules still on the left, or all moved over to the right, so what does it mean to assert that configurations (5.1) or (5.3) are just as likely as any others?

To resolve this let's start small by considering a gas of just two molecules. The possible states are described by bit strings 00, 01, 10, and 11. Of the four possibilities, one (00) corresponds to both molecules on the left, one (11) to both on the right, and two (01 and 10) have one molecule in either half. Extend this analysis to a four-molecule gas: of the 16 possibilities, only two (0000 and 1111) have all the molecules in just half of the vessel, 8 have a 3:1 split (e.g. 0111, 0100), and 6 have equal numbers in either half (e.g. 0101, 1001). There is a well-known number pattern called *Pascal's triangle* which describes how things go as the number of molecules in the gas is built up:

$$\begin{array}{ccccccccccccc} &&&&&&1&&&&&& \\ &&&&&1&&1&&&&& \\ &&&&1&&2&&1&&&& \\ &&&1&&3&&3&&1&&& \\ &&1&&4&&6&&4&&1&& \\ &1&&5&&10&&10&&5&&1& \\ 1&&6&&15&&20&&15&&6&&1 \end{array} \tag{5.6}$$

Each row corresponds to the number of molecules in the gas, starting from $n = 0$ at the top and increasing by one with each downward step. If you look closely you'll see that the pattern is generated by starting each row with 1 and then forming the next entry from the sum of the two numbers immediately above. Each entry corresponds to the number of configurations with the same number of molecules in the right-hand half; you should recognize the $n = 2$ and $n = 4$ cases just described in the third and fifth rows of the triangle. The mathematical formula which generates the entries in Pascal's triangle, and hence for the number of ways of finding r molecules on the right in an n-molecule gas, is

$${}_nC_r = \frac{n!}{(n-r)!r!}, \tag{5.7}$$

where the factorial symbol $n! \equiv n \times (n-1) \times (n-2) \times \cdots \times 2 \times 1$.[6] These are known as *combinatoric factors* and have the property (check the sums of the entries in each row of (5.6))

$$\sum_{r=0 \text{ to } n}^{n} {}_nC_{nr} = 2^n. \tag{5.8}$$

They can be derived as follows. Suppose we are given an initial string of 2^n 0s and asked to replace r of them with 1s, all possibilities being equally likely. There are n possible bit locations for the first 1, but then only $(n-1)$ for the second $(n-2)$ for the third, and so on down to $(n-r+1)$ for the rth. The number of ways to assemble this is then

$$n(n-1)(n-2)\ldots(n-r+1) = \frac{n!}{(n-r)!}. \tag{5.9}$$

However, since for a given string result it does not matter in which order we select the bits to become 1s, Equation (5.9) overcounts the number of possible combinations; focussing just on the 1s, we reason that there are r ways to identify the first, $(r-1)$ to identify the second, and so on down to the last remaining 1 which is uniquely identified. In order to correctly count combinations, therefore, we need to divide by a further factor of $r!$ to recover the result (5.7).

Figure 5.2 shows the result of using the Pascal triangle rule to count the relative preponderance of arrangements with r molecules to the right for four different choices of n, each time increasing the molecular population by a factor of 10. We've scaled the plots so that the area under each curve is the same, so the height of the curve can be interpreted as a relative likelihood for finding a particular fractional occupation r/n. For the least populous gas with $n = 10$, there are only 11 possible outcomes, clearly visible as discrete steps in the distribution. In this few-body regime, the distribution is still clearly a discrete-valued histogram; indeed, if you look closely, you'll also discern a step-like structure for $n = 100$. For larger n, the curve starts to resemble a continuous *probability distribution function*, which works such that the probability of finding an arrangement with a number of molecules to the right between r and $r + \delta$ is proportional to the area under the curve contained between two verticals at r/n, $(r+\delta)/n$. The most striking feature of Figure 5.2 is that

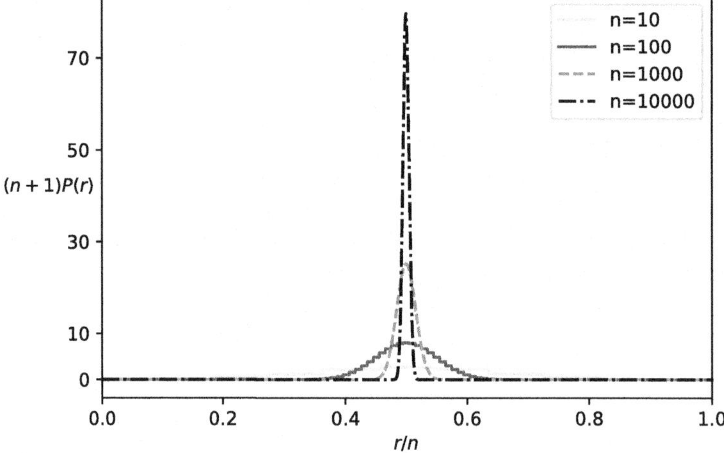

FIGURE 5.2 Binomial distribution (5.10) for molecular arrangements with $n = 10{,}100{,}1000{,}10000$.

as n increases, the curve becomes ever more sharply centred at $r = \frac{n}{2}$; here, "more sharply" means the central feature becomes narrower and more needle-like. As $n \to \infty$, therefore, although all molecule distributions are *a priori* equally likely, it is overwhelmingly likely that we'll observe the expected property that the gas is evenly spread over the two halves of the vessel and that departures from this expected behaviour become ever less likely. If the computer program which produced Figure 5.2 were capable of extending to the realistic case $n \sim O(N_A)$, the resulting curve would be a barely visible vertical line.[7]

It's also possible to obtain these results by pure calculation without resorting to graphs or computers. The basic starting point is the *binomial distribution* giving the probability $P(r)$ of finding a total of r instances of an event (in this case finding a molecule on the right) in a total population of n given the probability p of finding any single instance:

$$P(r) = {}_nC_r\, p^r (1-p)^{n-r}. \tag{5.10}$$

For an ideal gas, as already discussed, we're equally likely to find a molecule to the right or left, so $p = 1 - p = \frac{1}{2}$ and (5.10) reduces to

$$P(r) = \frac{1}{2^n}\, {}_nC_r. \tag{5.11}$$

Probability theory deals with "expected values" of different observables, which can be thought of as the average of all the outcomes we'd obtain were we to repeat an experiment many times. The simplest and most obvious observable to start with is r itself. Its expectation is given by

$$\langle r \rangle \equiv \sum_{r=0}^{n} rP(r) = \frac{1}{2^n} \sum_{r=1}^{n} r\ {}_nC_r, \tag{5.12}$$

where the first identity symbol defines what is meant by the expectation, and the second equality follows from (5.11) (the $r = 0$ term in the right-hand expression vanishes, so it is left out of the sum). In essence, to calculate an expectation, we multiply the value $f(r)$ of the observable in question found in a particular state of the gas by the probability $P(r)$ of occurrence of that state, then sum over all possible available states r. To evaluate (5.12), note that

$$\begin{aligned} r \times {}_nC_r &= \frac{n!}{(n-r)!(r-1)!} = \big(n-(r-1)\big)\frac{n!}{\big(n-(r-1)\big)!(r-1)!} \\ &= (n-r+1)\ {}_nC_{r-1}, \end{aligned} \tag{5.13}$$

so that

$$\langle r \rangle = \frac{1}{2^n}\sum_{r=1}^{n}(n-r+1)\ {}_nC_{r-1} = \frac{1}{2^n}\sum_{r=0}^{n-1}(n-r)\ {}_nC_r = \frac{1}{2^n}\sum_{r=0}^{n}(n-r)\ {}_nC_r,$$

$$\text{i.e.} \langle r \rangle = \langle n-r \rangle = \langle n \rangle - \langle r \rangle. \tag{5.14}$$

The second step relabels the terms in the sum $r \mapsto r-1$ and in the third step we reinstated the vanishing $r = n$ term in the sum. Now, n is constant for the gas sample so $\langle n \rangle = n$. We conclude

$$2\langle r\rangle = n \Rightarrow \frac{\langle r\rangle}{n} = \frac{1}{2}, \tag{5.15}$$

consistent with Figure 5.2.

In order to expose the narrowing of the distribution as $n \to \infty$ we need a measure of how far we expect a particular instance r to differ from the expectation $\langle r\rangle$. One's first thought might be to calculate the expectation of their difference, but unfortunately since $\langle r - \langle r\rangle\rangle \equiv \langle r\rangle - \langle r\rangle = 0$, this clearly won't do. It's far more useful to calculate the expectation of the difference squared, known as the *variance* σ^2, which as the mean over a set of positive definite quantities is guaranteed to be a positive quantity:

$$\sigma^2 = \left\langle \left(r - \langle r\rangle\right)^2 \right\rangle = \left\langle r^2 - 2r\langle r\rangle + \langle r\rangle^2 \right\rangle = \langle r^2\rangle - 2\langle r\rangle^2 + \langle r\rangle^2$$

$$= \langle r^2\rangle - \langle r\rangle^2, \tag{5.16}$$

or in words, "the mean of the square minus the square of the mean". To calculate the variance for our situation start with

$$\langle r^2\rangle \equiv \sum_{r=0}^{n} r^2 P(r) = \frac{1}{2^n}\sum_{r=0}^{n} r^2 \; {}_nC_r \tag{5.17}$$

and again use relation (5.13) to find

$$r^2 = \frac{1}{2^n}\sum_{r=0}^{n} r(n-r+1)\; {}_nC_{r-1} = \frac{1}{2^n}\sum_{r=0}^{n} (r+1)(n-r)\; {}_nC_r$$

$$= \langle (r+1)(n-r)\rangle, \tag{5.18}$$

where again some vanishing terms have been reinstated following relabelling $r \mapsto r+1$ in the sum in the second step. Since $\langle n\rangle = n$, $\langle nr\rangle = n\langle r\rangle$, we deduce

$$\langle r^2\rangle = \langle nr + n - r^2 - r\rangle = n\langle r\rangle + n - \langle r^2\rangle - \langle r\rangle, \tag{5.19}$$

$$\text{i.e,}\ \langle r^2\rangle = \frac{n^2}{4} + \frac{n}{4} \tag{5.20}$$

using the result $\langle r\rangle = \frac{n}{2}$. The variance is then

$$\sigma^2 = \langle r^2\rangle - \langle r\rangle^2 = \frac{n}{4}. \tag{5.21}$$

The square root of the variance is known as the *standard deviation* σ, and σ/n can be thought of as a characteristic measure of the width of the features in Figure 5.2 just as the root mean square speed $\sqrt{v^2}$ encountered in Chapter 3 characterized the molecular speed distribution in the gas. But as $n \to \infty$ this is incredibly small!:

$$\frac{\sigma}{n} = \frac{1}{2\sqrt{n}}. \tag{5.22}$$

The $1/\sqrt{n}$ behaviour of Equation (5.22) is characteristic of many systems described by probability theory. For $n \sim O(N_A)$ relevant for a macroscopic gas sample, we deduce the width of the peak in Figure 5.2 is 1 part in 10^{12} and for all practical purposes we are certain to observe $r = \frac{1}{2}$; indeed, it would be very hard to devise an experiment capable of measuring a discrepancy of this tiny magnitude.

5.2 BOLTZMANN'S STATISTICAL DEFINITION OF ENTROPY

For this simple example of Joule expansion, the statistical approach we have outlined yields physically reasonable answers in agreement with experiment. Is there any more to it than that? In 1877, in one of the boldest imaginative leaps in all of science, Ludwig Boltzmann answered this question in the affirmative, identifying the logarithm of the statistical weight Ω with the entropy S of classical thermodynamics in the celebrated equation

$$S = k_B \ln\Omega. \tag{5.23}$$

We've already seen that $\ln\Omega$ is an extensive function, a key property of S. In order to get its dimensionality correct the constant k_B, known as Boltzmann's constant, must have units JK^{-1}. To find the value of k_B use (5.23) to work out the entropy change during Joule expansion:

$$\Delta S = k_B \ln\Omega_{\text{after}} - k_B \ln\Omega_{\text{before}} = k_B \ln 2^{N_A} - k_B \ln 1 = k_B N_A \ln 2. \tag{5.24}$$

We've used the fact that the statistical weight *before* removal of the partition is unity (cf. 5.1), the statistical weight *after* is 2^{N_A} (5.4), and that $\ln 1 \equiv 0$. Compare (5.24) with the result $\Delta S = R\ln 2$ obtained for the Joule expansion using classical thermodynamics in Chapter 4.6, to conclude[8]

$$k_B = \frac{R}{N_A} \simeq 1.38 \times 10^{-23}\,\text{JK}^{-1}. \tag{5.25}$$

Equation (5.23) offers a bridge between the macroscopic world inhabited by gases, heatbaths, cylinders and pistons and the microworld of atoms, molecules and even subatomic particles, such as electrons and quarks. As we'll see, in thermal systems, the product $k_B T$ is a characteristic energy scale for processes involving such elementary constituents (5.23). It also ushers in a new way of thinking and working in physics, requiring the rules of probability theory. The Joule expansion example illustrates that when the number of component particles n is very large, then conclusions derived on the basis of probabilistic arguments become certainties governed by the classical thermodynamics developed in the first four chapters. In fact, $n \to \infty$ is often referred to as the *thermodynamic limit*. While we have developed the ideas in the context of a gas containing a definite number of particles, the method can be generalized to any physical degree of freedom of interest which is in some sense countable, notable examples being vacancies (i.e. missing atoms) in the regular arrangement of atoms in a crystal, the orientation of atomic-scale magnetic dipoles in a ferromagnetic material such as iron, or quantized sound wave excitations travelling through an

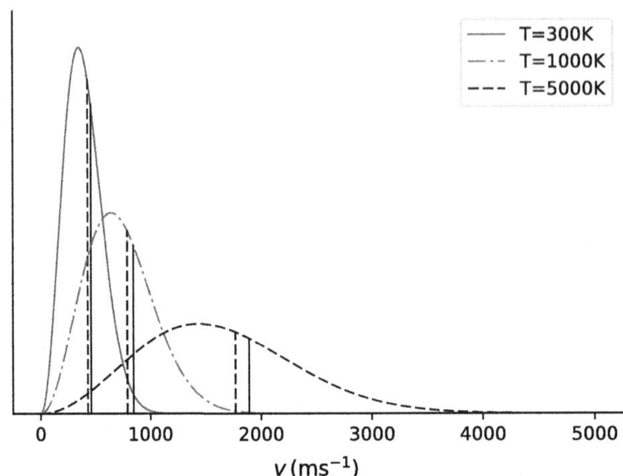

FIGURE 5.3 Distribution of molecular speeds for argon at three different temperatures.

elastic medium known as *phonons*. Boltzmann's leap has underpinned the development of a new and fertile area of theoretical physics known as *statistical thermodynamics*.

So far, however, evidence for (5.23) rests on a single, simple example. What claim does it have to be a universal law? Let's see how it works out for a different process: heating a gas at constant volume. As we saw in Chapter 3, raising a gas's temperature increases the speed v at which its component molecules fly around—it is therefore natural to think of describing the process as an expansion, not in physical space, but this time in the abstract space defined by the particle velocities, visualized as a three-dimensional plot with axes corresponding to the velocity components v_x, v_y, and v_z satisfying $v_x^2 + v_y^2 + v_z^2 = v^2$: every molecule can be represented as a point on this plot, with coordinates determined by its velocity at that instant.[9] The difficulty is that unlike the Joule expansion, where the physical space explored by the molecules is constrained by the rigid walls of a container, here the available space is not limited; instead, Figure 5.3 (cf. Figure 3.2) reminds us that the appropriate description uses a distribution, with no sharply defined maximum speed.[10] Nonetheless, as T increases, the range of the v-axis where the distribution has significant support grows. The volume in question has dimensions of velocity cubed ($\mathrm{m^3s^{-3}}$), so we need to identify a characteristic scale of this form. In another guerrilla physics move, assume the volume of v-space effectively explored by the molecular population scales as $\langle v^3 \rangle$, the mean cube of the molecular speeds. The cube root $\sqrt[3]{\langle v^3 \rangle}$ is shown by a full vertical line for the three distributions in Figure 5.3, while a dashed line denotes the root mean square speed $\sqrt{\langle v^2 \rangle}$.[11] Let's assume, therefore, that

$$\langle v^3 \rangle \simeq \langle v^2 \rangle^{\frac{3}{2}} = \left(\frac{3RT}{M} \right)^{\frac{3}{2}} = \left(\frac{3k_B T}{m} \right)^{\frac{3}{2}}, \tag{5.26}$$

where M is the mass of gas, m the mass of a single molecule and we've used the kinetic theory result (3.11) together with relation (5.25). By analogy with our reasoning for the Joule expansion, for a one mole sample at temperature T let's associate a statistical weight $\Omega \simeq \langle v^3 (T) \rangle^{N_A}$. The entropy change going from T_A to T_B is therefore (using $\ln x - \ln y = \ln(x/y)$ and $\ln x^y = y \ln x$)

$$\Delta S_{AB} = k_B \ln \Omega_B - k_B \ln \Omega_A = k_B \ln \left(\frac{T_B}{T_A} \right)^{\frac{3N_A}{2}} = \frac{3R}{2} \ln \left(\frac{T_B}{T_A} \right), \tag{5.27}$$

in perfect agreement with the classical thermodynamic result (4.19).

5.3 PURISM VERSUS PRAGMATISM

There are different ways to think about the statistical weight. In the Joule expansion example, we defined Ω as the total number of ways to partition the molecular population between the two halves of the vessel assuming all are equally probable. This is somewhat of a purist's approach,[12] which identifies the entropy increase as being entirely due to the removal of the partition separating the two halves of the container—once this is done, all subsequent states of the gas within the volume are equally likely, *including* the state where all molecules are still all found on the left-hand side. An important aspect of any state in thermal equilibrium is that it exhibits fluctuations; the equilibrium state of the gas includes all 2^{N_A} states, and we certainly anticipate encountering microstates with more molecules on the left-hand side than the right and *vice versa*. The extraordinary state ....00000000000000.... should therefore be regarded as an extreme fluctuation. Is this reasonable? Suppose the vessel has a linear dimension of 20 cm, and the mean molecular speed is $500\,\mathrm{ms}^{-1}$. A rough estimate for how frequently different microstates are explored by the gas is 500/0.2 =2500 s^{-1}. Working at this rate, the gas could explore $\sim 10^{21}$ microstates since the beginning of the universe some 15×10^{9} years ago. That's an astronomic number, but as we saw earlier, the total number of microstates 2^{N_A} is post-astronomic. Although the initial state is *a priori* equally likely as any other microstate, there's no sensible chance of ever re-encountering it—it's just not going to happen.

One feature of the purist approach is that the law of entropy increase (4.24) is rigorously observed—there's no sense in which thermal fluctuations of an isolated system can yield, even momentarily, a state of lower entropy. In statistical thermodynamics, however, very often a more pragmatic point of view is adopted, in which the statistical weight and hence entropy of a system is not determined *solely* by the external constraints on the system but depends in some way on the molecular arrangement. Unlike classical thermodynamics, application of the statistical approach requires some modelling of the microstates: in the Joule expansion, such a dependence could be parametrized by the fraction α of molecules on the right-hand side, with $0\le\alpha\le1$. The statistical weight is then[13]

$$\omega(\alpha) = {}_{N}C_{\alpha N} = \frac{N!}{(\alpha N)!((1-\alpha)N)!}. \tag{5.28}$$

To proceed we use an approximation for the factorial function valid for large N known as *Stirling's approximation:*

$$\ln N! \simeq N\ln N - N. \tag{5.29}$$

Figure 5.4 shows the Stirling approximation works well even for surprisingly low numbers: the error is already less than 4% for $N = 30$ and has shrunk to below 1% by $N = 100$. In equation (5.28) it yields

$$\ln\omega(\alpha) \simeq N\ln N - \alpha N\ln\alpha N - (1-\alpha)N\ln(1-\alpha)N. \tag{5.30}$$

The most probable state is given by maximizing (5.30):

$$\frac{d\ln\omega(\alpha)}{d\alpha} = -N\ln\alpha N - N + N\ln(1-\alpha)N + N = N\ln\left(\frac{1-\alpha}{\alpha}\right) = 0$$

$$\Rightarrow \alpha = \frac{1}{2}, \tag{5.31}$$

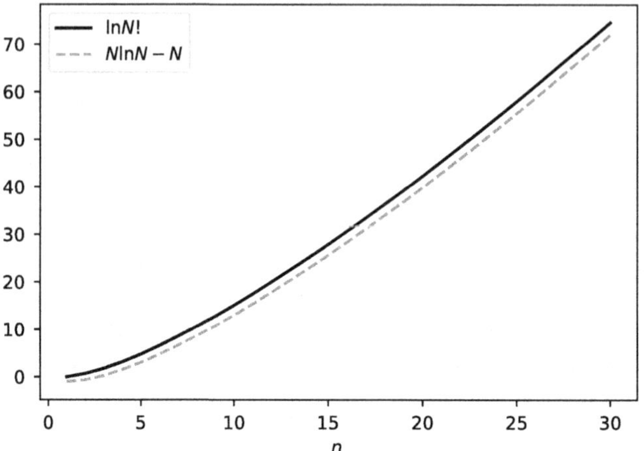

FIGURE 5.4 Stirling's approximation evaluated for $N \leq 30$.

where we used (MA.12c) and (MA.6e). We've recovered the previous result obtained using the exact binomial distribution. Maximizing entropy $S(\alpha) = k_B \ln\omega(\alpha)$ therefore yields an equilibrium condition $\langle\alpha\rangle = \frac{1}{2}$. This viewpoint[14] specifies the equilibrium state of an isolated system as that which maximizes the entropy $S(\alpha)$, and characterizes it in terms of expectations $\langle\cdots\rangle$ calculated by integrating a probability distribution $P(\alpha)$ which we'll go on to specify more precisely in subsequent paragraphs; for now we note this way of characterizing equilibrium is clearly consistent with the Clausius statement that entropy tends towards a maximum.

5.4 BOLTZMANN DISTRIBUTION, FREE ENERGY, AND PARTITION FUNCTION

For an isolated system in thermal equilibrium entropy is a function of internal energy and volume: $S = S(U,V)$ (cf. Equation (4.17)). Consider two systems in thermal and mechanical contact which can exchange internal energy and volume with each other but remain isolated from the environment as a whole. For the combined system it is clear that

$$U = U_1 + U_2;\ V = V_1 + V_2. \tag{5.32}$$

When counting microstates of the combined system, we identify a distinct state for every possible combination of a state from subsystem 1 with one from subsystem 2, implying the statistical weight is obtained as the product $\Omega_1\Omega_2$ of the weights of the component subsystems. Since $\ln xy \equiv \ln x + \ln y$ Boltzmann's definition (5.23) then yields

$$S = S_1 + S_2. \tag{5.33}$$

We can find equilibrium conditions from the requirement that S is maximized, e.g., with respect to U_1 while keeping V_1, V_2 fixed:

$$\left.\frac{\partial S}{\partial U_1}\right|_{V_1} = \left.\frac{\partial S_1}{\partial U_1}\right|_{V_1} + \left.\frac{\partial S_2}{\partial U_2}\right|_{V_2} \left.\frac{\partial U_2}{\partial U_1}\right|_{V_1,V_2} = 0. \tag{5.34}$$

Since U is constant $dU_2 = -dU_1$ and it follows immediately that a condition for equilibrium is

$$\left.\frac{\partial S_1}{\partial U_1}\right|_{V_1} = \left.\frac{\partial S_2}{\partial U_2}\right|_{V_2}, \tag{5.35}$$

that is, the two subsystems are at the same temperature T defined in general by

$$\frac{1}{T} = \left.\frac{\partial S}{\partial U}\right|_{V}, \tag{5.36}$$

consistent with the Zeroth Law. Repeat the analysis this time minimizing with respect to V_1 at fixed U_1, U_2 to conclude that the systems must also be at the same pressure defined by

$$p = T\left.\frac{\partial S}{\partial V}\right|_{U}. \tag{5.37}$$

Next consider the case where $U_1, V_1 \gg U_2, V_2$; in this case we can think of system 1 as a heatbath with entropy $S_1(U_1) = S_1(U - U_2)$ whose temperature $T_1 = T$ can be considered constant, i.e., thermal contact with a much smaller system 2 will have negligible impact on T. The quantities U_2, V_2 by contrast are still expected to fluctuate as a consequence of continual exchange of energy and volume with the heatbath. We can develop an expression for the probability P of observing a particular value U_2 using a statistical weight argument for the microstates r of the heatbath, which assumes that while they are countable in principle the heatbath is sufficiently large that S_1 is a smooth function of $U_1 = U - U_2$:

$$P(U_2) \propto \Omega_1(U - U_2) \propto \exp\left(\frac{S_1(U - U_2)}{k_B}\right), \tag{5.38}$$

where we have used Boltzmann's relation $S = k_B \ln\Omega$. (5.23). Since $U_2 \ll U$ we can estimate S_1 using a Taylor expansion[15]:

$$S_1(U - U_2) = S_1(U) - U_2\frac{\partial S_1(U)}{\partial U} + \frac{1}{2}U_2^2\frac{\partial^2 S_1(U)}{\partial U^2} + \cdots \tag{5.39}$$

i.e.

$$P(U_2) \propto \exp\left(-\frac{U_2}{k_B T} + \frac{U_2^2}{2k_B}\frac{\partial T^{-1}}{\partial U} + \cdots\right). \tag{5.40}$$

The second term in brackets depends on the change in heatbath temperature as a consequence of thermal contact with the smaller system, and is negligible in the heatbath's thermodynamic limit.[16] We conclude the probability P_r of observing a general microstate labelled r is proportional to $e^{-U_r/k_B T}$, a combination known as the *Boltzmann factor*.

To correctly normalize P_r we must divide by the sum of the Boltzmann factors over all available microstates, known as the *partition function*:

$$Z = \sum_r \exp\left(-\frac{U_r}{k_B T}\right). \tag{5.41}$$

For an observable quantity A taking a value A_r in the rth microstate, the thermal expectation value is then given by

$$\langle A \rangle = \frac{1}{Z}\sum_r A_r \exp\left(-\frac{U_r}{k_B T}\right). \tag{5.42}$$

The partition function in the denominator defined by (5.41) ensures $\langle 1 \rangle \equiv 1$. Equation (5.42) for the expected value of an observable at a specified temperature T is one of the workhorse equations of statistical thermodynamics. The probability distribution

$$p_r = Z^{-1} e^{-U_r / k_B T} \tag{5.43}$$

is known as the *Boltzmann* or *canonical* distribution; note the switch from upper-case P_r to lower-case p_r now there's no longer any danger of confusion with pressure. In order to use (5.41,42) we need a model of the microstates in which both U_r and A_r can be specified or calculated.

The final piece of formalism we will need is the extension of the definition of entropy to non-isolated systems. Consider an ensemble[17] of n identical systems in thermal equilibrium, each having microstates labelled by r and occurring with probability p_r. For large n the number of systems found in the rth microstate will be $n_r = np_r$. For the ensemble as a whole the statistical weight is then[18]

$$\Omega = \frac{n!}{n_1!n_2!\ldots n_r!\ldots}; \tag{5.44}$$

we arrive at the entropy

$$S = k_B \ln\Omega = k_B \ln(n!) - k_B \sum_{r=1}^{n} \ln(n_r!) = k_B\left[n\ln n - n - \sum_r n_r \ln n_r + \sum_r n_r\right]$$

$$= k_B\left[n\ln n - \sum_r np_r \ln np_r\right] = k_B n\left[\ln n - \sum_r p_r \ln n - \sum_r p_r \ln p_r\right] = -nk_B \sum_r p_r \ln p_r. \tag{5.45}$$

where both Stirling's approximation and the normalization conditions $\sum_r n_r = n$, $\sum_r p_r = 1$ have been used. Since S must be extensive, we surmise the result for a single system

$$S = -k_B \sum_r p_r \ln p_r. \tag{5.46}$$

Don't be anxious about the minus sign: since $p_r \leq 1$, $\ln p_r$ is guaranteed to be a negative-valued quantity. The definition (5.46) is known as the Gibbs definition of entropy[19]; we've moved from an expression (5.23) depending on counting microstates to one phrased instead in terms of the probabilities of their occurrence. Next, substitute the Boltzmann distribution (5.43) in (5.46):

$$S = -k_B \sum_r \left(Z^{-1}e^{-U_r/k_BT}\right)\left(-\ln Z - \frac{U_r}{k_BT}\right) = k_B \ln Z + \frac{\langle U\rangle}{T}, \tag{5.47}$$

where we have exploited the fact that $\ln Z$ is constant independent of r. For a sufficiently large system we can ignore fluctuations and replace $\langle U\rangle$ with U, to define a new quantity:

$$F = U - TS = -k_B T \ln Z. \tag{5.48}$$

Like (5.23), this is a relation bridging micro- and macroworlds. The extensive quantity F has the same units as U and is called the *Helmholtz Free Energy*.[20] It can be expressed either in terms of the state functions U, T and S of classical thermodynamics,[21] or in terms of a sum over microstates via the logarithm of the partition function. Formally, for a system in contact with a heatbath and therefore having constant temperature T, F has a similar relation with respect to Z as entropy S does with Ω for an isolated system. Systems in equilibrium at constant T correspond to the *minima* of the free energy F (it's down to the minus sign…). The condition for equilibrium is essentially a competition between states which tend to minimize U and those which maximize S; as T rises, the entropy must increase. Very often, analysis of systems in contact with a heatbath is much more relevant for physically realizable situations encountered in experiments and real life. For this reason, many textbooks switch focus from discussing S and Ω to F and Z somewhere about this point, and in particular, we will find an important role for it in Chapter 13.

5.5 THE MAXWELL-BOLTZMANN DISTRIBUTION

After this big chunk of formalism it will come as something of a relief to actually calculate something. Let's return to the distribution of molecular speeds in an ideal gas. Considering a single molecule in thermal equilibrium with the rest of the gas, and ignoring rotational motion (which doesn't depend on speed v), we have that the energy U_r is $\frac{1}{2}mv^2$ and hence the probability of having velocity $\vec{v}$ is proportional to the Boltzmann factor $e^{-mv^2/2k_BT}$, i.e.,

$$P(\vec{v}) = A\exp\left(-\frac{mv^2}{2k_BT}\right), \tag{5.49}$$

where $v = |\vec{v}|$ and A is a normalization constant to be determined. In order to find A we need to sum over all possible microstates, i.e. all possible velocities, keeping in mind that the velocity distribution is isotropic and three-dimensional. Three-dimensional geometry dictates that the number of velocity microstates corresponding to speeds between v and $v + dv$ is proportional to the volume of a spherical shell centred at the origin with radius v and thickness dv, which in turn equals the surface area $4\pi v^2 \times dv$. We conclude $P(v) = 4\pi v^2 P(\vec{v})$, and hence the normalization condition for P considered as a probability distribution is

$$\int d^3\vec{v}P(\vec{v}) = 4\pi A\int_0^\infty v^2 \exp\left(-\frac{mv^2}{2k_BT}\right)dv = 1, \tag{5.50}$$

where in the second step we substituted (5.49) for $P(\vec{v})$. The integral in (5.50) is a standard one but may not be immediately familiar to you,[22] so here's the answer:

$$A\left(\frac{2\pi k_BT}{m}\right)^{\frac{3}{2}} = 1 \Rightarrow A = \left(\frac{m}{2\pi k_BT}\right)^{\frac{3}{2}}. \tag{5.51}$$

The final answer for the Maxwell-Boltzmann speed distribution is

$$P_{MB}(v) = 4\pi v^2 P(\vec{v}) = 4\pi\left(\frac{m}{2\pi k_BT}\right)^{\frac{3}{2}} v^2 e^{-\frac{mv^2}{2k_BT}}. \tag{5.52}$$

This was used to generate the plots in Figures 3.2 and 5.3, and also to calculate expectations such as[23]

$$\langle v^2\rangle = \int dv v^2 P_{MB}(v) = 4\pi\left(\frac{m}{2\pi k_BT}\right)^{\frac{3}{2}} \int_0^\infty v^4 e^{-\frac{mv^2}{2k_BT}} dv = \frac{3k_BT}{m}. \tag{5.53}$$

The v^2 factor in $P_{MB}(v)$ makes the probability of finding very slow molecules vanishingly small, while the exponential factor $e^{-mv^2/2k_BT}$ ensures very high velocities are also suppressed, although as advertised in Chapter 3 there is technically no upper limit. In general the Boltzmann factor suppresses the likelihood of finding states with $U \gg k_BT$, so that there is a sense in which k_BT is a typical energy for any process at an atomic scale.

In this chapter, we've had a second glimpse at the link between the macroworld described by classical thermodynamics and the microworld addressed using statistical techniques, and in so doing developed a powerful formalism encapsulated in relations, such as (5.41, 42) and (5.48), which furnish a cornerstone for the study of complicated "many-body" systems and continue to be used in many active areas of research to the present day. The underlying key concept that at a fundamental level matter consists of discrete countable entities, is sometimes called *atomism.* The idea can be traced back to the 5th century BCE through the teachings of Leucippus and Democritus, and the list of prominent atomists over the intervening years includes names like Descartes, Boyle, Bacon, Galileo, and Newton. Empirical evidence for atomism, however, was slow to accumulate, probably due to the requirement for refined experimental techniques. Two key contributions are due to John Dalton, whose analysis of the weights of products of chemical reactions supported the idea that the elements consist of identical units called atoms, and Albert Einstein,[24] who in 1905 analysed the observation through a microscope of the random motion of pollen grains in suspension in terms of many many collisions with individual yet invisible water molecules.

Boltzmann's visionary work naturally stood firmly within the atomic paradigm, but his views and teachings were not accepted by some prominent members of the scientific establishment towards the close of the 19th century, and he had to spend time and energy defending his theories, probably at some cost to his mental health. In 1906 he was eventually forced to resign his university position in Vienna due to ill health. He took his own life in September of that year, while on vacation with his family near Trieste. He is buried in the Viennese *Zentralfriedhof,* where his imposing gravestone

FIGURE 5.5 Boltzmann's tomb (Photo courtesy Ian Ford).

bears the famous formula (5.23)[25] (Figure 5.5). It is of course inappropriate to speculate too closely on what drove Boltzmann to his final desperate action; none of us is privy to another's intimate thoughts. Nonetheless, for scientists today, Boltzmann's grave has the status of the tomb of a fallen warrior.

NOTES

1 The technical term for such a distribution is *isotropic*.

2 This microstate is a very crude representation of the gas, focussing on one particular aspect, namely which half of the vessel the molecule occupies; this is precisely the question the Joule expansion poses us once the partition is removed.

3 Some thought needs to be given as to the order in which to visit the molecules since if we started from the leftmost and worked our way to the right we'd generate a string where all the 0 s came before the 1 s. One possibility is order the molecules starting from the topmost surface of the vessel and work downwards; another is to pick the next molecule at random and then conceptually "remove it from the box" so that it isn't picked twice.

4 Just as the position of a numeral in a conventional decimal number stands for a power of 10, e.g. $361 = 3 \times 10^2 + 6\times10^1 + 1\times10^0$.

5 Omega (Ω) literally means "big O". It has a sibling omicron (o) which is just "little O". Omicron doesn't get much air-time in science and mathematics, but unfortunately rose to prominence in 2022 as the label for a particularly infectious strain of SARS-CoV-2.

6 Sometimes the definition $0! = 1$ is also needed.

7 As we'll calculate in a moment, on the vertical scale of Figure 5.2, this line would extend upwards as far as $\sqrt{N_A} \sim 10^{12}$.

8 In 2019 Boltzmann's constant was defined to be *exactly* $1.380649\times10^{-23}\,\mathrm{JK}^{-1}$, so that henceforth this also became the base calibration for the thermodynamic temperature scale.

9 In advanced applications it's more usual to work in "momentum space" (p-space) in this context, but since momentum $p = mv$ for gas molecules, our approach is completely equivalent.

10 Figure 5.3 is drawn for argon, a monatomic gas with $C_V = \frac{3}{2}R$. The treatment of polyatomic gases is complicated by the need to include rotational degrees of freedom.
11 For the Maxwell-Boltzmann distribution depicted in Figure 5.3 the ratios $\langle v\rangle : \sqrt{\langle v^2\rangle} : \sqrt[3]{\langle v^3\rangle} = 1:1.085:1.162$ independent of temperature or molecular mass.
12 This point of view is strongly advocated in Pippard's book.
13 We've used lower-case ω to emphasize that the two weights are distinct, related by $\int d\alpha\omega(\alpha) = \Omega$.
14 You may be getting anxious about the appearance of more than one "viewpoint" even at this relatively early stage of the book. In practice this presents no problem, due to the large number of degrees of freedom in the thermodynamic limit. The precise definition of $\omega(\alpha)$ requires us to specify a coarse-graining scale through the width of the element $d\alpha$ in the sum $\Sigma_\alpha \omega(\alpha)d\alpha = \Omega$. Suppose we chose $d\alpha$ large enough so that effectively only one value α_0 contributed to the sum, i.e. $\Omega = \omega(\alpha_0)d\alpha$. To calculate S we need $\ln\Omega = \ln\omega(\alpha_0) + \ln(d\alpha)$. As extensive quantities both $\ln\Omega$ and $\ln\omega$ are of order N, the number of degrees of freedom. We learned from our analysis of the binomial distribution that $d\alpha \sim \sqrt{N}$ so that $\ln(d\alpha) \sim \frac{1}{2}\ln N$. For $N \sim O(N_A)$ this correction is therefore utterly negligible. Purists and pragmatists alike agree on the numerical value of entropy.
15 Any smooth function $f(x)$ can be expanded around the point $x = a$ via the series $f(x) = f(a) + (x-a)\partial f/\partial x|_{x=a} + \cdots + [(x-a)^n/n!]\partial^n f/\partial x^n|_{x=a} + \cdots$. To derive (5.40) we also need (MA.12b).
16 For instance, if the heatbath is an ideal gas, the second term is suppressed with respect to the first by a factor $U_2/2U$.
17 The word *ensemble* is frequently used in statistical mechanics—it stands for a collection of identical systems, each in general occupying a different microstate.
18 A factor of $(\Omega')^n$, where Ω' is the statistical weight associated with a single isolated subsystem, cancels top and bottom in this equation.
19 Josiah Willard Gibbs (1839–1903) made seminal contributions to the development of statistical mechanics, as well as giving the field its name.
20 Hermann von Helmholtz (1821–1894) was a physicist and physician who also made notable contributions to electromagnetic theory.
21 The corresponding differential relation is $dF = dU - TdS - SdT = -SdT - pdV$. Note that to use this relation, for the first time we are required to calculate absolute entropy rather than just entropy changes ΔS. Very often this subtlety is brushed over by choosing to parametrize the microstates empirically.
22 It belongs to the family of so-called "Gaussian" integrals: $\int_0^\infty x^2 e^{-\alpha x^2} = \sqrt{\pi}/(4\alpha^{3/2})$.
23 The next one up is $\int_0^\infty x^4 e^{-\alpha x^2} = 3/(8\alpha^2)\sqrt{(\pi/\alpha)}$.
24 Albert Einstein (1879–1955) needs no introduction from us; suffice it to say his reputation today would still be enviable based on his contributions to statistical thermodynamics alone.
25 The "W" in Boltzmann's version stands for *Wahrscheinlichkeit*, the German for probability.

Part II

6 The Quantum Realm

In the 1920s, physics was shaken by the quantum revolution. Experimental data could not be reconciled with the older classical paradigms. A new view was required, along with a radical revision to concepts that had seemed completely secure. Arguably, the quantum revolution still rumbles on, although the aftershocks are somewhat more muted now. While we have a complete mathematical grasp of quantum theory, its interpretation remains open to debate (although not that many working physicists engage with the debate, taking a more "shut up and calculate" approach). It is entirely possible that the ambiguity over interpretation is a signal that we are still not in possession of the correct fundamental concepts. That would also chime with the difficulties inherent with reconciling the quantum and relativistic pictures. Modern theoretical developments towards a quantum gravity may also shed some light on entropy, or at least information. Consequently, for a variety of reasons, we need to engage with the quantum description of the microworld and to see how that impacts on our understanding of entropy.

6.1 BACKGROUND

In Part I, we developed a reasonably thorough account of the kinetic theory of gases, which allows thermodynamic investigations at the macroscopic level to be underpinned by the microscopic behaviour of atoms and molecules. The price for this increased sophistication and understanding is the practical limitation of keeping track of an astronomic number of particles and their positions and velocities, which forces us to use statistical techniques. On the other hand, this allowed the state function entropy to be defined in terms of statistical weight.

If it were understood only at the macroscopic level, entropy would have an important role as a "bookkeeping tally", but its nature, lacking the immediate "tactile" experience that other state function (temperature, pressure, volume, etc.) have, would be somewhat mysterious.

Put more simply, we would not really know what we were playing with.

The statistical approach necessary in the kinetic theory has, as a by-product, given a way of understanding the nature of entropy; we start to see it as a measure of microscopic possibilities leading to the same macroscopic state. It could be couched as a measure of our ignorance regarding the exact microscopic state or perhaps as an indication of our lack of ability to prepare a system in a precise microscopic state.

The discussion, however, is somewhat moot. After all, the kinetic theory of gases, no matter how successful, can only be a staging post on the way to something more fundamental. At the level of atoms and molecules, *quantum theory* has to be the correct theoretical domain. We must develop a quantum approach to the ideal gas and thermodynamics in general. This will also allow us to cast further illumination on the nature of entropy.

DOI: 10.1201/9781003121053-9

Some of the arguments presented in this chapter echo, or re-jig, arguments covered in Part I; however, now they will now be presented in an expanded quantum context.

6.2 SAUCE FOR GEESE...

In 1924, a young PhD student, Louis de Broglie,[1] was pondering the issue of quantized radiation. The earlier work of Planck[2] and Einstein[3] provided tantalizing and paradoxical hints that there is a particle-like aspect to electromagnetic radiation. Physicists had believed the matter to be settled, since the results obtained by Thomas Young[4] (1801) could only be interpreted by granting light a wave nature. The 1920s did not simply bring a clash of theoretical approaches; contrasting experiments were delivering data that in one case could only be understood by a wave model, whereas equally valid experiments pointed to particle-like aspects.

In the early part of the 20th century, physicists were learning to live with a double-think (complementary) approach, aided by the existence of compact relationships that mapped one conceptual framework into another. Firstly, the expression:

$$p = h/\lambda \tag{6.1}$$

relates the momentum, p, of the particle aspect of light to the wavelength, λ, of the wave aspect via *Planck's constant h* (6.63×10^{-34} Js) which first appeared in the expression for black body radiation spectra (Section 9.4.2), as derived by Planck in 1900. It also features in the second double-think equation:

$$E = hf \tag{6.2}$$

connecting the energy, E, of the radiation particle, the *photon*, with the frequency, f, of the wave.

While the rest of the physics community was trying to figure out how light could be *both* a wave and a particle in different circumstances, de Broglie had a rather mischievous thought. After all, if a wave such as light could in some weird way also be a particle, was it possible that a particle, such as an electron, might display a wave nature in the right conditions? In his thesis, de Broglie proposed generalizing the relationships (6.1) and (6.2) to apply to electrons and other particles.

His examiners were somewhat taken aback by this suggestion, so they sent a copy of the thesis to Einstein for his comments. With Einstein's approval, de Broglie gained his degree, and pocketed a Nobel Prize in 1929 along with it.

Experimental confirmation of an electron's wave nature came rapidly, but with it, a growing and increasingly urgent question: at what scale is the wave aspect of matter completely suppressed? People, cars and cricket balls do not display a wave nature. It would be unnerving, to say the least, to observe humans diffracting on passing through a doorway. For a typical male of 70 kg mass, moving at a comfortable walking speed, say $1.4\ \text{ms}^{-1}$, their wavelength would be:

$$\lambda = h/p = h/p = 6.63\times10^{-34}/(70\times1.4) = 6.77\times10^{-36}\,\text{m}$$

which is far too small to be picked up in any currently conceivable experimental device.

However, as technology has improved, the wave-aspects of increasingly large systems have been demonstrated. The current record stands at molecules with ~2000 atoms, with a wavelength $\sim 6\times10^{-14}\,\text{m}$, and it is likely that the limit will be raised further.

Clearly we need to consider the wave nature of the atoms in a typical gas, if we are to fully understand their thermodynamics. The rms molecular speed of Argon at room temperature ($\sim 25^{\circ}\text{C}$) is $\sim 430\ \text{ms}^{-1}$ and as its mass[5] is $\sim 40\ u = 6.64\times10^{-26}\,\text{kg}$, this gives a de Broglie wavelength of:

$$\lambda = h/p = 6.63\times10^{-34}/\left(6.64\times10^{-26}\times430\right) = 2.32\times10^{-11}\,\text{m}$$

which is comparable to atomic dimensions, confirming that we must see if the wave nature of argon has any notable effect on how the gas behaves.

6.3 PARTICLE IN A BOX

One of the simplest introductory problems in quantum mechanics is the *particle in a box.*

Consider a single particle (soon to become an argon molecule) held inside a rectangular box with impenetrable walls. There is nothing else within the box. We assume that the particle's motion within the box is completely free – i.e., that there are no forces acting on the particle. The plan now is to figure out the quantum wave of this particle.

A standard progressive wave is mathematically described by a function of the form:

$$y(x,t) = A\sin\left(\frac{2\pi}{\lambda}x - 2\pi ft\right) \tag{6.3}$$

where y is the wave's displacement, λ its wavelength, f the frequency and A the wave's amplitude. It would be equally valid to write:

$$y(x,t) = A\cos\left(\frac{2\pi}{\lambda}x - 2\pi ft + \Phi\right) \tag{6.4}$$

by introducing a phase, Φ, that can be adjusted to describe the wave correctly.

In quantum theory, however, things are not quite so simple. The particle's wave nature is expressed via a *quantum amplitude*, which must be represented by a *complex number*, $\psi = a + ib = Re^{i\theta}$. The *absolute square* of this amplitude, $\psi^*\psi$, gives the relative probability of finding the particle at location x at time t. This is an extremely important principle, known as the *Born Rule*, which has two consequences:

- amplitudes are not directly observable, since complex numbers cannot be used to express physical quantities[6] in the material world;
- the amplitudes ψ and $\psi' = e^{i\vartheta}\psi$ are physically indistinguishable as $(\psi')^*(\psi') = e^{-i\vartheta}\psi^*$ $e^{i\vartheta}\psi = \psi^*\psi$.

The most elegant way of generalizing our progressive wave equation to cater for complex values is to write it in the form:

$$\psi(x,t) = \text{A}\left\{\cos\left(\frac{2\pi}{\lambda}x - 2\pi ft\right) + i\sin\left(\frac{2\pi}{\lambda}x - 2\pi ft\right)\right\} = Ae^{i\left(\frac{2\pi}{\lambda}x - 2\pi ft\right)} \tag{6.5}$$

This expression can be somewhat tidied by inserting the de Broglie relationships (6.1) and (6.2):

$$\psi(x,t) = Ae^{i\left(\frac{2\pi}{\lambda}x - 2\pi ft\right)} = Ae^{i\frac{2\pi}{h}(px - Et)} = Ae^{i(px - Et)/\hbar} \tag{6.6}$$

and introducing the *reduced Planck's constant*, $\hbar = h/2\pi$.

However, this is still not quite right. If this were a situation in classical physics, equation (6.6) would represent a wave propagating in the direction of increasing x. With this quantum wave, however, that is not the case. If we take the absolute square:

$$\psi(x,t)\,\psi^*(x,t) = Ae^{i(px-Et)/\hbar}A^*e^{-i(px-Et)/\hbar} = AA^*$$

i.e., the same value at any location, and independent of time. To model a particle in motion, an artful combination of these *free particle solutions* needs to be constructed called a *wave packet.*

In any case, as there is no constraint other than the dimensions of the box, our particle can, in principle, have either positive or negative momentum. Both need to be represented. One of the rules of quantum theory is that every possibility must be included in the state unless they can be specifically excluded in some manner. We need both:

$$\psi_+(x,t) = A_+e^{i(px-Et)/\hbar} \qquad \psi_-(x,t) = A_-e^{i(-px-Et)/\hbar} \tag{6.7}$$

to give:

$$\Psi(x,t) = \psi_+ + \psi_- = A_+e^{i(px-Et)/\hbar} + A_-e^{i(-px-Et)/\hbar} \tag{6.8}$$

Next, we need to consider how the walls of the box affect the quantum amplitude within. At first glance, the walls would appear to be irrelevant, as we have specified that the particle is free inside the box, i.e., that no forces act within the contained volume. However, if the box has impenetrable walls (which is impossible in practice but a good starting model for a theoretical investigation), the quantum amplitude cannot exist outside of the box. Suppose that the box is of length L, so that it lies between $0 \le x \le L$. Our first thought is that:

$$\Psi(x,t) = A_+e^{i\frac{1}{\hbar}(px-Et)} + A_-e^{i\frac{1}{\hbar}(-px-Et)} \qquad 0 < x < L$$

$$\Psi(x,t) = 0 \qquad x < -w, x > L + w \tag{6.9}$$

where w is the width of the energy barrier. While correct, this ignores the region walls themselves.

To model walls that cannot be crossed, physicists generally think of them as a potential energy barrier of finite thickness, but infinite "height" (Figure 6.1)—in which case, any particle would have to start with infinite KE to cross the region the walls occupy. This is clearly never going to happen, making the walls a forbidden zone.

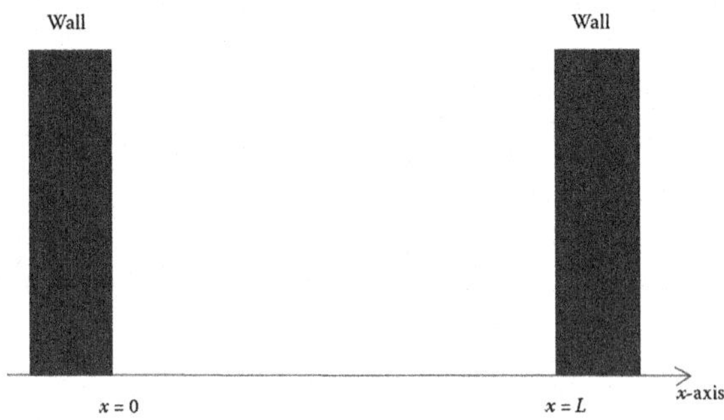

FIGURE 6.1 A box with impenetrable walls is modelled by a span of the x-axis bounded by regions where the potential energy is infinite.

The potential energy barrier is taken to be "sharp": there is no gradation, no matter how steep, between the point where the potential is zero and where it is infinite. So, the quantum amplitude must be zero at the leading edge of each barrier. Otherwise, there would be some probability of finding the particle at this point, requiring it to have infinite KE.

First, let's look at what happens when $x = 0$:

$$\Psi(0,t) = A_+ e^{-i\frac{1}{\hbar}Et} + A_- e^{-i\frac{1}{\hbar}Et} = 0 \tag{6.10}$$

giving $A_+ = -A_-$.

Now we move to $x = L$, incorporating $A_+ = -A_-$:

$$\begin{aligned}\Psi(L,t) &= A_+ e^{-i\frac{1}{\hbar}Et} \left\{ e^{i\frac{1}{\hbar}pL} - e^{-i\frac{1}{\hbar}pL} \right\} \\ &= A_+ e^{-i\frac{1}{\hbar}Et} \left\{ \left(\cos\left(\frac{1}{\hbar}pL\right) + i\sin\left(\frac{1}{\hbar}pL\right)\right) - \left(\cos\left(\frac{1}{\hbar}pL\right) - i\sin\left(\frac{1}{\hbar}pL\right)\right) \right\} \\ &= 2iA_+ e^{-i\frac{1}{\hbar}Et} \sin\left(\frac{1}{\hbar}pL\right)\end{aligned} \tag{6.11}$$

Setting $A = 2iA_+$, we next equate $\Psi(L,t)$ to zero, so that:

$$\Psi(L,t) = Ae^{-i\frac{1}{\hbar}Et} \sin\left(\frac{1}{\hbar}pL\right) = 0 \tag{6.12}$$

From here, we could consider $A = 0$, which would satisfy the equality, but results in $\Psi(x,t) = 0$ for *all* $0 \le x \le L$ and at *all times*. In essence, we get $\Psi(L,t) = 0$ at the cost of removing the particle from the box altogether.

More sensibly, we can leave $A \neq 0$, but set $\sin\left(\frac{1}{\hbar}pL\right) = 0$. From the properties of the $\sin(x)$ function (Figure 6.2), we see that $\sin(x) = 0$ when $x = n\pi$, n being an integer.

We are effectively imposing the constraint:

$$\frac{1}{\hbar}pL = n\pi \tag{6.13}$$

with integer n. This is our first example of a *quantum number*.

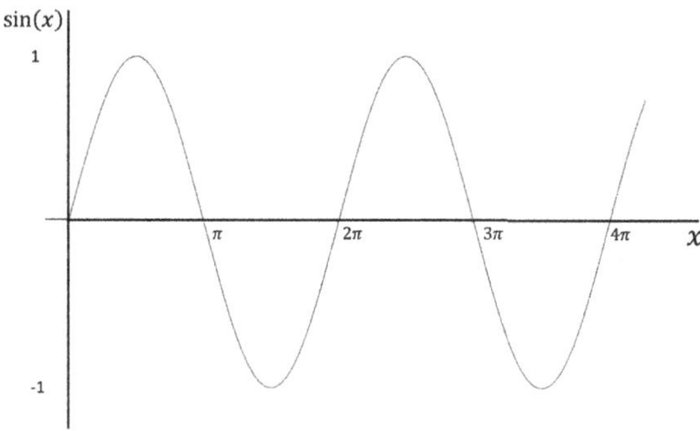

FIGURE 6.2 The function $\sin x$ has values equal to zero periodically, when $x = n\pi$.

This means that the momentum of the particle within our box is *quantized* to specific values:

$$p_n = \frac{\hbar n\pi}{L} \tag{6.14}$$

which impacts on the wavelengths of the de Broglie wave:

$$\lambda_n = \frac{h}{p_n} = \frac{hL}{\hbar n\pi} = \frac{2\pi hL}{hn\pi} = \frac{2L}{n} \tag{6.15}$$

Finally, we have a set of amplitudes for the particle in a box:

$$\Psi_n(x,t) = Ae^{-i\frac{1}{\hbar}E_n t}\sin\left(\frac{1}{\hbar}p_n x\right) \tag{6.16}$$

each of which is an example of a *wave function*—a term frequently used for an amplitude function which is continuous in time and space. Many physicists use the term "amplitude" to denote situations where there only a finite number of discrete possibilities that we are modelling (e.g., the spin components that will come up in Chapter 8). However, the terms "wave function" and "amplitude" are broadly interchangeable, and we will use them in that manner.

There is no loss of generality in taking $n > 0$, as a negative value would mean the opposite direction of the momentum. The case of $n = 0$ can also be eliminated as this would render $\Psi_0(x,t) = 0$, once again effectively removing the particle from the box.

In any practical situation, the value of n will be exceptionally large. For our argon molecule at room temperature, we calculated a de Broglie wavelength of 2.32×10^{-11} m. If this molecule were in a room-sized box with, $L = 4\,\text{m}$, its quantum number at that wavelength would be:

$$n = \frac{2L}{\lambda_n} = \frac{8}{2.32 \times 10^{-11}} = 3.44 \times 10^{11}$$

6.3.1 Normalization

We've got one more task that we need to sort out – finding an appropriate value for A in the wave function. We know from the Born rule that the absolute square of the amplitude gives a probability. We also know that the particle is certainly located *somewhere* $0 \le x \le L$ so:

$$I = \int_0^L \Psi_n^* \Psi_n \, dx = A^2 \int_0^L \sin^2\left(\frac{1}{\hbar}p_n x\right) dx = 1 \tag{6.17}$$

Using the standard trigonometric relationship:

$$\sin^2\vartheta = \frac{1}{2}(1 - \cos 2\vartheta)$$

$$I = \frac{A^2}{2}\int_0^L dx - \frac{A^2}{2}\int_0^L \cos\left(\frac{2}{\hbar}p_n x\right) dx = \frac{A^2}{2}L = 1 \tag{6.18}$$

We know that the $\cos\left(\frac{2}{\hbar}p_n x\right)$ integral will become a multiple of $\sin\left(\frac{2}{\hbar}p_n x\right)$, which will evaluate to zero at both $x = 0, x = L$ due to the way in which the wave function has been set up (the boundary conditions). That means:

$$A = \sqrt{\frac{2}{L}} \qquad \Psi_n(x,t) = \sqrt{\frac{2}{L}}e^{-i\frac{1}{\hbar}E_n t}\sin\left(\frac{1}{\hbar}p_n x\right) \tag{6.19}$$

6.3.2 Energy Quantization

Realistically, we must consider a particle to be free to move in the *y* and *z* directions as well. Our box should be three-dimensional. As directions in space are arbitrary, we expect solutions along each axis to be the same, leading to an overall amplitude:

$$\Psi(x,y,z,t) = \sqrt{\frac{8}{V}}e^{-i\frac{1}{\hbar}Et}\sin\left(\frac{1}{\hbar}p_x x\right)\sin\left(\frac{1}{\hbar}p_y y\right)\sin\left(\frac{1}{\hbar}p_z z\right) \tag{6.20}$$

or, with the boundary conditions imposed:

$$\Psi_{n_x,n_y,n_z}(x,y,z,t) = \sqrt{\frac{8}{V}}e^{-i\frac{1}{\hbar}E_{n_x,n_y,n_z}t}\sin\left(\frac{n_x\pi}{L}x\right)\sin\left(\frac{n_y\pi}{W}y\right)\sin\left(\frac{n_z\pi}{H}z\right) \tag{6.21}$$

Note that the *three* quantum numbers, (n_x, n_y, n_z) are independent of each other but that the energy, E_{n_x,n_y,n_z} is dependent on all of them together.

As the particle is non-interacting within the confines of the box, the only energy we need consider is kinetic. For any gases that we are likely to encounter in rooms, the speeds involved are very much less than the speed of light, so we can use $KE = p^2/2m$, giving:

$$E = \frac{1}{2m}\left(p_x^2 + p_y^2 + p_z^2\right) = \frac{\hbar^2\pi^2}{2m}\left(\frac{n_x^2}{L^2} + \frac{n_y^2}{W^2} + \frac{n_z^2}{H^2}\right) \tag{6.22}$$

With a cubical box of side *L*:

$$E = \frac{\hbar^2\pi^2}{2mL^2}\left(n_x^2 + n_y^2 + n_z^2\right) \tag{6.23}$$

and by writing $n^2 = n_x^2 + n_y^2 + n_z^2$, we obtain:

$$E_n = \frac{n^2\hbar^2\pi^2}{2mL^2} \tag{6.24}$$

As a general quantum feature *spatially localized systems have quantized energies*. It is common to refer to these as *energy levels*. In this case, the *n*th energy level, E_n, has energy $E_n = n^2\hbar^2\pi^2/2mL^2$. If we go up a level, $n \to n+1$, the energy increases by ΔE:

$$\Delta E = \frac{(n+1)^2\hbar^2\pi^2}{2mL^2} - \frac{n^2\hbar^2\pi^2}{2mL^2} = \frac{\hbar^2\pi^2}{2mL^2}\left(n^2 + 2n + 1 - n^2\right) = \frac{\hbar^2\pi^2(2n+1)}{2mL^2} \tag{6.25}$$

As n will be enormous in practical situations, $2n+1 \sim 2n$, meaning:

$$\Delta E = \frac{2n\hbar^2\pi^2}{2mL^2} \tag{6.26}$$

For a box of $L = 4\,\text{m}$, containing our argon with $m = 6.64\times10^{-26}$ kg, we obtain:

$$\Delta E = \frac{2\times3.44\times10^{11}\times\left(1.05\times10^{-34}\right)^2\times\left(3.14\right)^2}{2\times6.64\times10^{-26}\times16} = 3.52\times10^{-32}\ \text{J}$$

whereas E_n is:

$$E_n = \frac{n^2\hbar^2\pi^2}{2mL^2} = \frac{\left(3.44\times10^{11}\right)^2\times\left(1.05\times10^{-34}\right)^2\times\left(3.14\right)^2}{2\times6.64\times10^{-26}\times16} = 6.05\times10^{-21}\ \text{J}$$

This tells us that the energy gap between realistic quantum states is so small, we may as well take the argon atom's energy to be continuous.

6.3.3 Degeneracy

Equation (6.23) gives the energy levels for a 3D box:

$$E_n = \frac{\hbar^2\pi^2}{2mL^2}\left\{n_x^2 + n_y^2 + n_z^2\right\}$$

Up to now, we have assumed that each quantum state has a distinct and unique energy level. In fact, if we look at the three separate quantum numbers, the states $\left\{n_x, n_y, n_z\right\} = \{1,0,0\};\{0,1,0\};\{0,0,1\}$ will have the same energy and are hence *degenerate*. There are many more combinations where that is also the case.

Even if the box is not cubical (6.23):

$$E_{n_x,n_y,n_z} = \frac{\hbar\pi^2}{2m}\left(\frac{n_x^2}{L^2} + \frac{n_y^2}{W^2} + \frac{n_z^2}{H^2}\right)$$

degeneracy is possible whenever $n_x^2/L^2 = n_y^2/W^2 = n_z^2/H^2$.

6.4 UNCERTAINTY

The wave function for particle in a box (6.19):

$$\Psi_n(x,t) = \sqrt{\frac{2}{L}}e^{-i\frac{1}{\hbar}E_n t}\sin\left(\frac{1}{\hbar}p_n x\right)$$

allows us to calculate the probability of finding the particle within a region of the box. It forms a probability density such that $(0 \le x \le L)$:

$$\text{Prob}\left(x \rightarrow x+\Delta x\right) = \frac{2}{L}\int_{x}^{x+\Delta x} \Psi_n^* \Psi_\Delta \, dx \tag{6.27}$$

In more realistic situations, we use a combination of wave functions to construct a wave packet, which is more localized in space. For example, something like:[7]

$$\Phi\left(x,\, t\right) = \sum_{n=n_0-\Delta n}^{n_0+\Delta n} K\cos^2\left[\frac{\left(n-n_0\right)\pi}{2\left(\Delta n+1\right)}\right] e^{-i\left(n-n_0\right)\pi/2} \sin\left(\frac{n\pi x}{L}\right) e^{-\frac{i}{\hbar}E_n t} \tag{6.28}$$

(where $n_0 - \Delta n \leq n \leq n_0 + \Delta n$, $\Delta n \ll n_0$ and K is a normalization constant) gives a smooth, peaked probability distribution within the box (Figure 6.3).

In this instance, the most likely place to find the particle would be at the maximum $\langle x \rangle$, but there is a high probability of finding it between the broad shoulders of the peak. Typically, we measure the width of the peak by starting from the variance (5.16), which measures the average square distance from the mean:

$$\sigma^2\left(x\right) = \left\langle x^2 \right\rangle - \left\langle x \right\rangle^2$$

and then moving to the standard deviation:

$$\Delta x = \sigma = \sqrt{\left\langle x^2 \right\rangle - \left\langle x \right\rangle^2} \tag{6.29}$$

Statistically, the region bounded by $\langle x \rangle - \Delta x \leq x \leq \langle x \rangle + \Delta x$ covers roughly 68% of the population. In quantum theory, this is known as the uncertainty in the measurement. Repeated experiments using the same particles and boxes would find the particle within the uncertainty region ~ 68% of the time,

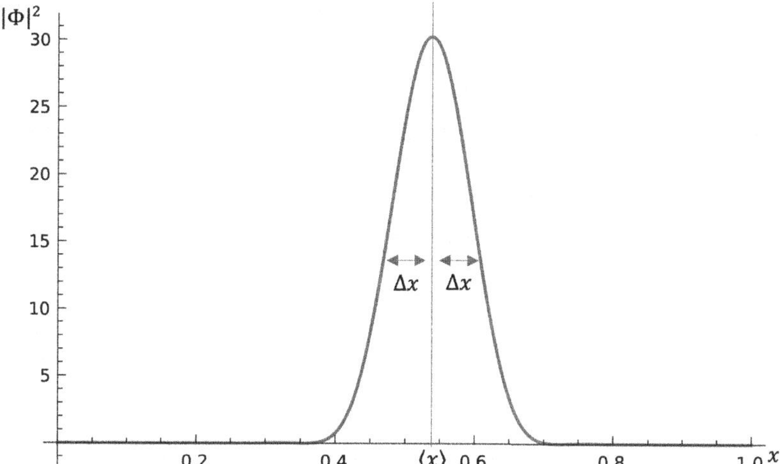

FIGURE 6.3 A wave packet is a combination of wave functions leading to a localized probability distribution for a particle's position. In this rendering of the probability distribution, t = 4s. Here, n_0 = 100 and Δn = 10. The walls of the box are not displayed but lie at $x = 0$ and $x = 1$. (Image produced by Mathematica code kindly provided by Michael Seifert, Associate Professor of Physics, Connecticut College.)

neglecting the standard experimental variations that take place. This is an inherent quantum limitation on the measurement, not a technological or methodological issue.

If instead of scrutinizing position we looked to measure the momentum of the particle, we would get a similar range of results, clustered around $\langle p \rangle$ with 68% 68% of results $\langle p \rangle - \Delta p \le p \le \langle p \rangle + \Delta p$. Again, this uncertainty is inherent and a reflection of the underlying probabilistic nature of the quantum world.

In 1927, Heisenberg published a relationship between these two measures:

$$\Delta x \Delta p \ge \hbar/2 \tag{6.30}$$

known universally as the *uncertainty principle*. This innocent looking inequality expresses a subtle but amazing link between an amplitude designed to yield a probability distribution over position $\Psi(x)$ and one that gives momentum $\varphi(p)$. The two amplitudes are mathematically linked.[8] If the peak of one is narrowed, the other broadens, the connection being (6.30). Any attempt to localize the particle destroys information regarding its momentum, and vice versa.

We will discuss the impact of this uncertainty relationship on our ability to count microstates in Section 7.1.2 and we will investigate its origin more in 11.1.5.

6.5 FIRST LAW REVISITED

A system, such as a gas, of $\mathcal{N}$ non-interacting particles shoehorned into a box would, by quantum theory, have a collection of energy levels, E_n. As the particles are non-interacting, finding one of them in a specific level, E_K, is a matter of indifference to the others. This is an important assumption. As a consequence, the energy levels are freely populated and their occupation, or otherwise, has no effect on the energy values themselves. If the system settles down into some pattern of occupancy, we can take the probability of finding a particle in energy state E_n to be p_n. If we now have a very large number of identical systems prepared in the same way and exposed to the same conditions,[9] the average energy across the systems will be:

$$\langle E \rangle = \sum_n p_n E_n \tag{6.31}$$

As a working hypothesis, to be developed further in Chapter 7, we take this average energy to be the energy we would observe in nature were we to measure an example system. In other words:

$$U = \sum_n p_n E_n \tag{6.32}$$

As a matter of simple approximation theory, we can write:

$$dU = \sum_n p_n dE_n + \sum_n E_n dp_n \tag{6.33}$$

Let's consider each of these terms separately.

The first denotes changes in the value of energy for each level, while the probability of finding a particle in that level remains the same. Taking our formula for the energy levels in a box (6.23):

$$E_n = \frac{n^2\hbar^2}{2mL^2}\left(n_x^2 + n_y^2 + n_z^2\right) \tag{6.34}$$

we can see that the only macroscopic variables involved are the dimensions of the box. In fact, other than the quantum numbers the rest of the terms are various fundamental constants. This justifies us writing:

$$dE_n = \frac{\partial E_n}{\partial V} dV \tag{6.35}$$

using V as the volume that the system occupies. Inserting this into the first term of (6.33):

$$dU\big|_{p_n} = \sum_n p_n dE_n = \sum_n p_n \frac{\partial E_n}{\partial V} dV = \frac{\partial}{\partial V}\left(\sum_n p_n E_n\right) dV = \frac{\partial \langle E_n \rangle}{\partial V} dV = \frac{\partial U}{\partial V} dV \tag{6.36}$$

From Chapter 2, we have (2.10) which, with $dQ = 0$, becomes:

$$\mathcal{P} = -\frac{\partial U}{\partial V}\bigg|_S \tag{6.37}$$

so, we feel warranted in writing:

$$dU\big|_{p_n} = -\mathcal{P}dV \tag{6.38}$$

and making the tentative identification of the first term with the work done on the system.

Unpicking this argument suggests that a (gentle, reversible, adiabatic) expansion of the gas results in the energy levels changing without disrupting the probability of finding particles in each level. This is the effect of allowing our gas to do work, and by inference also the effect of allowing work to be done on the gas.

However, it would be wrong to read too much into this argument. There are some hidden assumptions that would need to be justified more fully for a real system.

For example, we are assuming that the particles are reasonably distributed across a range of available states. That will be the case if they gently interact/collide with one another, and provided we give the system time to settle down. However, one of our primary assumptions has been that the particles are non-interacting. Hence, there is some conflict here.

We have simply shown that to some level of approximation, work in a gentle case adjusts the energy levels, not the distribution of particles. In more realistic (rough handling) cases, we would expect the approximations to break down and both the energy levels and their occupancy to change.

Now to the second term:

$$dU\big|_V = \sum_n E_n dp_n \tag{6.39}$$

Using the Gibbs entropy, as it involves probabilities (5.46):

$$\mathcal{S}_G = -k_B \sum_{n=1}^{\mathbb{N}} p_n \ln\left(p_n\right)$$

we find the differential form:

$$dS_G = -k_B \sum_{n=1}^{\mathbb{N}} \left(dp_n \ln(p_n) + \frac{p_n}{p_n} dp_n \right) = -k_B \sum_{n=1}^{\mathbb{N}} \left(\ln(p_n) + 1 \right) dp_n$$

$$= -k_B \left\{ \sum_{n=1}^{\mathbb{N}} \ln(p_n) dp_n + \sum_{n=1}^{\mathbb{N}} dp_n \right\} \tag{6.40}$$

In this equation, the second term can be struck out, as it is zero. Given any normalized probability distribution, $\sum p_n = 1$, so that $\sum dp_n = 0$. Hence:

$$dS_G = -k_B \sum_{n=1}^{\mathbb{N}} \ln(p_n) dp_n \tag{6.41}$$

One probability distribution that we have on the table is the Boltzmann distribution (5.43):

$$p_n = \frac{1}{Z} e^{-E_n / k_B T}$$

if we use this in (6.41):

$$dS_G = k_B \sum_{n=1}^{\mathbb{N}} \left(\ln(Z) + \frac{E_n}{k_B T} \right) dp_n$$

$$= k_B \left\{ \ln(Z) \cancel{\sum_{n=1}^{\mathbb{N}} dp_n} + \frac{1}{k_B T} \sum_{n=1}^{\mathbb{N}} E_n dp_n \right\} = \frac{1}{T} \sum_{n=1}^{\mathbb{N}} E_n dp_n \tag{6.42}$$

making:

$$T dS_G = \sum_{n=1}^{\mathbb{N}} E_n dp_n \tag{6.43}$$

Putting this into (6.39):

$$dU|_V = \sum_{n=1}^{\mathbb{N}} E_n dp_n = T dS_G \tag{6.44}$$

showing that the second term is the energy transfer via heating.

This demonstration hinges on adopting the Boltzmann distribution as the probability distribution involved and we have implicitly connected the Gibbs entropy with the thermodynamic function of state $dQ = TdS$. These two points are connected in a way that will be explored fully in the next chapter.

While this seems an intuitive and evident approach, there are subtleties involved. The Boltzmann distribution of (5.43) is defining the probabilities for a collection of microstates. This maps well into particle states if we have a single particle. However, we are clearly in a multi-particle regime, so that the microstates of the system overall must be related to the situations of each separate particle. Dealing with this would be a complex and lengthy diversion at this point. The reader is referred to Appendix PA1 to see how this resolves, at their leisure. Suffice to say, our intuitive approach turns out to be justified in the limit of large numbers, as we have specified.

6.6 SUMMARY

In this introductory chapter, we have established some basic aspects of the quantum world, such as the energy quantization that results from localizing a system to a region of space. We have started to build an understanding of amplitudes and the way in which they describe quantum objects. All of this provides foundations for the rest of this part of the book. Significantly, in the next chapter, we will use energy quantization to help distinguish microstates of a system.

NOTES

1 Louis Victor Pierre Raymond, 7th Duc de Broglie (1882–1987), Nobel Prize in Physics 1929.
2 Max Karl Ernst Ludwig Planck (1858–1947), Nobel Prize in Physics 1918.
3 Albert Einstein (1879–1955), Nobel Prize in Physics 1921.
4 Thomas Young FRS (1773–1829).
5 Where u is the atomic mass unit, introduced in Section 3.2.
6 To clarify the terminology, an *imaginary number* is any multiple of the square root of -1, i.e., $5i$, $8.9i$, ib are all imaginary numbers. A *complex number* take the form $a + ib$. Technically, from this perspective, the number a is also a complex number with $b = 0$. In that sense, measurements of physical variables do use complex numbers, but the key point is that there are no imaginary components to any measurements that we make.
7 Don't worry about where this comes from, we are not going to be using this wave packet. It is included here simply to illustrate the kind of construction that can be done.
8 One is the *Fourier transform* of the other.
9 This is the idea of an *ensemble*, first mentioned in Section 5.4 and discussed in detail in the next chapter.

7 Ensembles, Energy Levels, and Microstates

In the previous chapter, we set up a basic framework for exploring thermodynamics in a quantum regime. We have seen that constrained particles "in a box" have quantized energies (energy levels) specified by a quantum number that helps us to index microstates. Under appropriate and limited circumstances, we can model work done by a change in the energy values of the different levels while heating the system shuffles the pack in terms of the probability for occupying the levels. In this chapter, we proceed to fill in some vital details. The idea of an ensemble is formalized, and two of the key examples explored. We also derive the Boltzmann distribution in two complementary ways. Finally, we turn to the quantum version of an idea gas, and in the process derive a formula for the entropy. Our first task, however, is to discuss the issues pertinent to our ability to count microstates.

7.1 COUNTABILITY

In Chapter 5, we developed Boltzmann's statistical interpretation of entropy by considering the Joule expansion of a gas. It was then, *prima facie*, easy to define a statistical weight for the system by characterizing the set of microstates via the position of the molecules on one side of our partitioned box or the other. Consequently, for $\mathbb{N}$ particles, $\Omega = 2^{\mathbb{N}}$. However, even here, we are glossing over some issues:

- Exactly how do we define the boundary between the two sides of the box?
- Is the partition between the sides infinitesimally thin?
- If not, what are we to do with the few molecules within an arbitrarily small distance of that divide?
- Do we take the molecules themselves to have size?
- In which case, what fraction of the molecule must be over the divide to be counted as being on the other side?

These rather nit-picking examples point to a wider and more significant issue. Classical systems are described by properties that are generally continuous in nature. Specifically, each particle within the system has a location and momentum specified by a collection of six numbers: $\{x, y, z, p_x, p_y, p_z\}$. For a system of $\mathcal{N}$ particles, we imagine an abstract $6\mathcal{N}$- dimensional space, *phase space*, where each particle occupies a unique[1] point. The specific system microstate, defined by the individual particle states, forms a volume in phase space shaded out by the individual points. Trivially, within any one volume, it does not matter which particle is at which point.

Given that there are many different microstates compatible with a given macrostate, we can envisage a volume of phase space corresponding to the macrostate and spanning all the appropriate microstates. We know that the Boltzmann entropy relates to the number of microstates sitting under

DOI: 10.1201/9781003121053-10

TABLE 7.1
A Hypothetical Matching of Real Numbers (Right-Hand Column) to Integers (Left-Hand Column)

1	⇔	0	.	1	2	3	7	5	6	3	4	...
2	⇔	0	.	3	3	3	4	5	7	2	1	...
3	⇔	0	.	2	6	5	4	9	7	7	4	...
						etc						

TABLE 7.2

1	⇔	0	.	1	2	3	7	5	6	3	4	...
2	⇔	0	.	3	3	3	4	5	7	2	1	...
3		0	.	2	6	5	4	9	7	7	4	...
						etc						

a macrostate, so we can glimpse a connection between the phase space volume and the entropy, especially if we can count the microstates within that volume. However, when it comes to counting the number of microstates equivalent to a given microstate, we run into a profound issue: continuous quantities are not countable.

7.1.1 Cantor's Diagonal Argument

The first person to explore the rigours of countability was Georg Cantor[2] in 1891. His starting point was the set of positive integers (the *natural numbers*) $\mathcal{N} = \{1,2,3,\ldots\}$. Any other collection of objects would be *countable* if a rule could be defined, mapping each member of the collection to a single unique member of the naturals. The total number of objects in the collection would be the largest natural needed to ensure each object in the collection was matched. A collection is then countable if and only if a rule of this nature can be found.

Crucially, Cantor was able to show that *no such rule can be created for the real, continuous, numbers*. In outline, his proof works as follows.

First, we accept that any member of the set of real numbers can be expressed as a continuous decimal expansion. Next, we imagine that we have discovered a rule that matches each real number to a unique integer (Table 7.1).

Now consider the following number, 0.246... where I have taken the first digit from the first number and added one, the second digit from the second number and added one, the third digit from the third number and added one, etc. (Table 7.2).

This new number, formed by taking digits along the diagonal of the table, is not found in the list and hence is not mapped to an integer.

In this way, *we can construct a number which is not present in the list of numbers matched to the naturals*.

Our new number can't be the first number in the list, as the first digit of our new number is different to the first digit of the number in the list (it's actually +1). Equally it can't be the second number in the list as its second digit is different. In fact, it can't be the *n*th number as the *n*th digit is also different. Hence, the number is not on the list[3]. If it is not in the list, then it is not mapped to one of the naturals, so it has not been counted. Yet, it is clearly a real number, and so we are forced to conclude that the reals are not countable by this rule. As we have not specified the matching rule that we are using, the argument applies to *all possible rules*, so the real numbers are not countable under any circumstances.

By extension then, *states dependent on physical variables expressed as real numbers are not countable*, and we have a problem…

7.1.2 Coarse Graining

In classical (i.e., non-quantum) statistical mechanics, the countability issue is avoided by *coarse graining* the range of possibilities[4].

Effectively, we pick a collection of points within phase space and define a position—momentum *cell* of volume $\delta x \delta y \delta z \delta p_x \delta p_y \delta p_z$ surrounding each point. A cell will be occupied if one of our specific particle states lies within its bounds. Furthermore, if two or more particles occupy the same cell, we declare that these particles are in the same state. Granulating (coarse graining) phase space into these volume cells allows the number of cells that are occupied to be countable, and hence the number of microstates, Ω. We have a way to calculate the entropy.

However, as the size of the cells and hence the scale of the course graining is arbitrary, different choices could well give rise to different measures of entropy. Fortunately, it can be shown that these different entropies will all be within an additive constant of each other. Effectively, this is footnote 14 of Chapter 5.

Quantum theory gives some credence to this approach. It could be argued that Heisenberg's Uncertainty Principle (6.30):

$$\Delta p \Delta x \geq \hbar/2$$

divides phase space into region volumes $\sim \hbar^{3N}$ dependent on the number of particles in the system.

Energy quantization and state superposition also come into play. For example, if we have a system with energy E, then we can construct Ω by counting the number of energy states with the range $E \to E + \delta E$ (taking care to account for degenerate energy states as well). Once again, the specific value of entropy will be dependent on the choice of δE, but we hope that in the transition between microscopic and macroscopic descriptions, all such choices converge to give us a single value for entropy. This is known as the *thermodynamic limit*.

7.1.3 The Thermodynamic Limit

The notion of a thermodynamic limit is fundamental when you analyse a system statistically. We justify the use of statistical techniques by arguing that various approximations and probabilities are washed out as the number of particles in the system grows.

In Section 5.1 we reasoned that the predicted spread of values for a physical variable tightens $\sim 1/\sqrt{\mathbb{N}}$. Hence, in this limit, probabilities effectively become certainties. This calculation is done specifically for the ideal gas in Section 9.2.2.

Formally, the thermodynamic limit is approached by allowing the number of particles, $\mathbb{N}$, and the volume, V, of the modelled system to grow without limit, subject to $\mathbb{N}/V = \text{constant}$, i.e., that the number density of particles remains the same. Applying these limits to our statistical calculations should mean that the results converge on the observed macroscopic information. In this way, any aspects of our calculations that depend on the size of the system or any boundary conditions that may have been imposed at the edges drop out. Essentially, we reduce all the microscopic fluctuations in our calculated variables to insignificance compared with the macroscopic values.

However, it is worth noting that there are many significant aspects of system behaviour, such as phase changes, that only occur in the thermodynamic limit.

7.2 TERMINOLOGY

Before we can explore the relation between statistical mechanics and quantum theory, it is sensible to agree on some basic terminology:

Macrostate: the macrostate of a system is defined by values of the relevant thermodynamic variables (functions of state), such as pressure, volume, temperature, etc.

Microstate: this is a complete description of the system at the atomic scale. We will take a slightly simplified approach in this chapter and define the microstate by specifying the energy level for each particle in the system. This will need to be finessed when we deal with the quantum theory of identical particles (Chapters 8 and 9)

Single particle state: the state of an individual particle within the system. In quantum mechanics, this would be the wave function of the particle

Multi-particle state: a quantum state constructed by combining single particle states. If the particles are weakly interacting or non-interacting, then the microstates of the system would be the collection of multi-particle states made by combining single-particle states for every particle in the system. Otherwise, the Schrödinger equation would have to be solved to find the microstates of the system. In practice this would be an exceptionally difficult task for even the smallest reasonably sized multi-particle system

7.3 ENTROPIES

After the discussion in Section 1, we have three distinct approaches to entropy on the table.

First, we have the *thermodynamic entropy*, $\mathcal{S}_T$, (Chapter 4), shown to be a state function and enshrined in (4.17):

$$\Delta U = T\Delta\mathcal{S}_T - \mathcal{P}\Delta V$$

We will also refer to this as the *Clausius entropy* or occasionally the *phenomenological entropy* to emphasize that this slant on entropy has come from a high-altitude view of physics, using macroscopic properties and investigating their relationships. This entropy is certainly "real" as it is a function of state, but without a link to the microscopic behaviour of the system, we are left wondering exactly what it is that we are manipulating.

That question was partially answered in Chapter 5 where we introduced *the Boltzmann entropy*, $\mathcal{S}_B$, as the logarithm of statistical weight, Ω, which is a count of the number of microstates that are consistent with the given macrostate (5.23):

$$\mathcal{S}_B = k_B \ln(\Omega)$$

More generally, the macrostate of a system determines the volume of phase space populated by the appropriate microstates. The Boltzmann entropy follows from the count of microstates found within that volume. Performing that count is not easy and, as we have seen, typically requires coarse graining. There is also an evident limitation to this approach. It is entirely possible that the different microstates occur with different probabilities. In which case, it seems unreasonable that a very low probability state would be weighted equally to one of its highly probable siblings.

For such cases we need a more general form of entropy, which is the *Gibbs formula* (5.46):

$$\mathcal{S}_G = -k_B \sum_{n=1}^{\mathbb{N}} p_n \ln(p_n)$$

(where $\mathbb{N}$ is the number of microstates of the system consistent with the macrostate, and n is the index used to distinguish microstates.)

The value of (5.46) depends on the probability distribution, $\{p_n\}$, in play. This makes the Gibbs expression a *functional*, i.e., a measure that depends on a choice of function. We can think of it as

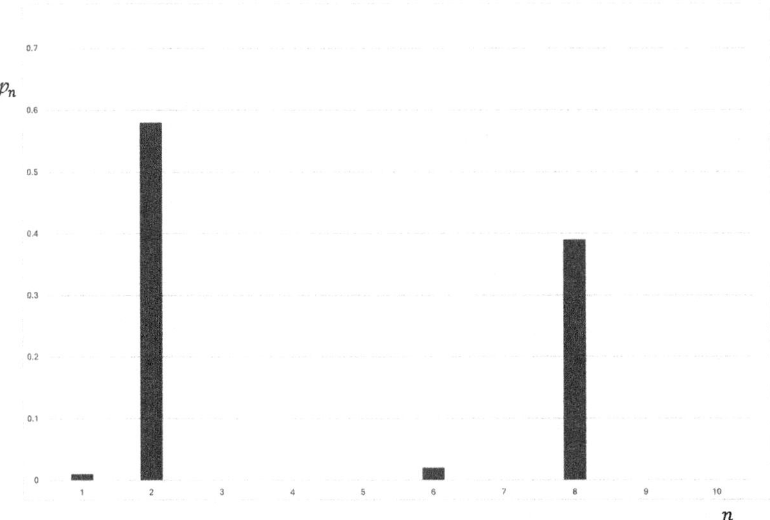

FIGURE 7.1 A probability distribution that is sharply peaked at two values.

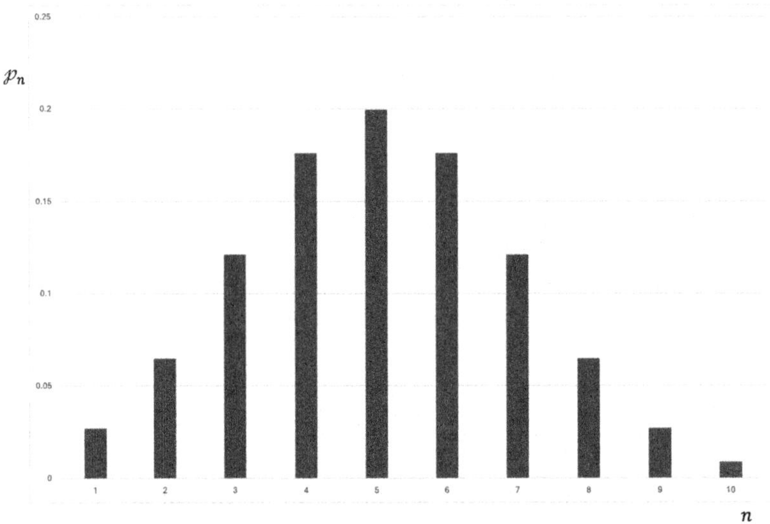

FIGURE 7.2 A discrete normal distribution of probabilities.

a "rating" or "score" for a particular probability distribution. This in turn reflects on our ability to know things about the system. Consider a probability distribution that is sharply peaked for certain values, Figure 7.1:

If this represented the microstates of a system, we would be pretty sure that we were going to encounter examples in states 2 and 8 more than anything else. The Gibbs functional for this distribution evaluates as $0.81k_B$. On the other hand, with a distribution like Figure 7.2 (which is a discrete version of the normal distribution's bell-curve), we are far less certain about what we will encounter. The Gibbs functional in this case is $2.0k_B$. Finally, with a completely flat distribution, Figure 7.3, we really don't know what we are going to find at any time and the Gibbs functional is $2.3k_B$.

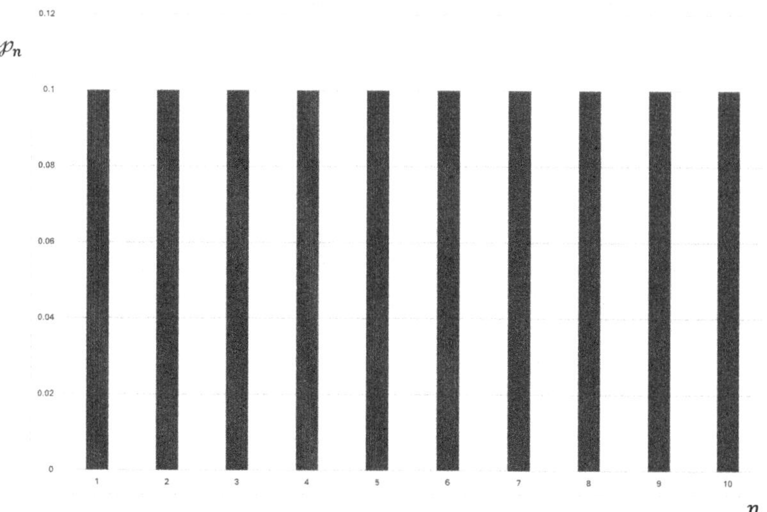

FIGURE 7.3 A flat probability distribution where each microstate is equally likely.

We propose *to take the Gibbs functional as a measure of our ignorance when we encounter a system.* Finding a probability distribution across microstates that is consistent with the macroscopic state of the system and which maximizes the Gibbs functional will a key aim. This probability distribution will be the one that we are most likely to encounter in nature for that system in that state when we know the least about what is happening. We further suggest that the value of the *Gibbs functional* over this distribution will be the *Gibbs entropy* and *equal to the thermodynamic entropy*. The rest of this section will be, in part, a justification for these proposals.

Once we have the probability distribution that maximizes the Gibbs functional, we can extract everything else of thermodynamic interest via the partition function, as we will see.

However, before we can move to carry out these determinations for different situations, we need a few more fundamental ideas from statistical mechanics.

7.4 ENSEMBLES

One of the most important concepts in statistical thermodynamics is the *ensemble*: a collection of identical systems prepared for our inspection in the same manner. This first came up in Section 5.4, but we now wish to place the idea on a more formal footing.

The ensemble could be a set of systems given to us all at the same time, or it could be the same system subjected, Sisyphus-like, to our experimentation over and over again, provided we have a mechanism for restoring it back to its original unblemished state. Either way, the ensemble provides a context for investigating the thermodynamic behaviour of a system and comparing it with the real world.

When we set out to prepare systems for membership, the best we can do is to pin down an individual macrostate. We do not have the level of fine control needed to select microstates. As a result, across the ensemble each microstate that is compatible with the macrostate will be represented. This is what gives the ensemble its power as a theoretical tool. If we subject each system to the same experiment, we expect all possible behaviours to arise. The more common behaviours will appear in many of the systems, and the less likely ones will be represented fewer times. We could say that the ensemble will draw a probability distribution for us.

Obviously, the more systems we have in our collection, the better this "simulation" will be, so in principle we like to have an infinite ensemble, or at the very least a large number which we can project to infinity. In a sense, this is taking the thermodynamic limit.

We will be interested in three types of ensemble[5]:

- The **microcanonical ensemble**: here, every system is completely isolated, shut off from any thermal or mechanical interaction with its surroundings or other members of the herd. When the systems are prepared, they will have a certain energy and contain a given number of particles. Those values are then fixed
Definition: $\mathcal{N}, V, T, U$ fixed
- The **canonical ensemble**: in this case, the systems are allowed thermal contact with a reservoir, either identical reservoirs, one for each system, the same reservoir for all systems or even, with a large ensemble, the herd of systems can act as a thermal reservoir for each individual
Definition: $\mathcal{N}, V, T$ fixed, $U = \langle E_n \rangle$
- The **grand canonical ensemble**: the systems are allowed to exchange energy and/or particles with reservoir(s)
Definition: V, T fixed $U = \langle E_n \rangle$, $\mathcal{N} = \langle \mathbb{N}_n \rangle$

Crucially, we will look for a probability distribution across the ensemble, which maximizes the Gibbs functional. Once we have that probability distribution, *the observed value of any macroscopic variable is given by the ensemble average of that variable*. For example, the observed energy of a system, U, is:

$$U = \left\langle E_n \right\rangle = \sum_{n=1}^{\mathbb{N}} p_n E_n \tag{7.1}$$

With that task in mind, we need a mathematical tool for maximizing the functions of more than one variable, especially when they are subject to constraints.

7.4.1 Looking for Functionals

Before we get to the techniques we need for maximizing, it's illuminating to look at the Gibbs functional in abstract.

Imagine we wanted to construct some functional, $\mathcal{F}(\{p_n\})$ on a probability distribution $\{p_n\}$. We need this functional to reflect the amount of information we could have about a system where the microstates are governed by $\{p_n\}$. We also want the functional to be relatable to entropy, under the right circumstances. Consequently our functional must be:

1. *Additive*—the entropy of two independent systems, considered as a whole, is the sum of the separate entropies $S(AB) = S(A) + S(B)$
2. *Continuous*—whatever function we use, it needs to be well behaved mathematically, so it can be differentiated and manipulated in other ways
3. *Maximized*—the functional should come to a *maximum* value when we know the *least* about the system, i.e., when the probability distribution over macrostates is flat
4. *Minimized*—the functional should be a *minimum* if we know *exactly which state the system should is in*, i.e., that the probability distribution has $p_n = 1, n = K; p_n = 0, n \neq K$

Probabilities for independent systems are multiplicative, but entropy is additive *so our functional must be based on a logarithm*, which is the continuous well-behaved mathematical object that converts multiplication into addition. As a start we take $\mathcal{F}(\{p_n\}) = K\log(p_n)$. The base we choose for the log is arbitrary, so for mathematical elegance we use the natural log. Our conditions 1 and 4 imply that the functional *increases* when states become *less probable*, so the constant of proportionality must be negative; $\mathcal{F}(\{p_n\}) = -K\ln(p_n)$. If we want this to converge on the thermodynamic entropy, we ought to take a clue from (7.1), and look to the ensemble average:

$$\langle \mathcal{F}(\{p_n\}) \rangle = -K\sum_n p_n \ln(p_n) \tag{7.2}$$

Finally, for consistency with the thermodynamic entropy, we should pick $K = k_B$ and we have:

$$\mathcal{S}_G = -k_B\sum_n p_n \ln(p_n) \tag{7.3}$$

Note that $\mathcal{S}_G = 0$ for the situation described in 4 above.

7.5 MAXIMIZING

We will frequently need to find the maximum value of a function of several variables, such as $\mathcal{F}(x_1, x_2, \ldots x_n)$. Generally, this is done by first building the differential form:

$$d\mathcal{F} = \frac{\partial \mathcal{F}}{\partial x_1}dx_1 + \frac{\partial \mathcal{F}}{\partial x_2}dx_2 + \ldots + \frac{\partial \mathcal{F}}{\partial x_n}dx_n = \sum_{i=1}^{n}\frac{\partial \mathcal{F}}{\partial x_i}dx_i \tag{7.4}$$

To identify a maximum, or minimum for that matter, we set this equal to zero, as we are looking for a "spot" where the function is hardly changing no matter which "direction" we move in:

$$\sum_{i=1}^{n}\frac{\partial \mathcal{F}}{\partial x_i}dx_i = 0 \tag{7.5}$$

If this is to be a local maximum (or minimum) then for each of the variables:

$$\frac{\partial \mathcal{F}}{\partial x_i} = 0 \tag{7.6}$$

Normally, the various x_i, and hence the dx_i are independent of one another, and so can take any value irrespective of the values of the others. Being independent, we can't reply on a conspiracy between them to find values that cause terms in (7.5) to cancel each other out. In which case, the only way to make (7.5) balance is if $\partial\mathcal{F}/\partial x_i = 0$ for each x_i.

However, in some situations, the variables are not independent. Instead, they are governed by a *constraint* that connects their values. For example, the number of particles sitting in different energy levels are subject to a simple constraint: they must add up to the total number of particles in the system. A constraint is expressed via a parallel function of the variables, such as $\mathcal{G}(x_1, x_2, \ldots x_n) = 0$. With one constraint in place, the first $(n-1)$ variables are free to adopt any values, but then the last, x_n, is fixed by having to comply with $\mathcal{G}(x_1, x_2, \ldots x_n) = 0$. Two constraints, $\mathcal{G} = 0$, and $\mathcal{H} = 0$ means that $(n-2)$ variables are free, with the values of the last two nailed by the constraints. Of course, the order of the variables is not important. Simply that the number of free variables is the total number minus the count of constraints in place.

So, with any collection of constraints governing values of variables, we cannot make the jump from (7.5):

$$\sum_{i=1}^{n} \frac{\partial \mathcal{F}}{\partial x_i} \Delta x_i = 0$$

to $\frac{\partial \mathcal{F}}{\partial x_i} = 0$ for each x_i.

The technique needed to fix this issue was invented by Lagrange.[6] He suggested writing:

$$\mathcal{F}' = \mathcal{F} + A\mathcal{G} + B\mathcal{H} \tag{7.7}$$

where A and B are *Lagrange multipliers* (constants that will be determined later), and proceeding to maximize $\mathcal{F}'$ instead. We can always do this, as $\mathcal{G} = 0$, and $\mathcal{H} = 0$, so $\mathcal{F}' = \mathcal{F}$. Now when we construct the differential form we have:

$$d\mathcal{F}' = \sum_{i=1}^{n} \frac{\partial \mathcal{F}}{\partial x_i} dx_i + A\sum_{i=1}^{n} \frac{\partial \mathcal{G}}{\partial x_i} dx_i + B\sum_{i=1}^{n} \frac{\partial \mathcal{H}}{\partial x_i} dx_i$$

$$= \sum_{i=1}^{n} \frac{\partial}{\partial x_i} (\mathcal{F} + A\mathcal{G} + B\mathcal{H}) dx_i = 0 \tag{7.8}$$

which we set to zero to maximize, as shown.

For the first $(n-2)$ variables we select, their independence allows us to conclude:

$$\frac{\partial}{\partial x_i} (\mathcal{F} + A\mathcal{G} + B\mathcal{H}) = 0 \tag{7.9}$$

which we can solve to maximize/minimize $\mathcal{F}'$ with respect to those variables. This is the same as maximizing/minimizing $\mathcal{F}$ as the constraints do not add to the function.

For the remaining two variables, which are not so free as their colleagues, we have the Lagrange multipliers to help. *We find values of A and B that force:*

$$\frac{\partial}{\partial x_i} (\mathcal{F} + A\mathcal{G} + B\mathcal{H}) = 0 \tag{7.10}$$

irrespective of the dx_i. We are then free to universally apply (7.9) and proceed to our maximum/minimum.

Lagrange's elegant solution will allow us to work with different sorts of ensemble and extract thermodynamic information.

7.6 THE MICROCANONICAL ENSEMBLE

The microcanonical ensemble seems to represent a rather facile situation where nothing much can happen, but in fact, we can make some important deductions. Also, it gives us an interesting test case for the techniques we will apply to other forms of ensemble.

A specific microcanonical ensemble will contain isolated systems in the same macrostate, and so all possible compatible microstates will be present, if we have enough examples in the ensemble.

At first, it is tempting to think that having every system completely isolated would result in no constraints for us to worry about. After all, we have prepared the systems to correspond to selected values of energy, E, and a number of particles, $\mathcal{N}$. The systems are then isolated, so the volume does not change. There seem to be no external constraints acting. However, this is not true. Our aim is to maximize the Gibbs functional, leading to a probability distribution, which if it is to be well behaved must be normalized. So, we seek to maximize (5.46):

$$\mathcal{S}_G = -k_B \sum_{n=1}^{\mathbb{N}} p_n \ln(p_n)$$

subject to the *normalization constraint*:

$$\sum_{n=1}^{\mathbb{N}} p_n = 1 \qquad \sum_{n=1}^{\mathbb{N}} p_n - 1 = 0 \tag{7.11}$$

Following Lagrange's advice we first build:

$$\mathcal{S}_G' = -k_B \sum_{n=1}^{\mathbb{N}} p_n \ln(p_n) + A\left(\sum_{n=1}^{\mathbb{N}} p_n - 1\right) \tag{7.12}$$

The differential form is (using (6.41)):

$$d\mathcal{S}_G' = -k_B \sum_{n=1}^{\mathbb{N}} \ln(p_n)\, dp_n + A \sum_{n=1}^{\mathbb{N}} dp_n$$

$$= -k_B \sum_{n=1}^{\mathbb{N}} \left(\ln(p_n) - A\right) dp_n \tag{7.13}$$

which we set equal to zero and, comfortable that the constraint is dealt with, deduce:

$$\ln(p_n) - A = 0 \tag{7.14}$$

for each p_n. A deft rearrangement yields:

$$p_n = e^A \tag{7.15}$$

Now we need to find the value of A, which we can do from the normalization constraint. After all:

$$\sum_{n=1}^{\mathbb{N}} p_n = \sum_{n=1}^{\mathbb{N}} e^A = \mathbb{N}e^A = 1 \tag{7.16}$$

So:

$$e^A = \frac{1}{\mathbb{N}} \qquad A = -\ln(\mathbb{N}) \tag{7.17}$$

In fact, we don't need A in this case. The important point is the left-hand side of (7.17), which tell us that $p_n = 1/\mathbb{N}$ for all n. Crucially, in the microcanonical ensemble *each microstate is equally likely*.

One of the most important assumptions behind statistical mechanics is the *ergodic hypothesis* (due to Boltzmann and Maxwell). In broad terms, this claims that any point in phase space can be reached from any other point with the same energy. Specifically, a single isolated system will, over sufficient time, wander across all points in the phase space of the whole system consistent with the energy we gave it to start with. If we have an ensemble of systems, then each point in phase space will be equally represented, with a large enough herd in the ensemble. This makes *the time-averaged behaviour of a single system equal to the ensemble average behaviour of the herd*.

Given a collection of systems meandering through phase space, or equally exploring all microstates, we have very little information about the ensemble, *which is why the Gibbs functional is maximized in this situation*. We now know the probability distribution that this produces, so we can determine the relevant value of the Gibbs functional:

$$\mathcal{S}_G = -k_B \sum_{n=1}^{\mathbb{N}} p_n \ln(p_n) = k_B \sum_{n=1}^{\mathbb{N}} \frac{1}{\mathbb{N}} \ln(\mathbb{N}) = k_B \ln(\mathbb{N}) \tag{7.18}$$

which is the Boltzmann entropy. This establishes the more universal applicability of the Gibbs functional, and that the Boltzmann entropy is only useful in contexts where the microstates are all equally likely, as we mentioned in Section 7.3.

To summarize, the ergodic hypothesis (that an isolated system will wander through all its microstates over time) means that we have no specific information about where the system is at any time, which justifies our maximizing the Gibbs functional, which leads to equal probability microstates (consistent with the ergodic hypothesis) giving the Boltzmann entropy. *The Boltzmann entropy is the entropy of an ensemble of isolated systems*: the microcanonical ensemble. Now let's apply the microcanonical ensemble to a quantum gas of particles.

7.6.1 Energy Levels

The results of Chapter 6 show us that any system localized to a certain volume of space will have quantized energy levels. Our quantum gas is going to be such a system. As a result, the particles in the gas can only exist with certain specified discrete energy values. We will use drawings such as that shown in Figure 7.4 to facilitate imagining how such a system works.

Each horizontal line represents an energy level with a specific value, ε_n on some notional vertical scale representing energy. The horizontal length of the line has no physical meaning and is simply drawn to give room for further annotations.

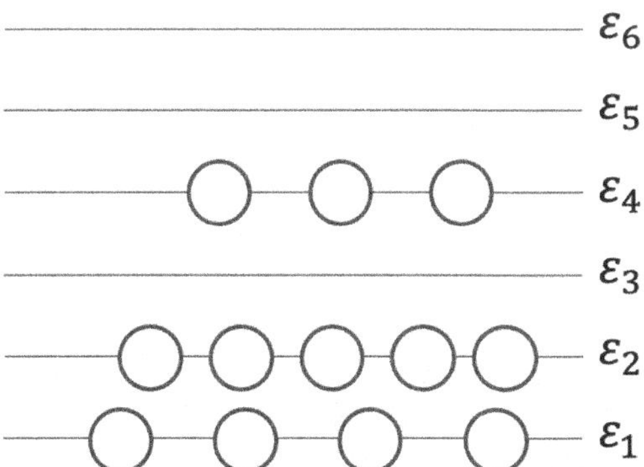

FIGURE 7.4 Notional energy levels for a localized system with particles distributed among the energy levels.

In Figure 7.4, we have displayed the first six energy levels for a system. They are shown as equally spaced, for convenience. In practice, equal spacing is unlikely, but it would add a significant complication to the argument at this stage.

The particles making up our system are inserted into the diagram as "blobs" sitting on the horizontal lines. Their horizontal positioning has no physical relevance.

In this simple situation, the single particle state of an individual particle is the energy level it is occupying. Our assumption that the particles are not interacting with each other means that the energy level adopted by any one particle is not influenced by what the others are doing.

We will use $\mathbb{N}_n$ to represent the number of particles in level n, and hence the number of particles with energy ε_n. The total number of particles is:

$$\mathcal{N} = \sum_1^{\mathbb{N}} \mathbb{N}_n = 12 \tag{7.19}$$

and the total energy of the system:

$$E = \sum_1^{\mathbb{N}} \mathbb{N}_n \varepsilon_n = 4 \times \varepsilon_1 + 5 \times \varepsilon_2 + 0 \times \varepsilon_3 + 3 \times \varepsilon_4 + \ldots \tag{7.20}$$

If we set $\varepsilon_1 = \epsilon$, $\varepsilon_2 = 2\epsilon$, and $\varepsilon_n = n\epsilon$, then $E = 26\epsilon$. The combination $\{E, \mathcal{N}\} = \{26\epsilon, 12\}$ defines the macrostate of the system.

In this case, the microstate is specified by an ordered set of numbers identifying the energy level occupied by each particle. If we take Figure 7.4 and enhance it by including particle labels, we get Figure 7.5, and we can specify this microstate by the ordered collection $\{4,2,1,4,2,1,4,2,1,2,2,1\}$.

Clearly, this is not the only way that the particles can be arranged in this fashion. A few alternatives are shown in Figure 7.6. Each of these represents a distinct microstate leading to the same macrostate, $\{26\epsilon, 12\}$.

In fact, the number of microstates following this pattern is given by:

$$\Omega_{\text{PT}} = \frac{\mathcal{N}!}{\mathbb{N}_1!\mathbb{N}_2!\ldots\mathbb{N}_{\mathbb{N}}!} \tag{7.21}$$

The numerator, $\mathcal{N}!$, is the number of ways of picking $\mathcal{N}$ objects out of a bag. This vastly overcounts the number of microstates following this pattern, as the order of particles on each horizontal row has

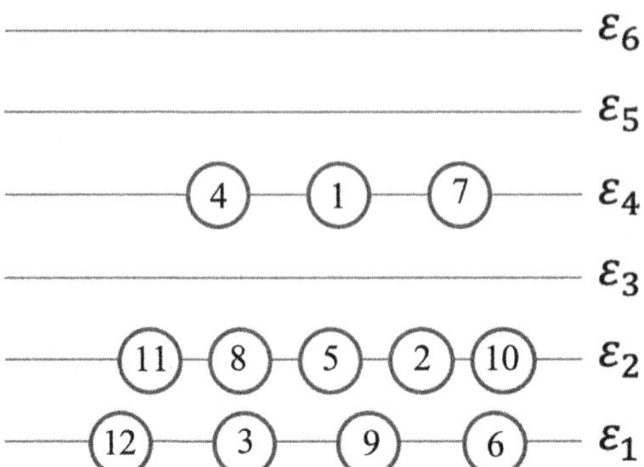

FIGURE 7.5 In this figure the particles in each energy level are identified by a numerical label.

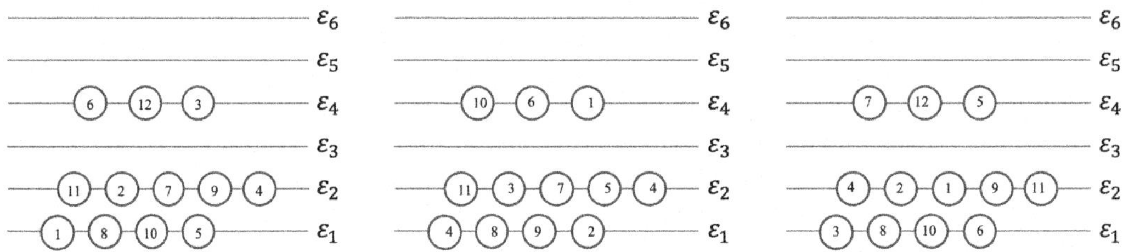

FIGURE 7.6 Different ways in which particles can be distributed following the same pattern as Figure 7.4.

no physical relevance. If there are $\mathbb{N}_n$ particles in row n, they can be ordered $\mathbb{N}_n!$ ways so we need to divide by that factor to eliminate the over counting. This suggests that there are Ω_{PT} microstates following this pattern in the energy levels and leading to that macrostate $\{26\epsilon, 12\}$.

However, that is not quite the end of the story. It is very likely that *a different pattern of particles across energy levels would lead to the same macrostate*. To see this, we're going to simplify things a little further and reduce to 6 particles.

Table 7.3 shows the different patterns leading to the macrostate $\{15\epsilon, 6\}$. Here we have extended the number of energy levels up to 10. These are the rows at the top of the table. The number of particles in each energy level is indicated by values in the rows.

The columns labelled A–Z are the 26 different patterns across energy levels leading to the macrostate. For each pattern, we have also calculated the number of microstates that adhere to that pattern. For example, there are 120 distinct microstates corresponding to pattern G. In total, there are 2002 microstates consistent with the macrostate.

If we were to construct a microcanonical ensemble of systems (each being a "thin" gas of 6 particles) and ensured that they were all in the macrostate $\{15\epsilon, 6\}$, then all the 2002 microstates would be equally likely and found an equal number of times across the ensemble.[7] However, if we were to pick out a system at random, there is a good chance that it would be in one of the patterns N, O, P, Q, V as all of these have 180 microstates.

With this data, we can calculate the *average occupation of the energy levels*. For example, if we take the first energy level, which has energy ϵ, then the average number of particles that we would find across the ensemble in this level is:

$$\langle \mathbb{N}_1 \rangle = \frac{\begin{array}{c} 5\times 6 + 4\times(4\times 30) + 3\times(3\times 120) + 3\times(3\times 60) + 3\times 20 + 2\times 60 + 4\times(2\times 180) \\ + 2\times 60 + 1\times 30 + 1\times 120 + 1\times 60 + 1\times 180 + 1\times 30 \end{array}}{2002}$$

$$= 2.143$$

TABLE 7.3
The Number of Diagrams (Labelled A–Z) and Number of Microstates with Six Particles and 15 Units of Energy

	Patterns																									
Energy Levels	**A**	**B**	**C**	**D**	**E**	**F**	**G**	**H**	**I**	**J**	**K**	**L**	**M**	**N**	**O**	**P**	**Q**	**R**	**S**	**T**	**U**	**V**	**W**	**X**	**Y**	**Z**
10	1																									
9		1																								
8			1						1																	
7				1		1							1													
6					1		1			1				1					1							
5					1			1			2				1		1			1				1		
4				1			1	1				3			1	2		1			2	1			1	
3			1			1		1		2				1		1	2	3		1		2	4		1	3
2		1				1	1		2		1		3	2	2	1	1		4	3	3	2	1	5	4	3
1	5	4	4	4	4	3	3	3	3	3	3	3	2	2	2	2	2	2	1	1	1	1	1			
Microstates per pattern	6	30	30	30	30	120	120	120	60	60	60	20	60	180	180	180	180	60	30	120	60	180	30	6	30	20
Total Particles	6	6	6	6	6	6	6	6	6	6	6	6	6	6	6	6	6	6	6	6	6	6	6	6	6	6
Total Energy	15	15	15	15	15	15	15	15	15	15	15	15	15	15	15	15	15	15	15	15	15	15	15	15	15	15

TABLE 7.4
The average occupation number for energy levels 1–10

Energy Level	⟨N⟩
10	0.003
9	0.015
8	0.045
7	0.105
6	0.210
5	0.378
4	0.629
3	0.989
2	1.484
1	2.143

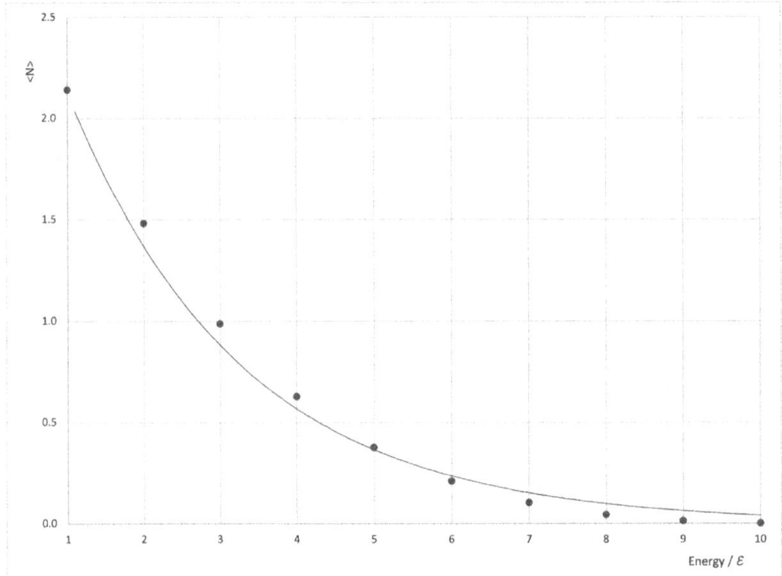

FIGURE 7.7 The average occupation number plotted as a function of energy.

Note: The line is an exponential fit.

Repeating the same calculation for each level in turn produces the data in Table 7.4, which is then plotted in Figure 7.7. The line shown to guide the eye happens to be a fit[8] of the form $\langle \mathbb{N} \rangle = Ae^{-kx}$ ...

7.6.2 Pattern Recognition

An interesting possibility suggests itself. What if one of the patterns dominated the statistics as it had so many microstates adhering to it? It is very possible that allowing the number of particles in the system to grow to a realistic level, one pattern would emerge as having a vast number of microstates in its collection. We could then *ignore all the other patterns*. To all intents and purposes, this one pattern would be found so many times across the ensemble that all the averages and other calculations could be done *just for that pattern*, as the others would only give negligible contributions.

To see if this is the case, we need to find the pattern with the largest number of microstates. In other words, we must maximize (7.21):

$$\Omega_{PT} = \frac{\mathcal{N}!}{\mathbb{N}_1!\mathbb{N}_2!\ldots\mathbb{N}_{\mathbb{N}}!}$$

subject to the constraints:

$$\sum_1^k \mathbb{N}_n\varepsilon_n - \mathrm{U} = 0 \qquad \sum_1^k \mathbb{N}_n - \mathcal{N} = 0 \tag{7.22}$$

However, before we start on that process there is one more piece of physics that we ought to take care of.

7.6.3 Dealing with Degeneracy

In Section 6.3.3, we showed how different combinations of quantum numbers can lead to the same energy level, so various quantum states are degenerate. We need to take account of this in our deliberations. In essence, for every line that we drew across Figure 7.4 to indicate an energy level, there could be multiple quantum states with the same energy. To cater for such possibilities, we denote g_n as the number of degenerate quantum states with energy ε_n.

For the moment, and subject to later revision, we assume that our particles are distinguishable so that there is no restriction on the number of particles that can occupy each quantum state.[9] Hence, as we build our pattern by placing $\mathbb{N}_n$ particles one by one into a quantum state with energy ε_n, we have g_n choices each time. Our formula for the total number of microstates in each pattern becomes:

$$\Omega_{PT} = \frac{\mathcal{N}!g_1^{\mathbb{N}_1}g_2^{\mathbb{N}_2}\ldots g_i^{\mathbb{N}_i}\ldots g_k^{\mathbb{N}_{\mathbb{N}}}}{\mathbb{N}_1!\mathbb{N}_2!\ldots\mathbb{N}_i!\ldots\mathbb{N}_{\mathbb{N}}!} = \mathcal{N}!\prod_1^k \frac{g_n^{\mathbb{N}_n}}{\mathbb{N}_n!} \tag{7.23}$$

which we need to maximize.

7.6.4 To the Max

The first step in making this a tractable proposition is to maximize $\ln(\Omega_{PT})$ rather than Ω_{PT}:

$$\ln(\Omega_{PT}) = \ln\left(\frac{\mathcal{N}!g_1^{\mathbb{N}_1}g_2^{\mathbb{N}_2}\ldots g_i^{\mathbb{N}_i}\ldots g_k^{\mathbb{N}_{\mathbb{N}}}}{\mathbb{N}_1!\mathbb{N}_2!\ldots\mathbb{N}_i!\ldots\mathbb{N}_{\mathbb{N}}!}\right) = \ln(\mathcal{N}!) - \sum_1^{\mathbb{N}}\ln(\mathbb{N}_n!) + \sum_1^{\mathbb{N}}\mathbb{N}_n\ln(g_n) \tag{7.24}$$

A convenient consequence of making this move is that we can now deploy Stirling's Approximation[10] (5.29) $\ln(n!) \approx n\ln(n) - n$ for sufficiently large values of n. Given that our task relates to macroscopic systems, we shall certainly fulfil that criterion for each of our $\mathbb{N}_n$. We now have:

$$\ln(\Omega_{PT}) = \mathcal{N}\ln(\mathcal{N}) - \mathcal{N} - \sum_1^{\mathbb{N}}\left(\mathbb{N}_n\ln(\mathbb{N}_n) - \mathbb{N}_n + \mathbb{N}_n\ln(g_n)\right) \tag{7.25}$$

The next step involves a mathematical fudge. To maximize, we need the differential form, which is only defined for continuous variables. However, as physicists we do not always follow the mathematical niceties. Instead, we wave our hands, claim that as each $\mathbb{N}_n$ is so large, the difference between $\mathbb{N}_n$ and $\mathbb{N}_n+1$ is so small compared to $\mathbb{N}_n$, we may as well call it continuous and press on (we have not forgotten about the constraints…):

$$d\left(\ln\left(\Omega_{\mathrm{PT}}\right)\right)=\sum_{1}^{\mathbb{N}}\frac{\partial}{\partial \mathbb{N}_n}\left(\mathcal{N}\ln\left(\mathcal{N}\right)-\mathcal{N}-\sum_{1}^{\mathbb{N}}\left(\mathbb{N}_n\ln\left(\mathbb{N}_n\right)-\mathbb{N}_n+\mathbb{N}_n\ln\left(g_n\right)\right)\right)d\mathbb{N}_n$$

$$=\sum_{1}^{\mathbb{N}}\left(-\frac{\partial\left(\mathbb{N}_n\ln\left(\mathbb{N}_n\right)\right)}{\partial \mathbb{N}_n}+\frac{\partial \mathbb{N}_n}{\partial \mathbb{N}_n}+\frac{\partial \mathbb{N}_n}{\partial \mathbb{N}_n}\ln\left(g_n\right)\right)d\mathbb{N}_n$$

$$\sum_{1}^{\mathbb{N}}\left(-\frac{\partial \mathbb{N}_n}{\partial \mathbb{N}_n}\ln\left(\mathbb{N}_n\right)-\mathbb{N}_n\frac{\partial \ln\left(\mathbb{N}_n\right)}{\partial \mathbb{N}_n}+1+\ln\left(g_n\right)\right)d\mathbb{N}_n$$

$$=-\sum_{1}^{\mathbb{N}}\left(\ln\left(\mathbb{N}_n\right)+\ln\left(g_n\right)\right)d\mathbb{N}_n \tag{7.26}$$

Now with the constraints added in:

$$d\left(\ln\left(\Omega_D\right)\right)=-\sum_{1}^{\mathbb{N}}\left(\ln\left(\mathbb{N}_n\right)+\ln\left(g_n\right)\right)d\mathbb{N}_n+A\sum_{1}^{\mathbb{N}}d\mathbb{N}_n+B\sum_{1}^{\mathbb{N}}d\mathbb{N}_n\varepsilon_n$$

$$=\sum_{1}^{\mathbb{N}}\left(-\ln\left(\mathbb{N}_n\right)+\ln\left(g_n\right)+A+B\varepsilon_n\right)d\mathbb{N}_n \tag{7.27}$$

This can be set to zero to maximize:

$$\Delta\left(\ln\left(\Omega_D\right)\right)=\sum_{1}^{\mathbb{N}}\left(-\ln\left(\mathbb{N}_n\right)+\ln\left(g_n\right)+A+B\varepsilon_n\right)d\mathbb{N}_n=0 \tag{7.28}$$

From which we obtain

$$-\ln\left(\mathbb{N}_n\right)+\ln\left(g_n\right)+A+B\varepsilon_n=0 \tag{7.29}$$

Hence:

$$\ln\left(\frac{\mathbb{N}_n}{g_n}\right)=A+B\varepsilon_n$$

$$\frac{\mathbb{N}_n}{g_n}=e^{A+B\varepsilon_n} \tag{7.30}$$

or equivalently:

$$\frac{\mathbb{N}_n}{g_n} = Ke^{B\varepsilon_n} \tag{7.31}$$

While we still have several mallards that require rectilinear arrangement before we can make a formal identification, our argument has evidently converged on the Boltzmann factor and so we expect to find that $B = -1/k_B T$.

7.6.5 Ponder Point

It is worth pausing to take in the significance of this results. We set out in a spirit of optimism to find the pattern that governed the largest number of microstates. Our result shows that such a pattern exists and can be constructed by populating the states according to (7.31), which is the Boltzmann distribution. However, we have not justified our other hope: that this pattern would have so many microstates under its wing, it would be pointless to include any others in calculations. Yet, the fact that we have converged on the Boltzmann factor, which we know applies for systems in nature justifies and confirms this hope!

One other aspect of this needs some further thought. Dividing (7.31) by the total number of particles, brings a probability (using $B = -1/k_B T$):

$$p_n = \frac{K}{\mathcal{N}} e^{-\varepsilon_n / k_B T} \tag{7.32}$$

i.e., the chance of finding a particle in a state with energy ε_n, at least once it has been properly normalized. However, we started this by constructing a microcanonical ensemble where each microstate is equally likely. What gives?

The Boltzmann distribution enshrined in (7.32) is *not* giving the probability of a *system microstate*, rather p_n *in this case* it is the probability of *a single particle state* for the particles within the system.

Finally, there is one more question. When this distribution was first derived in Chapter 5, the systems were in thermal contact with a reservoir. Here, the microcanonical ensemble is a collection of isolated systems. Once again, there is a danger of confusing the wood with the trees. The distribution we have found *applies to the single-particle states of any one particle within our system*. So, any one particle within any one system across the ensemble. From this perspective, we are treating the single particles as *tiny individual systems* within an "ensemble" (gas), which is *our larger system* represented multiple times across the ensemble. Within the gas, *the other particles are acting as the thermal reservoir for our individual.*

We will develop this result and use it later in the chapter. For now, we are going to re-enforce our understanding of the Boltzmann distribution by switching to the canonical ensemble.

7.7 THE CANONICAL ENSEMBLE

Many times, we do not know the energy of a system. Sometimes, we don't even care. If the system is in thermal contact with a reservoir, then provided a sufficient time has elapsed to ensure that the situation has reached equilibrium, we do know that the system and the reservoir will be the same *temperature*. When $\mathcal{N}, V, T$ are fixed but E is allowed to fluctuate, we have the canonical ensemble.

Although we do not *fix* the energies of the systems across our ensemble, we do *constrain* them. The ensemble average energy E_n must equal our observed energy, U. This is just like the probability constraint that we put in place with the microcanonical ensemble. With that in mind, we modify (7.12) to include the energy constraint:

$$\mathcal{S}'_G = -k_B \sum_{n=1}^{\mathbb{N}} p_n \ln(p_n) + A\left(\sum_{n=1}^{\mathbb{N}} p_n - 1\right) + B\left(\sum_{n=1}^{\mathbb{N}} p_n E_n - U\right) \tag{7.33}$$

The differential form is then:

$$d\mathcal{S}'_G = -k_B \sum_{n=1}^{\mathbb{N}} \left(\ln(p_n) - A - BE_n\right) dp_n \tag{7.34}$$

Which we set equal to zero, hence deducing:

$$\ln(p_n) - A - BE_n = 0$$

$$\ln(p_n) = A + BE_n$$

$$p_n = e^{A+BE_n} = Ke^{BE_n} \tag{7.35}$$

and we have the Boltzmann distribution again. Arriving at the same distribution using the canonical ensemble confirms that the approach in Section 7.6 worked in part, as the other particles in the gas were effectively acting as a reservoir for the individual. Of course, we neglected interactions between the particles, so they would have to be included in a full formal demonstration.

7.7.1 The Last Brick in the Wall

The canonical ensemble constrains its member systems, so the ensemble average energy is equal to the observed energy (7.1):

$$U = \sum_{n=1}^{\mathbb{N}} p_n E_n$$

If we allow the observed internal energy to change, in an infinitesimal and reversible manner (6.33):

$$dU = \sum_{n=1}^{\mathbb{N}} p_n dE_n + \sum_{n=1}^{\mathbb{N}} dp_n E_n$$

as we explored in Chapter 6. If infinitesimal work is done, that shifts the energy in the levels, if energy flows via heating, then the level probability/occupation changes.

Taking the Gibbs functional in its differential form (modifying (7.13) by removing the constraint):

$$d\mathcal{S}_G = -k_B \sum_{n=1}^{\mathbb{N}} \ln(p_n) dp_n$$

and inserting the Boltzmann distribution in the form that we have gets us:

$$d\mathcal{S}_G = -k_B \sum_{n=1}^{\mathbb{N}} \left(\ln(K) + BE_n\right) dp_n = -k_B \left\{ \sum_{n=1}^{\mathbb{N}} \ln(K) dp_n + B\sum_{n=1}^{\mathbb{N}} E_n dp_n \right\} = -k_B B \,\Delta U|_V$$

the first term dropping out as $\sum dp_n = 0$. We can now refer to 4.16:

$$dS = \frac{dQ}{T}$$

and make the identification $B = -1/k_B T$.

As a passing thought, the ability to make this identification also justifies our taking the maximum value of the Gibbs functional, which after all is what lead to the Boltzmann distribution, as being equivalent to the thermodynamic entropy.

7.7.2 A Nice Aside

We're now going to find an expression which we will need to use in a later chapter.

We start by considering:

$$\frac{\partial}{\partial T}\left(e^{-E_n/k_BT}\right) = \frac{E_n e^{-E_n/k_BT}}{k_B T^2} = Z\frac{E_n p_n}{k_B T^2}$$

Now we pluck (5.47) from Chapter 5 (amending the notation slightly):

$$\mathcal{S}_G = \frac{1}{T}\langle E\rangle + k_B \ln(Z)$$

and play:

$$\langle E\rangle = \sum_{n=1}^{\mathbb{N}} E_n p_n = \frac{k_B T^2}{Z}\sum_{n=1}^{\mathbb{N}} \frac{\partial}{\partial T}\left(e^{-E_n/k_BT}\right) = \frac{k_B T^2}{Z}\frac{\partial}{\partial T}\sum_{n=1}^{\mathbb{N}} e^{-E_n/k_BT} = \frac{k_B T^2}{Z}\frac{\partial Z}{\partial T}$$

Hence:

$$\mathcal{S}_G = \frac{k_B T}{Z}\frac{\partial Z}{\partial T} + k_B \ln(Z) = k_B \frac{\partial}{\partial T}\left(T\ln(Z)\right) \tag{7.36}$$

This elegant and useful expression re-enforces the utility of the partition function in calculations as we suggested at the end of Section 7.3

For the moment, this closes our discussion of different ensembles. However, we will be taking a very close look at the grand canonical ensemble in Chapter 9.

7.8 CLOSELY SPACED ENERGY LEVELS

Arguably, there is a significant flaw in our reasoning thus far. We have regarded the number of particles in an energy level, $\mathbb{N}_n$, as being pseudo-continuous. Indeed, the energy values ε_n have indirectly been treated in a similar fashion. Our initial explorations lent themselves to this, as we were considering very simple systems with a small number of particles and distinct energy levels. In practice, a typical macroscopic system would contain a vast number of particles and energy levels that are significantly crowded together.

To make the transition to a slightly more plausible case, the degeneracy factor, g_i, morphs into a *density of states*, $g(\varepsilon)$, whereby the number of states between ε and $\varepsilon + d\varepsilon$ is $g(\varepsilon)d\varepsilon$:

$$d\mathbb{N} = g(\varepsilon)d\varepsilon \tag{7.37}$$

or, more formally[11]:

$$g(\varepsilon) = \frac{d\mathbb{N}}{d\varepsilon} \tag{7.38}$$

In which case, rather than $\mathbb{N}_n$ we now have:

$$d\mathcal{N} = Kg(\varepsilon)e^{-\varepsilon/k_BT}d\varepsilon \tag{7.39}$$

This leads us to:

$$\mathcal{N} = \int_0^\infty Kg(\varepsilon)e^{-\varepsilon/k_BT}d\varepsilon \tag{7.40}$$

Hence:

$$K = \frac{\mathcal{N}}{\int_0^\infty g(\varepsilon)e^{-\varepsilon/k_BT}d\varepsilon} \tag{7.41}$$

and we have determined the second Lagrange constant. Note that the integral in this expression is the partition function for this continuous situation:

$$Z = \int_0^\infty g(\varepsilon)e^{-\varepsilon/k_BT}d\varepsilon \tag{7.42}$$

Once we know the density of states for a given system, computing the fraction of particles to be found between ε and $\varepsilon + \Delta\varepsilon$ becomes a case of evaluating:

$$\frac{\Delta\mathcal{N}}{\mathcal{N}} = \frac{1}{Z}\int_{\varepsilon}^{\varepsilon+\Delta\varepsilon} g(\varepsilon)e^{-\varepsilon/k_BT}d\varepsilon \tag{7.43}$$

7.8.1 Modelling an Ideal Gas

To see how this can be followed through in practice, we model a gas as being a collection of particles constrained within a box and with quantum states appropriate to that situation. This will represent an ideal gas as we are not including any interaction forces between the particles. Hence, the quantum state of each particle is that of a single particle alone in the same box.

One option would be to use this approach and calculate the velocity distribution of particles within the gas and, hence, produce the Maxwell–Boltzmann distribution from Chapter 5. This is a standard result which is easily found in a range of texts.

To ring the changes, we'll calculate the entropy of the ideal gas from our quantum model.

With a 3D box of size L, there is degeneracy among the quantum states as various combinations of $\{n_x, n_y, n_z\}$ will lead to the same energy (Section 6.3.3):

$$E\left(n_x, n_y, n_z\right) = \frac{\pi^2 \hbar^2}{2mL^2}\left(n_x^2 + n_y^2 + n_z^2\right)$$

So, trivially, $\{n_x, n_y, n_z\} = \{1,0,0\} = \{0,1,0\} = \{0,0,1\}$ are degenerate, for example.

If we take $n^2 = n_x^2 + n_y^2 + n_z^2$ and treat n as a pseudo-continuous variable, then it acts like the radius of a sphere in a configuration space with axes $\{n_x, n_y, n_z\}$. As the individual quantum numbers are all positive, we are only interested in the "top right" quadrant which is 1/8 of the total sphere, i.e., $V = \frac{1}{6}\pi n^3$. This volume represents the number of states, $\mathbb{N}$, that exist up to energy E where:

$$E = \frac{\pi^2 \hbar^2 n^2}{2mL^2} \qquad n = \sqrt{\frac{2mL^2 E}{\pi^2 \hbar^2}} = \frac{L}{\pi \hbar}\sqrt{2mE} \tag{7.44}$$

Putting this together we obtain:

$$\mathbb{N} = \frac{1}{6}\pi n^3 = \frac{L^3}{6\pi^2 \hbar^3}\left(2mE\right)^{3/2} \tag{7.45}$$

The density of states, $g\left(E\right) = d\mathbb{N}/dE$, so:

$$g\left(E\right) = \frac{L^3}{6\pi^2 \hbar^3}\left(2m\right)^{3/2}\frac{3}{2}E^{1/2} = \frac{L^3}{4\pi^2 \hbar^3}\left(2m\right)^{3/2}E^{1/2} \tag{7.46}$$

Having completed a set of opening moves, we can now turn to the entropy using (7.36) as promised:

$$\mathcal{S}_G = k_B \frac{\partial}{\partial T}\left\{T \ln\left(\mathcal{Z}\right)\right\}$$

However, as we are dealing with a gas of many non-interacting particles, we need to be sure and use the multi-particle partition function from Appendix PA1, i.e.:

$$\mathcal{Z} = \frac{Z^{\mathcal{N}}}{\mathcal{N}!}$$

First, we need to evaluate the single particle partition function (7.42):

$$Z = \int_0^\infty g(\varepsilon) e^{-\varepsilon/k_B T} d\varepsilon = \frac{L^3}{4\pi^2\hbar^3}(2m)^{3/2} \int_0^\infty E^{1/2} e^{-E/k_B T} dE$$

If we set $u = E/k_B T$ and hence $E = k_B Tu$, $dE = k_B Tdu$, and this becomes a standard form:

$$Z = \frac{L^3}{4\pi^2\hbar^3}(2m)^{3/2}(k_B T)^{3/2} \int_o^\infty u^{1/2} e^{-u} du = \frac{L^3}{4\pi^2\hbar^3}(2m)^{3/2}(k_B T)^{3/2}\left[\frac{\sqrt{\pi}}{2}\right]$$

$$= L^3 \frac{(2mkT)^{3/2}}{8\pi^{3/2}\hbar^3} = \frac{L^3}{\hbar^3}\left(\frac{mk_B T}{2\pi}\right)^{3/2} \tag{7.47}$$

Next, we insert this result into (7.36) to yield:

$$\mathcal{S}_G = k_B \frac{\partial}{\partial T}\{T\ln(Z^{\mathcal{N}}/\mathcal{N}!)\}$$

$$= k_B \frac{\partial}{\partial T}\left\{T\mathcal{N}\ln\left(\frac{L^3}{\hbar^3}\left(\frac{mk_B T}{2\pi}\right)^{3/2}\right) - T\ln(\mathcal{N}!)\right\} \tag{7.48}$$

There is clearly a role for Stirling's Approximation, which gets us to:

$$\mathcal{S}_G = k_B \frac{\partial}{\partial T}\left\{T\mathcal{N}\ln\left(\frac{L^3}{\hbar^3}\left(\frac{mk_B T}{2\pi}\right)^{3/2}\right) - T\mathcal{N}\ln(\mathcal{N}) + T\mathcal{N}\right\} \tag{7.49}$$

It is now a matter of tidying up and performing the differentiation. The first step is to combine the two logarithmic terms:

$$\mathcal{S}_G = k_B \frac{\partial}{\partial T}\left\{T\mathcal{N}\ln\left(\frac{L^3}{\mathcal{N}\hbar^3}\left(\frac{mk_B T}{2\pi}\right)^{3/2}\right) + T\mathcal{N}\right\} \tag{7.50}$$

Now we differentiate using the product rule:

$$\mathcal{S}_G = k_B\left\{\mathcal{N}\ln\left(\frac{L^3}{\mathcal{N}\hbar^3}\left(\frac{mk_B T}{2\pi}\right)^{3/2}\right) + \mathcal{N}T\frac{\hbar^3}{L^3}\left(\frac{mk_B T}{2\pi}\right)^{-3/2}\frac{3L^3}{2\hbar^3}\left(\frac{mk}{2\pi}\right)^{3/2} T^{1/2} + \mathcal{N}\right\}$$

$$= k_B\left\{\mathcal{N}\ln\left(\frac{L^3}{\mathcal{N}\hbar^3}\left(\frac{mk_B T}{2\pi}\right)^{3/2}\right) + \frac{3}{2}\mathcal{N}TT^{-3/2}T^{1/2} + \mathcal{N}\right\}$$

$$= \mathcal{N}k_B \left\{ \ln\left(\frac{L^3}{\mathcal{N}\hbar^3} \left(\frac{mk_BT}{2\pi} \right)^{3/2} \right) + \frac{5}{2} \right\} \tag{7.51}$$

The last step is to render this result, which is known as the *Sackur–Tetrode equation*, in a form that will be of most use to us later, in Chapter 15.

As the volume of the box is L^3, the mass density of particles within the box is:

$$\rho = \frac{\mathcal{N}m}{L^3} \tag{7.52}$$

so that $\mathcal{N} = \rho L^3/m$. This we insert to produce:

$$\mathcal{S}_G = \mathcal{N}k_B \left\{ \frac{3}{2}\ln\left(\frac{2\pi m^{5/3}k_BT}{\rho^{2/3}h^2} \right) + \frac{5}{2} \right\} \tag{7.53}$$

7.8.2 *h* Marks the Spot

It's a bit surprizing to see Planck's constant in the middle of an expression for the entropy of an ideal gas, which you would have thought would be entirely classical.

Clearly, the derivation relied on quantum arguments to model the gas as a collection of non-interacting particles in a box with suitably constrained wave functions. However, since the ideal gas is a macroscopic system, we might have expected (hoped!) that all the specifically quantum aspects would have cancelled out.

If we decompose (7.51) by splitting the log into two terms, we get:

$$\mathcal{S}_G = \mathcal{N}k_B \left\{ \ln\left(\frac{L^3}{\mathcal{N}\hbar^3} \right) + \frac{3}{2}\ln\left(\frac{mk_BT}{2\pi} \right) + \frac{5}{2} \right\}$$

Following the script from Section 7.1.3, and taking thermodynamic limit by allowed $\mathcal{N}$ and V to grow, subject to $\mathcal{N}/V$ remaining constant we see that the first term is constant.

As a result, the change in entropy as we shift from temperature T_A to T_B is:

$$\Delta\mathcal{S}_G = \frac{3}{2}Nk_B\ln\left(\frac{mk_BT_A}{2\pi} \right) - \frac{3}{2}Nk_B\ln\left(\frac{mk_BT_B}{2\pi} \right)$$

$$= \frac{3}{2}Nk_B\ln\left(\frac{T_A}{T_B} \right) = \frac{3}{2}nR\ln\left(\frac{T_A}{T_B} \right) \tag{7.54}$$

In perfect agreement with (4.19) from Chapter 4. Once again, we see that the thermodynamic limit washes all sins…

NOTES

1 If two particles cannot occupy the same physical point in space…
2 Georg Ferdinand Ludwig Philipp Cantor (1845–1918), German mathematician.
3 The complete rule to apply is to take the digit if it is 0–8 and add one, but if the digit is 9, then replace that with 0.
4 In footnote 15 of Chapter 5, the relation between ω and Ω, the $d\alpha$ used in the expansion was a form of coarse graining. Here we progress to a more general consideration.
5 There is also the *Gibbs ensemble*, where volume is allowed to vary, but we will not have call on that version.
6 Joseph Louis Lagrange (1736–1813), Italian Mathematician and considered one of the greatest mathematical talents to have ever lived.
7 Always assuming the number of systems in the ensemble is very much more than 2002.
8 Admittedly, this is not the greatest of fits, but the idea here is to hint at what will happen when the number of particles is very much larger.
9 This becomes a major issue in Chapters 8 and 9.
10 We introduced this approximation in Chapter 5, section Y.
11 An alternative definition which is often used is $g(\epsilon)=\frac{1}{V}\frac{dN}{dE}$ where V is the volume of the system. Normalizing to the volume is convenient in solid state physics, for example, where you are not so much interested in the total number of particles or states, but the number per unit volume of material.

8 Identical Particles

In quantum theory, and especially quantum field theory, the nature of quantum states is transformed if the particles are identical. In fact, the impact of particles (or better quanta) being identical is more profound in quantum theory than in classical mechanics. Their identity has direct experimental consequences over and above those related to the occupation of states. Over the next couple of chapters, we will explore the quantum theory of identical particles to draw out the impact on the entropy of a collection of such particles.

8.1 SPIN

The transition from classical mechanics to quantum theory requires a recasting of familiar physical properties to conform to the behaviours observed in quantum experiments. This challenged the imaginations of the early pioneers in the 1920s, but they were at least guided by a principle to work with: that the quantum versions of the physical properties had to smoothly merge into the classical understanding under the right conditions.

However, there is one physical property, revealed by experiment which has no classical equivalent. The *spin* of a fundamental particle is a quantum property that does not map over into a classical equivalent. Confusingly, the name "spin" creates the understandable but false impression that we are describing the rotation of a particle about some axis, rather like a toy top or gyroscope. In fact, spin is best viewed as a symmetry property of the wave function.

In Section 6.3, we introduced the Born rule, which relates the complex square of an amplitude to an appropriate probability. For example, the probability of finding a particle in a region $x \to x + \Delta x$ is given by:

$$P(x \to x + \Delta x) \approx \psi^*(x)\,\psi(x)\,\Delta x \tag{8.1}$$

or more correctly:

$$P(x \to x + \Delta x) = \int_x^{x+\Delta x} \psi^*(x)\,\psi(x)\,dx \tag{8.2}$$

At the same time, we saw that any two wave functions are physically indistinguishable if there is only a phase factor between them, i.e., ψ and $\psi' = e^{i\vartheta}\psi$.

If we rotate a particle through an angle θ about some axis (conventionally the z axis), then we can illustrate the impact on the respective wave function by:

$$\psi' = \hat{\mathcal{R}}(\theta)\,\psi \tag{8.3}$$

DOI: 10.1201/9781003121053-11

where $\hat{\mathcal{R}}(\theta)$ is an appropriate rotation operator. Although common sense is not a reliable guide when it comes to the quantum world, we would guess that a full rotation of 2π will not have any physical impact on the particle. However, it is inaccurate to deduce from this that:

$$\hat{\mathcal{R}}(2\pi)\,\psi = \psi \tag{8.4}$$

due to the phase factor alluded to earlier. It turns out that:

$$\hat{\mathcal{R}}(\vartheta)\,\psi = e^{im\vartheta}\,\psi \tag{8.5}$$

where m is either a half-integer (i.e., $1/2, 3/2, 5/2$, etc.) or an integer (i.e., 0, 1, 2, 3…). Given the smallest half-integer case, a rotation of 2π produces:

$$\hat{\mathcal{R}}(2\pi)\,\psi = e^{i\pi}\,\psi = -\psi \tag{8.6}$$

inverting the phase of the wave function. It would take a full 4π revolution to restore the original wave function with the same phase. In fact, this holds for any half-integer value of m. Particles (quanta) that are correctly described by wave functions which have half-integer values of m are referred to as *fermions*, while the integer m wave functions denote *bosons*. The respective rotational symmetries of the quanta are known by their *spin values*, the case of $m = 1/2$ being a spin 1/2 fermion, for example, an electron.

As it turns out, the mathematical structures surrounding these wave functions have the same properties as those used to describe situations where angular momentum is involved, e.g., electrons in orbit around a nucleus. This naturally leads to the suspicion that the spin of a particle is a reference to an internal angular momentum brought about by the particle rotating about some axis. As already suggested, this model cannot be sustained in practice for a range of reasons, not the least of which is that the angular momentum ascribed to the spin is $\propto \hbar$.[1] Hence, if we (in our imaginations) make the transition from quantum behaviour to classical behaviour by letting $\hbar \to 0$, the spin angular momentum will vanish. There is no classical equivalent to spin.

Nevertheless, spin is a crucially important aspect of quantum reality and impacts very directly on the properties of systems.

8.2 QUANTUM IDENTICAL PARTICLES

States that contain more than one quantum of the same species have specific *symmetry properties* which are tied to their spin via an important theorem. Consider a general wave function for two particles, $\psi(\boldsymbol{r}_1, \boldsymbol{r}_2)$, where we are using $\boldsymbol{r}_1, \boldsymbol{r}_2$ as abbreviated forms for the coordinates of the two particles (i.e., $\boldsymbol{r}_1 = (x_1, y_1, z_1)$). We consider a convenient operator, $\hat{\mathcal{P}}$, which acts on the wave function and swaps the particles about, so:

$$\hat{\mathcal{P}}\psi(\boldsymbol{r}_1, \boldsymbol{r}_2) = e^{i\delta}\,\psi(\boldsymbol{r}_2, \boldsymbol{r}_1) \tag{8.7}$$

If the particles are identical, then this exchange can have no observable consequences, hence the wave function after the exchange can only differ from the original by some fixed phase factor, $e^{i\delta}$, as shown. Applying the operator again must put everything back the way it was:

$$\hat{\mathcal{P}}\psi(\boldsymbol{r}_2, \boldsymbol{r}_1) = e^{2i\delta}\,\psi(\boldsymbol{r}_1, \boldsymbol{r}_2) \tag{8.8}$$

implying that $e^{2i\delta} = 1$, hence $e^{i\delta} = \pm 1$. As is typical in nature, both possibilities are exploited.
Boson states are *symmetric* under the exchange of two quanta:

$$\psi_B(\boldsymbol{r}_1, \boldsymbol{r}_2) = +\psi_B(\boldsymbol{r}_2, \boldsymbol{r}_1) \tag{8.9}$$

whereas fermion states are *antisymmetric* under a similar exchange:

$$\psi_F(\boldsymbol{r}_1, \boldsymbol{r}_2) = -\psi_F(\boldsymbol{r}_2, \boldsymbol{r}_1) \tag{8.10}$$

Note that these symmetry properties also apply to states involving more than two quanta, when an exchange of *any* selected pair, e.g., quanta 1 and 3, takes place:

$$\psi_B(\boldsymbol{r}_1, \boldsymbol{r}_2, \boldsymbol{r}_3, \ldots, \boldsymbol{r}_n) = +\psi_B(\boldsymbol{r}_3, \boldsymbol{r}_2, \boldsymbol{r}_1, \ldots, \boldsymbol{r}_n)$$

$$\psi_F(\boldsymbol{r}_1, \boldsymbol{r}_2, \boldsymbol{r}_3, \ldots, \boldsymbol{r}_n) = -\psi_F(\boldsymbol{r}_3, \boldsymbol{r}_2, \boldsymbol{r}_1, \ldots, \boldsymbol{r}_n) \tag{8.11}$$

In quantum theory, multi-quanta states are constructed by combining single quanta states.[2] However, the recipe used must conform to the overall symmetry required by the nature of the quanta. For example, if we set out to build a two-quanta boson state, by gluing together the single particle states ψ and φ, the result must be symmetrical; for example:

$$\psi_B(\boldsymbol{r}_1, \boldsymbol{r}_2) = \varphi(\boldsymbol{r}_1)\,\varphi(\boldsymbol{r}_2) \tag{8.12}$$

$$\psi_B(\boldsymbol{r}_1, \boldsymbol{r}_2) = \psi(\boldsymbol{r}_1)\,\psi(\boldsymbol{r}_2) \tag{8.13}$$

$$\psi_B(\boldsymbol{r}_1, \boldsymbol{r}_2) = \frac{1}{\sqrt{2}}\left(\varphi(\boldsymbol{r}_1)\,\psi(\boldsymbol{r}_2) + \psi(\boldsymbol{r}_1)\,\varphi(\boldsymbol{r}_2)\right) \tag{8.14}$$

In the case of a two-quanta fermion state, only the following assembly has the requisite anti-symmetry:

$$\psi_F(\boldsymbol{r}_1, \boldsymbol{r}_2) = \frac{1}{\sqrt{2}}\left(\varphi(\boldsymbol{r}_1)\,\psi(\boldsymbol{r}_2) - \psi(\boldsymbol{r}_1)\,\varphi(\boldsymbol{r}_2)\right) = -\frac{1}{\sqrt{2}}\left(\varphi(\boldsymbol{r}_2)\,\psi(\boldsymbol{r}_1) - \psi(\boldsymbol{r}_2)\,\varphi(\boldsymbol{r}_1)\right) \tag{8.15}$$

which has very significant consequences. Should the two states be the same, i.e., $\psi = \varphi$, *then the overall state vanishes*. In other words, *no two identical fermions can occupy the same quantum state*. This powerful constraint on fermion states is known as the *Pauli exclusion principle* and has significant physical consequences.

8.3 THE FERMI-DIRAC DISTRIBUTION

The overall state of a particle is not simply a matter of co-ordinates, $\boldsymbol{r}, t$, the spin of the particle is also relevant:

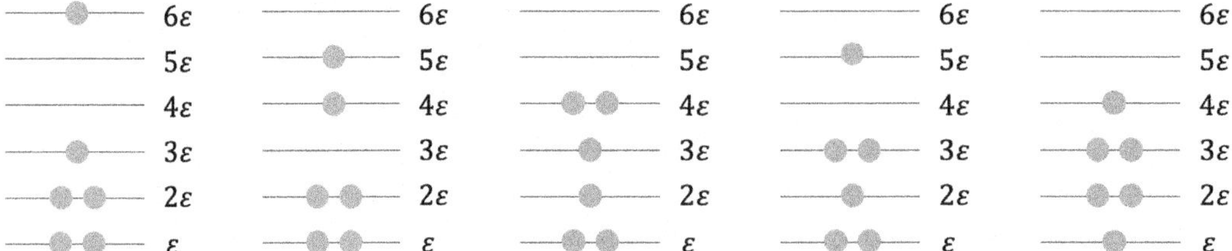

FIGURE 8.1 Possible ways of distributing six identical fermions so that the total energy is 15ε but there are at most two electrons in each state.

TABLE 8.1
Comparison Data between the Boltzmann Occupancy Calculated in Chapter 7, Table 7.5, and the Fermion Occupancy Calculated

5 Level	$\langle N \rangle$ Boltzmann	$\langle N \rangle$ Fermions
10	0.003	0
9	0.015	0
8	0.045	0
7	0.105	0
6	0.210	0.2
5	0.378	0.4
4	0.629	0.8
3	0.989	1.2
2	1.484	1.6
1	2.143	1.8

$$\Psi = \psi(\boldsymbol{r},t)\,\phi(\text{spin}) \tag{8.16}$$

In the case of electrons, which are spin-1/2 quanta, there are two possible spin states, which we call 'spin-up' and 'spin-down' depending on the spin component's direction along the z-axis.

As energy levels are generally determined by the spatial aspect of the overall state[3], we can slot at most two electrons, one in each spin state, into each energy level without falling foul of the Pauli exclusion principle. This has immediate significance for the distinction between states and patterns, as per Section 7.6.1. For example, Figure 8.1 itemises all the ways in which six electrons can be placed in energy levels so that the total energy of the system is 15 units, subject to the restriction that no more than two electrons can be in each state.[4] For the sake of simplicity, all the successive energy levels are separated by the same gap in energy, ε.

In the ground state, energy ε, the *average occupation* across the possibilities is:

$$\langle \mathrm{N}_1 \rangle = \frac{2+2+2+2+1}{5} = 1.8 \tag{8.17}$$

whereas for the first energy level, energy 2ε:

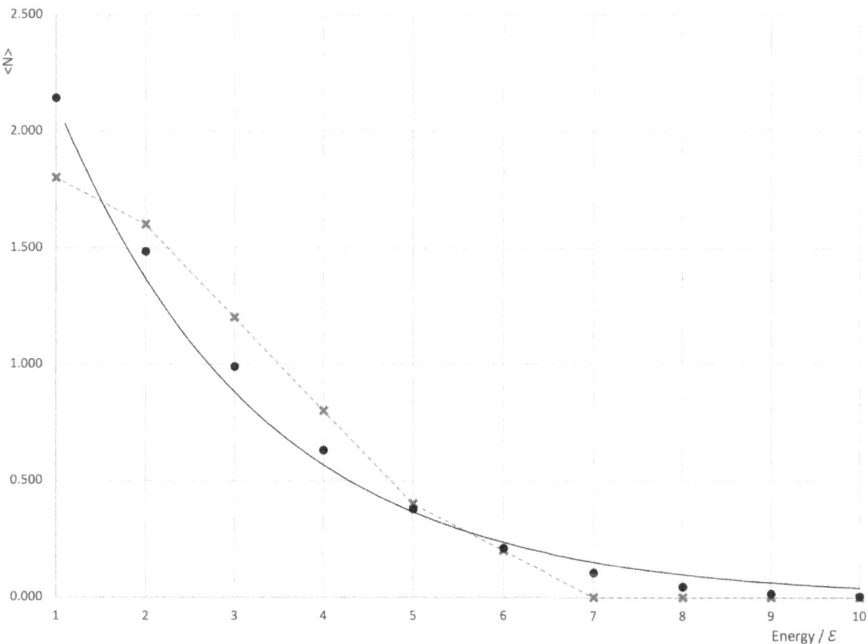

FIGURE 8.2 In this graph the filled circles are the data points from Table 7.2 which was used to calculate energy level occupancy according to Boltzmann statistics. The solid line is an exponential fit to that data. The open crosses are data points from the Fermi statistics shown in Table 8.2, with a dashed line to guide the eye.

$$\langle \mathbb{N}_2 \rangle = \frac{2+2+1+1+2}{5} = 1.6 \tag{8.18}$$

It is interesting to see how these average occupations stack up against those calculated in Section 7.6.1 for the Boltzmann statistics. This is shown in Table 8.1 and plotted in Figure 8.2.

The number of particles involved here is far too small for any binding conclusions, but we can see that the tendency for fermions to occupy the lower energy levels is greater than that for particles obeying Boltzmann statistics.

8.3.1 Deriving the Fermi-Dirac Statistics

Our next step is to see how fermion particle distributions differ from the Boltzmann case in general. The work was first done independently by Fermi[5] and Dirac[6] and published in 1926. In his classic text, *The Principles of Quantum Mechanics*, Dirac cites Fermi as the first person to analyse the anti-symmetric particle case and coins the term fermions for such quanta, in his honour.[7]

Applying fermion exclusion to our states and patterns means that when distributing $\mathbb{N}_n$ identical fermions among g_n degenerate[8] states of energy ε_n, we can only slot one particle into each of the g_n states: the value of g_n becomes the maximum number of particles that can exist with energy ε_n. So, the first time that you drop a fermion into a state of energy ε_n, you have g_n choices. However, when it comes to the second fermion, there are only $(g_n - 1)$ states available, due to the exclusion rule.

With the third, there are $(g_n - 2)$ choices, etc. In the end, the total number of choices for inserting $\mathbb{N}_n$ fermions into g_n states is:

$$g_n \times (g_n - 1) \times (g_n - 2) \ldots \times (g_n - (\mathbb{N}_n - 1)) = \frac{g_n!}{(g_n - \mathbb{N}_n)!} \tag{8.19}$$

Finally, accounting for the different identical permutations of the $\mathbb{N}_n$ particles, the total number of distinguishable ways of arranging fermions into the g_n states of the ε_n energy level is:

$$\omega_n = \frac{g_n!}{\mathbb{N}_n!(g_n - \mathbb{N}_n)!} \tag{8.20}$$

making the overall number of states:

$$\Omega_F = \prod_1^k \frac{g_n!}{\mathbb{N}_n!(g_n - \mathbb{N}_n)!} \tag{8.21}$$

and hence using Stirling's approximation:

$$\ln(\Omega_F) = \sum_1^k \{\ln(g_n!) - \ln(\mathbb{N}_n!) - \ln((g_n - \mathbb{N}_n)!)\}$$

$$= \sum_1^k \{g_n \ln(g_n) - g_n - \mathbb{N}_n \ln(\mathbb{N}_n) + \mathbb{N}_n - (g_n - \mathbb{N}_n)\ln(g_n - \mathbb{N}_n) + (g_n - \mathbb{N}_n)\}$$

$$= \sum_1^k \{g_n \ln(g_n) - \mathbb{N}_n \ln(\mathbb{N}_n) - (g_n - \mathbb{N}_n)\ln(g_n - \mathbb{N}_n)\} \tag{8.22}$$

Now we can proceed to maximize this quantity, using our Lagrange multipliers to factor in the constraints, as before:

$$\Delta(\log(\Omega_F)) = \sum_1^k \frac{\partial}{\partial \mathbb{N}_n}(g_n \ln(g_n) - \mathbb{N}_n \ln(\mathbb{N}_n) - (g_n - \mathbb{N}_n)\ln(g_n - \mathbb{N}_n))\Delta\mathbb{N}_n + \sum_1^k (A + B\varepsilon_n)\Delta\mathbb{N}_n$$

$$= \sum_1^k \left(\ln\left(\frac{g_n}{\mathbb{N}_n} - 1\right) + A + B\varepsilon_n\right)\Delta\mathbb{N}_n = 0 \tag{8.23}$$

Hence:

$$\ln\left(\frac{g_n}{\mathbb{N}_n} - 1\right) = -A - B\varepsilon_n \tag{8.24}$$

giving:

$$\frac{\mathbb{N}_n}{g_n} = \frac{1}{1 + e^{-A - B\varepsilon_n}} \tag{8.25}$$

As before, we take $B = -1/k_B T$, and set $K = e^{-A}$ as a normalization constant. This gives the *Fermi–Dirac distribution* (but in a non-standard form that we will update shortly):

$$\frac{\mathbb{N}_n}{g_n} = \frac{1}{1 + Ke^{\varepsilon_n/k_B T}} \tag{8.26}$$

The value of K is determined by the normalization condition:

$$\mathbb{N} = \sum_1^{\infty} \mathbb{N}_n = \sum_1^{\infty} \frac{g_n}{1 + Ke^{\varepsilon_n/k_B T}} \tag{8.27}$$

which can only be done if we have a specific function $g_n(\varepsilon_n)$ for the appropriate density of states.

8.3.2 Classical Limits

Clearly, this is a markedly different result to the Boltzmann distribution, but consider what happens if the occupation is much less than the number of states available at that energy. In other words, if $\mathbb{N}_n/g_n \ll 1$, in which case:

$$\frac{1}{1 + Ke^{\varepsilon_n/k_B T}} \ll 1 \tag{8.28}$$

making $1 + Ke^{\varepsilon_n/k_B T} \gg 1$. Consequently, we may as well adjust things as $1 + Ke^{\varepsilon_n/k_B T} \approx Ke^{\varepsilon_n/k_B T}$:

$$\frac{\mathbb{N}_n}{g_n} = \frac{1}{Ke^{\varepsilon_n/k_B T}} = \frac{1}{K} e^{-\varepsilon_n/k_B T} \tag{8.29}$$

which is the Boltzmann distribution, realized in a "dilute" situation, such as at high temperatures, with the particles distributed over a wide range of energies, or with a low concentration of particles.

8.3.3 Low Temperature and Non-dilute...

The opposite extreme would be a situation where the available states at each energy are close to being fully occupied. To investigate this case, it's best to tweak the Fermi distribution. If instead of setting $K = e^{-A}$, explicitly revealing the Lagrange multiplier A as a normalization factor, we write $A = \mu/kT$ then:

$$\frac{\mathbb{N}_n}{g_n} = \frac{1}{1 + e^{(\varepsilon_n - \mu)/k_B T}} \tag{8.30}$$

which is more conventional. A simple re-arrangement gives:

$$\mathbb{N}_n = \frac{g_n(\varepsilon_n)}{1+e^{(\varepsilon_n-\mu)/k_BT}} = \mathcal{F}(\varepsilon_n)g_n(\varepsilon_n) \tag{8.31}$$

with the *Fermi function* $\mathcal{F}(\varepsilon_n) = 1/(1+e^{(\varepsilon_n-\mu)/k_BT})$.

As $T \to 0$, the value of $e^{1/k_BT} \to \infty$. If $\varepsilon_n > \mu$, then $(\varepsilon_n - \mu) > 0$ and we have:

$$\frac{\mathbb{N}_n}{g_n} = \frac{1}{1+e^{(\varepsilon_n-\mu)/k_BT}} \to \frac{1}{1+e^{+\infty}} \to 0 \tag{8.32}$$

In other words, any states with energy greater than μ are left vacant.

On the other hand, should $\varepsilon_n \le \mu$, then $(\varepsilon_n - \mu) \le 0$ and:

$$\frac{\mathbb{N}_n}{g_n} = \frac{1}{1+e^{(\varepsilon_n-\mu)/k_BT}} \to \frac{1}{1+e^{-\infty}} \to 1 \tag{8.33}$$

In other words, any states with energy less than μ are fully occupied.
Summarizing:

- at low temperatures, $\mathbb{N}_n/g_n = 1$ up to $\varepsilon_n = \mu$,
- but for $\varepsilon_n > \mu$, $\mathbb{N}_n/g_n = 0$.

If we populate the available states at $T = 0$, were that possible, by filling them one fermion at a time until we have used up all $\mathcal{N}$ particles, we will have reached a maximum energy level, equal to μ. This reveals how μ is *implicitly acting as a normalization constant.* The value of μ must be determined, at least in part, by the number of fermions in the system. However, when working in solid-state physics, it is easier to deal with the number of fermions (electrons) per unit volume in the material.

In Chapter 9, we will extend the idea of μ into the *chemical potential* in classical thermodynamics. This will be further developed in Chapter 13.

8.3.4 Normalization Again

As a simple example, consider a case where $g_n(\varepsilon_n) = g$, i.e., a constant independent of energy. This is clearly a very crude approximation, but it allows us to illustrate the normalization process. We then have:

$$\mathcal{N} = \sum_1^\infty \mathbb{N}_n = \sum_1^\infty \frac{g}{1+e^{(\varepsilon_n-\mu)/k_BT}} \tag{8.34}$$

Next, we convert the sum to an integral, on the assumption that the energy levels are very closely spaced and hence impersonate a continuous span of energies:

$$\mathcal{N} = \int_0^\infty \frac{g\,d\varepsilon}{1+e^{(\varepsilon-\mu)/k_BT}} = g\int_0^\infty \frac{d\varepsilon}{1+e^{(\varepsilon-\mu)/k_BT}} \tag{8.35}$$

We set $u = e^{(\varepsilon-\mu)/k_BT}$ so that $du = \frac{1}{k_BT} e^{(\varepsilon-\mu)/k_BT} d\varepsilon$ and so:

$$\mathcal{N} = gk_BT \int_{\varepsilon=0}^{\varepsilon=\infty} \frac{du}{u(1+u)} = gk_BT \left\{ \int_{\varepsilon=0}^{\varepsilon=\infty} \frac{du}{u} - \int_{\varepsilon=0}^{\varepsilon=\infty} \frac{du}{1+u} \right\} = gk_BT \left[\ln\left(\frac{u}{1+u} \right) \right]_{\varepsilon=0}^{\varepsilon=\infty} \tag{8.36}$$

Hence:

$$\mathcal{N} = gk_BT \left[\ln\left(\frac{e^{(\varepsilon-\mu)/k_BT}}{1+e^{(\varepsilon-\mu)/k_BT}} \right) \right]_{\varepsilon=0}^{\varepsilon=\infty} = -gk_BT \left[\ln\left(\frac{e^{-\mu/k_BT}}{1+e^{-\mu/k_BT}} \right) \right] \tag{8.37}$$

Re-arranging:

$$\frac{\mathcal{N}}{gk_BT} = \ln\left(\frac{1+e^{-\mu/k_BT}}{e^{-\mu/k_BT}} \right)$$

$$e^{\mu/k_BT} + 1 = e^{\mathbb{N}/gk_BT} \tag{8.38}$$

so that finally:

$$\mu(\mathcal{N},T) = k_BT\ln\left(e^{\mathcal{N}/gk_BT} - 1\right) \tag{8.39}$$

revealing μ to be a function of temperature and the total number of fermions (or total number per unit volume, depending on how you want to normalize).

8.3.5 Degenerate Fermi Systems

It is convenient to define the *Fermi temperature*, $T_F = \mu(\mathcal{N},0)/k_B$. Here $\mu(\mathcal{N},0)$ is the limit of μ as the temperature tends to absolute zero, also known as the *Fermi energy*, ε_F, which is equal to the (energetically) highest occupied state in the material at $T = 0$.

As an example, consider a metal such as copper, which has a Fermi energy ~7 eV. The corresponding Fermi temperature is $\sim 8\times10^4$ K. Clearly it would be impossible to heat a sample of copper to anything like this temperature, without radically changing its physical structure. From this we conclude that Fermi statistics must *always* be a crucial aspect of election physics in solid copper.

Our formula for $\mu(\mathcal{N},T)$ can now be parameterized and so plotted (Figure 8.3)

$$\frac{\mu(\mathcal{N},T)}{\varepsilon_F} = \frac{T}{T_F}\ln\left(e^{\mathbb{N}T_F/g\varepsilon_FT} - 1\right) \tag{8.40}$$

Firstly, we see from Figure 8.3 that μ can go negative for sufficiently high temperatures.

Secondly, subject to the limitations of the model we are using, $\mu(\mathcal{N},T)$ remains reasonably constant up to $T/T_F \sim 0.5$. So again, for copper, with a Fermi temperature $\sim 8\times10^4$ K, we can consider μ to be independent of temperature for the physics of the solid material.

If we plot the occupation for copper, according to the Fermi–Dirac distribution, for a range of temperatures above 0 K, we get Figure 8.4. The sharp "step-like" distribution at 0 K is rounded off at higher temperatures, indicating that some electrons with $\varepsilon_n \leq \mu$ but $\varepsilon_n \sim \mu$ are thermally excited into higher, i.e., $\varepsilon_n > \mu$ levels.

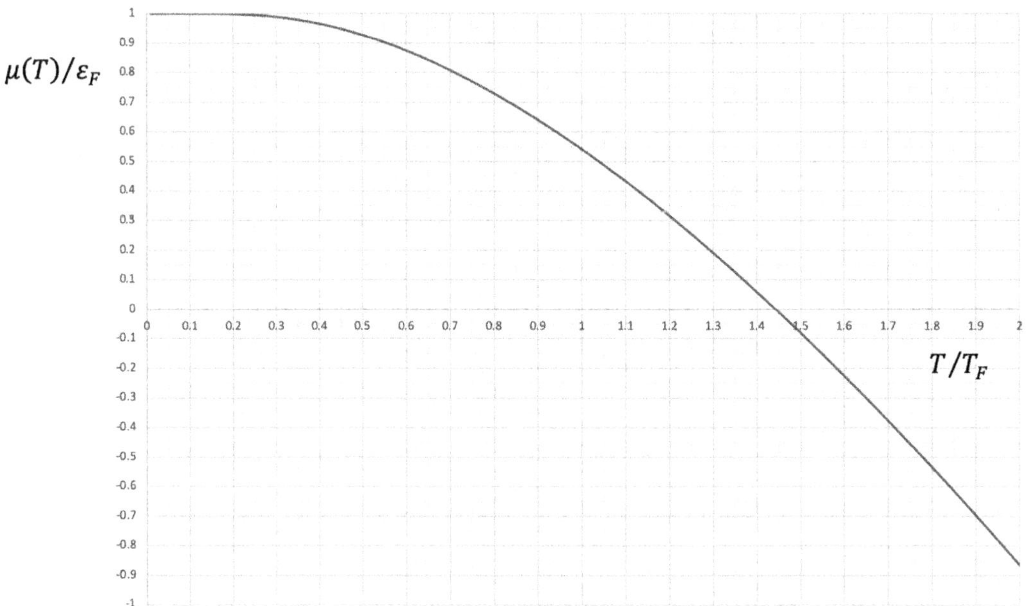

FIGURE 8.3 The variation of $\mu(\mathcal{N}, \mathrm{T})$ with temperature for the case of fixed g and $\mathbb{N}/g\mu_F = 1$.

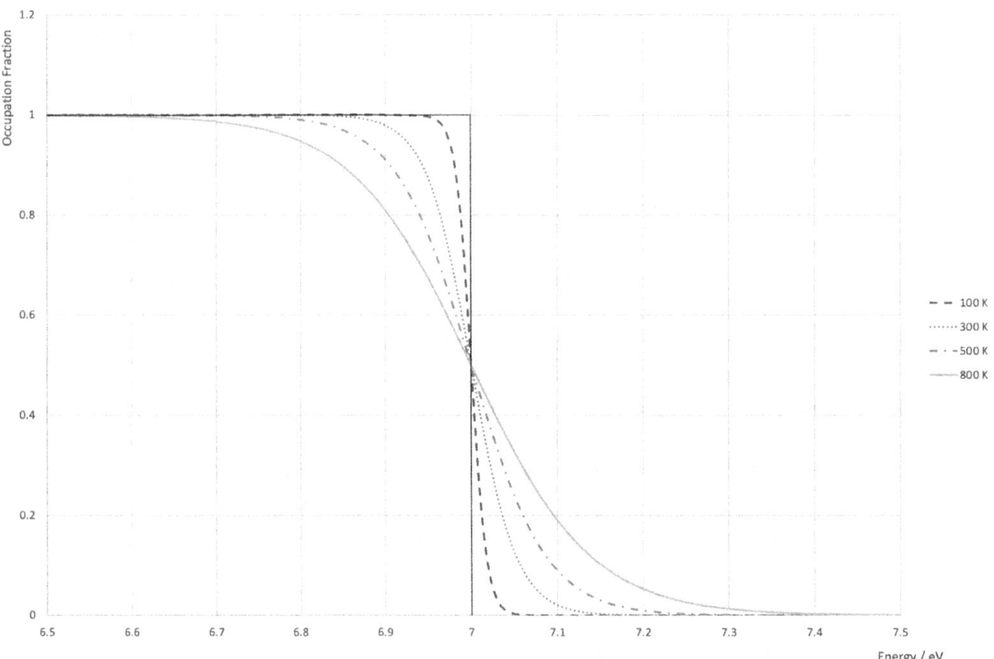

FIGURE 8.4 The relative occupation of states at various temperatures for Copper with a Fermi energy of 7 eV. The hypothetical occupation at 0 K is also shown as a guide. NB this diagram does not include the detail that μ changes with temperature.

However, even with the highest temperature plotted (800 K) the occupation only diverges from 1 for states that are very close to the Fermi level $|\varepsilon - \varepsilon_F| \sim 0.4$ eV. Equally, empty states only crop up on the energetic far side of ε_F.

At a room temperature of 300 K, typical thermal energies, ε_T:

$$\varepsilon_T \sim kT = \frac{\left(1.38\times10^{-23}\times300\right)}{\left(1.6\times10^{-19}\right)} \sim 0.026 \text{ eV}$$

So, unless the electrons are already in states that are very close to ε_F, they are not going to be thermally promoted into vacant states. Hence for temperatures $T \ll T_F$ there is a very large *Fermi Sea* of states filled with electrons which are too far, energetically speaking, from ε_F to take part in any thermal activity in the metal.

Systems like this, where the great majority of fermions in the system exist in states below the Fermi level, are referred to as *highly degenerate*, but note that this is a different use of the term, which in other contexts is used to refer to states of equal energy.

8.3.6 Higher Temperatures

It is not easy to come up with an analytical solution for $\mu(\mathcal{N},T)$ in any reasonably realistic situation where g is not constant. However, we can parametrize some data using information extracted from one of the classic texts on solid-state physics.[9] In Figure 8.5, we plot μ for a range of temperatures applied to a *Fermi gas* (i.e., a gas of fermions, such as electrons). This data is well modelled by a quadratic fit of the form $\mu(T) \approx \mu(0) - \alpha T^2$, with $\mu(0)$ being the value at 0 K, ~ 4.31 eV in this case.[10]

If we apply this parameterization of $\mu(T)$ to the Fermi distribution, we get Figure 8.6.

In this diagram, a solid horizontal line is drawn at an occupation fraction, $\mathbb{N}_n/g_n = 0.5$. This enables us to read off $\mu(T)$ at each temperature as the point where the distribution line crosses $\mathbb{N}_n/g_n = 0.5$.

Given (based on (8.30)):

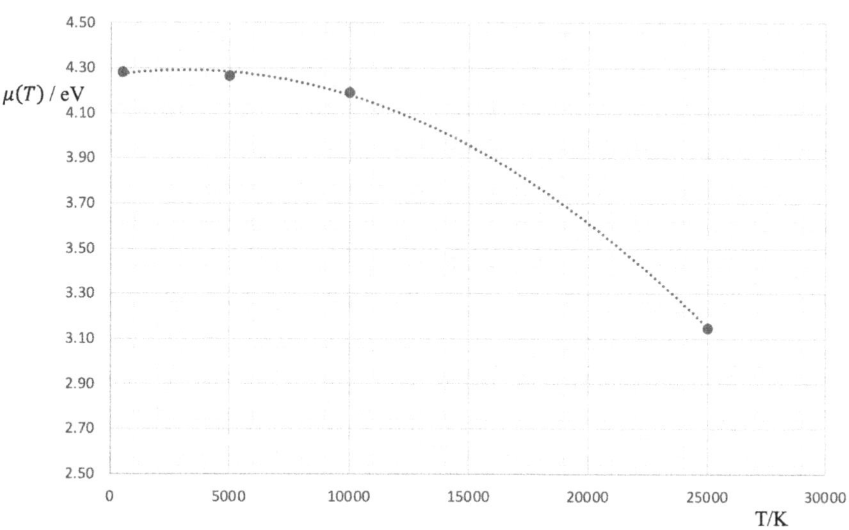

FIGURE 8.5 The variation of μ with temperature for a selection of temperatures applied to a fermi gas. The line shows the result of a quadratic fit of the form $\mu(\mathrm{T}) \approx \mu(0) - \alpha\mathrm{T}^2$.

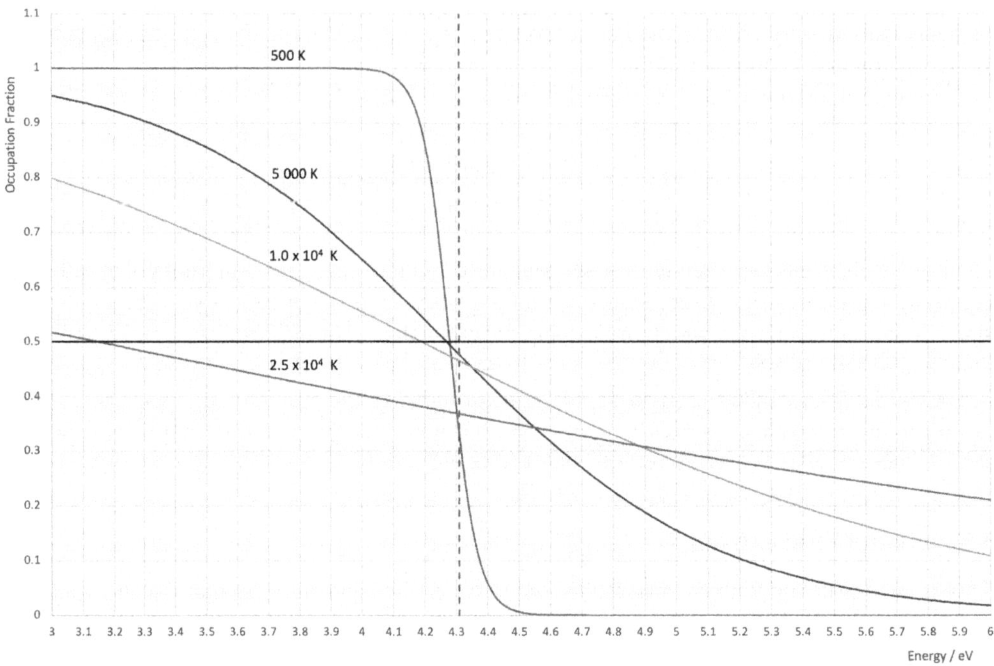

FIGURE 8.6 The Fermi distribution at a range of temperatures accounting for $\mu(\mathrm{T})$ with a fixed total number of particles. The vertical dotted line shows the value of ε_F which corresponds to a Fermi temperature of 50,000 K. The reduction of $\mu(T)$ with temperature can clearly be seen.

$$\frac{\mathrm{N}_n}{g_n} = \frac{1}{1+e^{(\varepsilon_n-\mu(T))/k_BT}} \tag{8.41}$$

whenever $\varepsilon_n = \mu(T)$, this becomes:

$$\frac{\mathrm{N}_n}{g_n} = \frac{1}{1+e^0} = \frac{1}{2} \tag{8.42}$$

Hence, while it is possible to equate $\mu(0)$ with the Fermi energy, ε_F, shown as the vertical dashed line on Figure 8.6, this is not true at other temperatures, but for many purposes the approximate equality is sufficient.

8.3.7 The Fermi Gas

Now, let's explore further the properties of degenerate systems. For example, the conduction electrons in a volume of metal. In the *free electron model*, it is assumed that the conduction electrons do not interact with each other. Their passage through the periodic potential of the metal lattice can be modelled by quantum theory, which shows that the electrons are essentially free to flow, but their interaction with the metal ions in the lattice has the effect of making their mass appear higher than that for electrons in a vacuum.[11] So, we can consider the conduction electrons as a gas of free particles within the box of the metal sample.

In Chapter 7, we derived the density of states for particles in a box, obtaining the result (7.46):

$$g(E)=\frac{d\mathcal{N}}{dE}=\frac{L^3}{6\pi^2\hbar^3}(2m)^{3/2}\frac{3}{2}E^{1/2}=\frac{L^3}{4\pi^2\hbar^3}(2m)^{3/2}E^{1/2}$$

with E being the maximum energy in the system.

The only modification we need to apply this to the context of non-interacting electrons in a metal is to remember that each electron has two spin states, so we double the density of states:

$$g(E)=\frac{L^3}{2\pi^2\hbar^3}(2m)^{3/2}E^{1/2} \tag{8.43}$$

The number of electrons, $\mathcal{N}$, sitting in states up to the Fermi level is:

$$\mathcal{N}=\int_0^{\varepsilon_F}\langle n(\varepsilon)\rangle g(\varepsilon)d\varepsilon \tag{8.44}$$

Taking $\langle n(\varepsilon)\rangle$ as the average occupation of states with energy ε. As we have explained, we can effectively model the situation at lower temperatures by:

$$\langle n(\varepsilon)\rangle=1, \varepsilon\leq\varepsilon_F \qquad \langle n(\varepsilon)\rangle=0, \varepsilon>\varepsilon_F \tag{8.45}$$

As a result:

$$\mathcal{N}=2\times\left(\frac{L^3}{4\pi^2\hbar^3}(2m)^{3/2}\right)\int_0^{\varepsilon_F}E^{1/2}dE=\frac{L^3}{2\pi^2\hbar^3}(2m)^{3/2}\left(\frac{2}{3}\varepsilon_F^{3/2}\right) \tag{8.46}$$

Rearranging gets us two options.

First there is;

$$\varepsilon_F=\left(\frac{3\pi^2\hbar^3\mathcal{N}}{L^3(2m)^{3/2}}\right)^{2/3}=\frac{\hbar^2}{2m}\left(\frac{3\pi^2\mathcal{N}}{L^3}\right)^{2/3} \tag{8.47}$$

Second, we have:

$$L^3=\frac{3\pi^2\hbar^3\mathcal{N}}{(2m)^{3/2}}\varepsilon_F^{-3/2} \tag{8.48}$$

which we can now use to refine our equation for the density of states:

$$\begin{aligned}g(E)&=\frac{2L^3}{6\pi^2\hbar^3}(2m)^{3/2}\frac{3}{2}E^{1/2}=2\left(\frac{(2m)^{3/2}}{4\pi^2\hbar^3}\right)\left(\frac{3\pi^2\hbar^3\mathcal{N}}{(2m)^{3/2}}\varepsilon_F^{-3/2}\right)E^{1/2}\\&=\left(\frac{3}{2}\mathcal{N}\varepsilon_F^{-3/2}\right)E^{1/2}\end{aligned} \tag{8.49}$$

Having obtained these intermediate results, we can proceed to calculate the total energy of electrons in this system:

$$E_{tot} = \int_0^{\varepsilon_F} \varepsilon g(\varepsilon) d\varepsilon = \frac{3}{2}\mathcal{N}\varepsilon_F^{-3/2}\int_0^{\varepsilon_F}\varepsilon^{3/2}d\varepsilon = \left(\frac{3}{2}\mathcal{N}\varepsilon_F^{-3/2}\right)\left(\frac{2}{5}\varepsilon_F^{5/2}\right) = \frac{3}{5}\mathcal{N}\varepsilon_F \tag{8.50}$$

giving an average energy per electron of:

$$\langle\varepsilon\rangle = \frac{3}{5}\varepsilon_F \tag{8.51}$$

It is worth comparing these results to the same outcomes for a standard ideal gas:

	Average energy per particle	**Total energy**
Ideal gas	$\frac{3}{2}kT \sim 0.039$ eV	$\frac{3}{2}\mathcal{N}kT$
Degenerate Fermi gas	$\frac{3}{5}\varepsilon_F \sim 4.2$ eV	$\frac{3}{5}\mathcal{N}\varepsilon_F$

Clearly, having to stack up electrons in their energy levels, according to the exclusion principle, leads to a much higher average energy per electron, which is not really surprising.

8.3.8 Degeneracy Pressure

As the total energy of electrons in a degenerate Fermi gas is:

$$E_{tot} = \frac{3}{5}\mathcal{N}\varepsilon_F = \frac{3}{5}\mathcal{N}\frac{\hbar^2}{2m}\left(\frac{3\pi^2\mathcal{N}}{L^3}\right)^{2/3} \tag{8.52}$$

we can calculate the pressure that these particles exert by using $\Delta U = -\mathcal{P}\Delta V$. In this context, $V = L^3$, so that:

$$E_{tot} = \frac{3\hbar^2}{10m}\mathcal{N}^{5/3}\left(3\pi^2\right)^{2/3}V^{-2/3} \tag{8.53}$$

resulting in:

$$\mathcal{P} = -\frac{\partial E_{tot}}{\partial V} = -\frac{\partial}{\partial V}\left(\frac{3\hbar^2}{10m}\mathcal{N}^{5/3}\left(3\pi^2\right)^{2/3}V^{-2/3}\right) = \left(\frac{3\hbar^2}{10m}\mathcal{N}^{5/3}\left(3\pi^2\right)^{2/3}\right)\left(\frac{2}{3}V^{-5/3}\right)$$

$$= \frac{\left(3\pi^2\right)^{2/3}\hbar^2}{5m}\left(\frac{\mathcal{N}}{V}\right)^{5/3} = \frac{2}{3}\frac{E_{tot}}{V} = \frac{2}{5}\frac{\mathcal{N}\varepsilon_F}{V} \tag{8.54}$$

This pressure will still be exerted as the temperature of the system drops towards absolute zero, whereas in a normal ideal gas, the pressure would also decline. Of course, this calculation must be refined should the upper energy levels result in relativistic speeds for the particles involved. The appropriate calculation modifies the pressure to be:

$$\mathcal{P}_r = \frac{\hbar c}{4}\left(3\pi^2\right)^{1/3}\left(\frac{\mathcal{N}}{V}\right)^{4/3} \tag{8.55}$$

In practice, the pressure estimation at 0 K, $\mathcal{P} = \frac{2}{5}\frac{\mathcal{N}\varepsilon_F}{V}$, remains viable up to temperatures comparable to the Fermi temperature of the material.

To estimate the size of this pressure in a metal, we return to copper with its Fermi energy ~ 7 eV and look up the relevant number density for its free electrons, $\mathcal{N}/V \sim 8.5\times10^{28}\ \text{m}^{-3}$. Our estimation for the pressure is then:

$$\mathcal{P} = \frac{2}{5}\times\left(8.5\times10^{28}\right)\times\left(7\times1.6\times10^{-19}\right) = 3.8\times10^{10}\ \text{Pa}$$

$$\sim 3.8\times10^{5}\ \text{atmos} \tag{8.56}$$

which is clearly an extraordinary figure. In a metal, we have an odd situation. While the conduction electrons are essentially free to flow through the metal, the enormous outwards pressure they exert is resisted by the forces that bind them to the lattice. The major factor causing the incompressibility of metals is the outward pressure exerted because of the Pauli exclusion principle. Given that we are working in the context of a degenerate Fermi gas, this is also known as the *degeneracy pressure*.[12]

Sometimes abbreviated references to this pressure makes it sound like it arises *directly* from the exclusion principle, as if that brings about some new kind of force to drive the pressure. In fact, degeneracy pressure is an *indirect* consequence of exclusion. Given the necessity to stack up fermions uniquely within states, the average energy of particles in a Fermi gas is much higher than the thermal energies seen in an ordinary gas. In other words, the particles are moving much more quickly than those within an ideal gas at the same temperature. That's why they exert excess pressure compared to an ideal gas.

One of the most well-documented examples of degeneracy pressure is the stability of *white dwarf stars*. An ordinary *main sequence star*, like our sun, exists in a balance between the tendency to collapse under the enormous self-gravitational pressure brought about by its mass and the outwards pressure of the super-heated ions within. In turn, the temperatures within the star's core are sustained by the energy released from nuclear fusion reactions. The exact reaction pathways exploited cannot be reproduced on Earth as they rely on some steps that are statistically highly unlikely[13] but nevertheless occur sufficiently often given the vast number of particles in play in the core. This makes the reaction balances very sensitive to the available density of fuel. In short, the reactions start to shut down well before all the fuel is consumed. In the latter stages of the lifecycles of stars with a mass comparable to the Sun's, the outer layers of the star are expelled in a relatively gentle process to form a region of diffuse, glowing gas known as a *planetary nebula*. This leaves the material comprising the core in the form of a compact remnant, typically around the size of the Earth, yet with a mass up to 1.4 times that of the Sun. This is a white dwarf, which is still emitting light as it cools down but no longer has any fusion reactions to replace the lost energy. The inward gravitational pressure is much too high to be resisted by outwards thermal pressure at these temperatures, yet the star will remain stable indefinitely.[14] The necessary outward force is provided by the degeneracy pressure of the Fermi gas of electrons within the star. In essence, any compression of the star increases the number density, $\mathcal{N}/V$, of the electrons, which in turn causes the degeneracy pressure to grow, resisting the compression.

Subrahmanyan Chandrasekhar[15] was the first person to demonstrate that degeneracy pressure could support a white dwarf, and in the process, also calculated the maximum mass of such remnants (1.4 solar masses as noted earlier) before gravity overcame even that outward pressure. Remnants of mass greater than this *Chandrasekhar limit* collapse straight through the white dwarf state and become *neutron stars*. As the name suggests, these fascinating objects are predominantly composed of neutrons, which being fermions exist within the structure as a Fermi gas and hence also exert an outwards degeneracy pressure supporting the star. While there is currently no calculated mass limit to such objects, one must exist: the heaviest observed neutron star is ~ 2.5 solar masses. Beyond this, the imploding stellar core cannot be supported by any known force and collapses through both white dwarf and neutron star phases to become a black hole. As discussed further in Chapter 16, at the centre of the black hole exists the *singularity*, which marks the boundary of our current understanding.

8.4 THE BOSE–EINSTEIN DISTRIBUTION

Now we turn to bosons. In Section 8.2, we saw that multi-quanta boson states are symmetric under the exchange of any two quanta. Consequently, unlike with fermions, there is no restriction on the number of identical bosons that can sit in any one quantum state. This leads to a rather different occupation distribution, which can be derived by following our familiar procedure.

As an indication of how this might go, we can revisit the scenario of six particles and 15 units of energy. We have already calculated occupancy for the Boltzmann and Fermion cases, now we need to look again at Table 7.3 and consider identical bosons instead.

In this instance, the number of microstates for each pattern is not relevant, as the bosons are identical particles. Consequently, the 180 microstates for pattern N could not be distinguished from each other—*they are effectively just the one state*.[16] In which case, calculating the average occupancy for, say, the 5th energy level across the table:

$$\langle \mathbb{N}_5 \rangle = \frac{1+1+2+1+1+1+1}{26} = 0.308 \tag{8.57}$$

Continuing to compute occupancies for the other energy levels and stacking them against the Boltzmann and Fermion cases, gives Table 8.2 and the plot in Figure 8.7.

Again, it would not be prudent to draw firm conclusions from such sparse data, but a slight excess of occupancy for the boson case compared to Boltzmann statistics can be seen. The excess becomes considerably more dramatic with a greater number of particles in play (see section 8.5).

8.4.1 DERIVING THE BOSE–EINSTEIN DISTRIBUTION

As a starting point, imagine a collection of g_n degenerate states with energy ε_n. This collection of states is blessed with $\mathbb{N}_n$ identical bosons, distributed in some fashion. We can picture this arrangement as a collection of "cells" bounded by "lines", Figure 8.8, which demark one degenerate state from another.

If we have g_n degenerate states, then our illustration must contain $g_n - 1$ vertical lines[17] and $\mathbb{N}_n$ blobs.

Imagine populating this diagram by picking blobs, to represent bosons, out of a bag, putting them into the diagram until we are satisfied that we have enough in the first state. We insert a line to mark the end of that state and the start of another. Then, we start picking bosons again, or not if the

TABLE 8.2
Average Occupancies for Boltzmann, Fermion and Boson Cases When Six Particles Have 15 Units of Energy

Energy Level	<N> Boltzmann	<N> Fermions	<N> Bosons
10	0.003	0	0.038
9	0.015	0	0.038
8	0.045	0	0.077
7	0.105	0	0.115
6	0.210	0.2	0.192
5	0.378	0.4	0.308
4	0.629	0.8	0.538
3	0.989	1.2	0.885
2	1.484	1.6	1.538
1	2.143	1.8	2.269

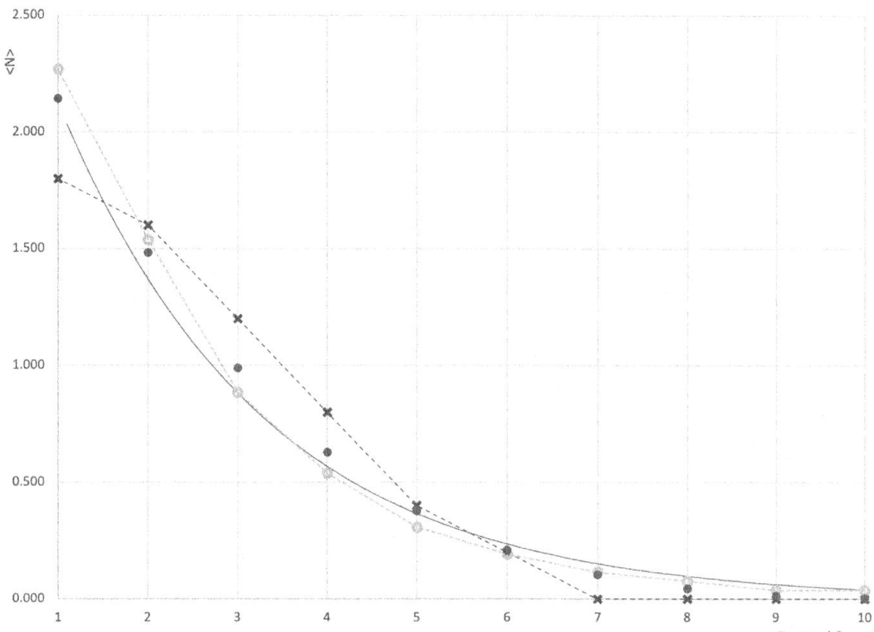

FIGURE 8.7 Comparison average occupation for Boltzmann, fermion and boson statistics where six particles have 15 units of energy.

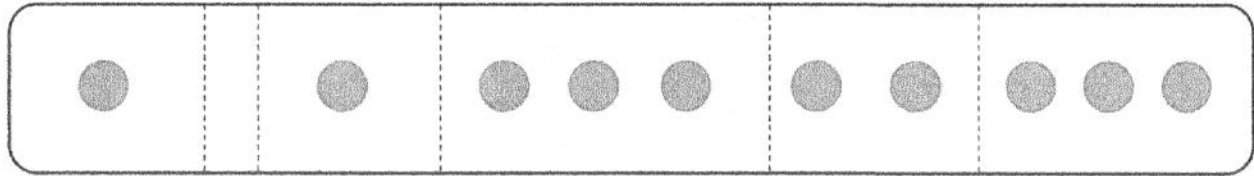

FIGURE 8.8 In this drawing, identical bosons are represented by grey blobs. These bosons have been distributed among six degenerate states, hence $g_n = 6$ and $\mathbb{N}_n = 10$. The "boundaries" between one state and another are the vertical dotted lines. The width of the "cell" containing bosons is of no physical significance and has simply been adjusted to the number of bosons in that state. Note that one of the degenerate states does not have any occupying bosons.

second state is empty (as per Figure 8.8). If we continue in this vein, we are effectively picking from $\mathbb{N}_n + (g_n - 1)$ objects in our bag. The number of ways of selecting objects is then $\{\mathbb{N}_n + (g_n - 1)\}!$. However, amongst our selections, the particles are identical (as a matter of physics), and the lines are identical (as a matter of discrimination), so we must compensate for $\mathbb{N}_n!$ and $(g_n - 1)!$ overcounting. Consequently, the number of arrangements for state n is:

$$\omega_n = \frac{\{\mathbb{N}_n + (g_n - 1)\}!}{\mathbb{N}_n!(g_n - 1)!} \tag{8.58}$$

and hence the overall number of arrangements across all states:

$$\Omega_B = \prod_1^k \frac{\{\mathbb{N}_n + (g_n - 1)\}!}{\mathbb{N}_n!(g_n - 1)!} \tag{8.59}$$

The argument now follows our well-trodden path. First, we take logs:

$$\ln(\Omega_B) = \sum_1^k \left\{\ln(\{\mathbb{N}_n + (g_n - 1)\}!) - \ln(\mathbb{N}_n!) - \ln((g_n - 1)!)\right\} \tag{8.60}$$

Then we apply Stirling's approximation:

$$\begin{aligned}\ln(\Omega_B) = \sum_1^k \Big\{&\{\mathbb{N}_n + (g_n - 1)\}\ln(\mathbb{N}_n + (g_n - 1)) - \cancel{\mathbb{N}_n} - \cancel{(g_n - 1)} - \mathbb{N}_n\ln(\mathbb{N}_n) \\ &\cancel{+\mathbb{N}_n} - (g_n - 1)\ln(g_n - 1) + \cancel{g_n - 1}\Big\}\end{aligned}$$

$$= \sum_1^k \left\{\{\mathbb{N}_n + (g_n - 1)\}\ln(\mathbb{N}_n + (g_n - 1)) - \mathbb{N}_n\ln(\mathbb{N}_n) - (g_n - 1)\ln(g_n - 1)\right\} \tag{8.61}$$

Now we maximize and add in the constraints, as before:

$$\begin{aligned}\Delta(\ln(\Omega_B)) = \sum_1^k \frac{\partial}{\partial \mathbb{N}_n}&\left\{\{\mathbb{N}_n + (g_n - 1)\}\ln(\mathbb{N}_n + (g_n - 1)) - \mathbb{N}_n\ln(\mathbb{N}_n) - (g_n - 1)\ln(g_n - 1)\right\} \\ &\Delta\mathbb{N}_n + \sum_1^k (A + B\varepsilon_n)\Delta\mathbb{N}_n\end{aligned}$$

$$= \sum_1^k \left\{\ln(\mathbb{N}_n + (g_n - 1)) + \frac{\mathbb{N}_n + (g_n - 1)}{\cancel{\mathbb{N}_n + (g_n - 1)}}\frac{\partial\cancel{(\mathbb{N}_n + (g_n - 1))}}{\partial \mathbb{N}_n} - \ln(\mathbb{N}_n) - \cancel{1}\right\}\Delta\mathbb{N}_n + \sum_1^k (A + B\varepsilon_n)\Delta\mathbb{N}_n$$

$$= \sum_1^k \left\{\ln\left(\frac{\mathbb{N}_n + (g_n - 1)}{\mathbb{N}_n}\right) + A + B\varepsilon_n\right\}\Delta\mathbb{N}_n = 0 \tag{8.62}$$

Hence:

$$\ln\left(\frac{\mathbb{N}_n+(g_n-1)}{\mathbb{N}_n}\right)=-(A+B\varepsilon_n) \tag{8.63}$$

so that:

$$\frac{\mathbb{N}_n}{(g_n-1)}=\frac{1}{e^{-(A+B\varepsilon_n)}-1} \tag{8.64}$$

Now we set $B=-1/k_BT$, and $K=e^{-A}$, resulting in:

$$\frac{\mathbb{N}_n}{(g_n-1)}=\frac{1}{Ke^{\varepsilon_n/k_BT}-1} \tag{8.65}$$

and showing B to be a normalization constant as in the Boltzmann and Fermi–Dirac cases. It is also conventional to assume that in any quantum level, n, there are a very large number of degenerate states, so that $g_n-1\approx g_n$, reducing our distribution to:

$$\frac{\mathbb{N}_n}{g_n}=\frac{1}{Ke^{\varepsilon_n/k_BT}-1} \tag{8.66}$$

If we are dealing with a dilute situation, $\mathbb{N}_n/g_n\ll 1$, then:

$$\frac{1}{Ke^{\varepsilon_n/k_BT}-1}\ll 1 \tag{8.67}$$

which is equivalent to saying $Ke^{\varepsilon_n/k_BT}-1\gg 1$, which in turn justifies the approximation $Ke^{\varepsilon_n/k_BT}-1\sim Ke^{\varepsilon_n/k_BT}$ converting our distribution to:

$$\frac{\mathbb{N}_n}{g_n}=\frac{1}{K}e^{-\varepsilon_n/k_BT} \tag{8.68}$$

establishing consistency with the Boltzmann distribution.

As with the Fermi–Dirac case, the Bose–Einstein distribution is more often quoted by recasting the Lagrange multiplier A into another form using the definition $A=\mu/k_BT$

$$\frac{\mathbb{N}_n}{g_n}=\frac{1}{e^{(\varepsilon_n-\mu)/k_BT}-1} \tag{8.69}$$

It is worth making a comparison between the Fermi–Dirac and Bose–Einstein distributions at this point:

Bose–Einstein	Fermi–Dirac
$\frac{\mathbb{N}_n}{g_n} = \frac{1}{e^{(\varepsilon_n - \mu(T))/k_B T} - 1}$	$\frac{\mathbb{N}_n}{g_n} = \frac{1}{e^{(\varepsilon_n - \mu(T))/k_B T} + 1}$

The functional form of the two is clearly very similar, but the difference of sign is crucial. As $e^x > 0$, in the Fermi–Dirac case it is clear that $\mathbb{N}_n \leq g_n$, hence there can be at most only 1 particle in each state, which is clearly a reflection of the exclusion principle. However, in the Bose–Einstein distribution, it is possible for $\mathbb{N}_n \gg g_n$, especially if $\varepsilon_n - \mu \ll k_B T$; clearly many bosons can exist in the same state, again as expected.

Another significant consequence of the negative sign in the Bose–Einstein distribution is the sign of μ. In Figure 8.3, we saw that $\mu(T) < 0$ at high enough temperatures. However, *that cannot be the case for bosons.* As (based on (8.69):

$$\frac{\mathbb{N}_n}{g_n} = \frac{1}{e^{(\varepsilon_n - \mu(T))/k_B T} - 1} \tag{8.70}$$

$e^{(\varepsilon_n - \mu(T))/k_B T} > 1$ if we wish to avoid the distribution blowing up. Taking the ground state as $\varepsilon_n = 0$, clearly $\mu(T) < 0$.

At this point, we will delay further comment on the physics of these distinct distributions until we have explored a different way of deriving their form, which we do in the next chapter. This alternative derivation will cast a helpful new light on the role of μ.

8.4.2 Consequences

Given the tendency for bosons to enjoy their mutual company (which is a consequence of $\mu < 0$ as developed further in the next chapter), one might reasonably ask why they don't all sit in the lowest energy. The answer is *thermal fluctuations*: some mechanism that can be classed as thermal supplying energy to particles and lifting them into higher energy states. This is why we have a boson *distribution*, after all. However, if we were to lower the temperature sufficiently so that typical thermal fluctuations ~ kT were less than the energy level gap between the ground and first exited states, then we could create a system where all the particles existed in the same quantum state. Research in this area has generated considerable interest in recent times. *Bose–Einstein condensates*,[18] where a very large fraction of the particles in a system all occupy the same quantum state, are of practical and theoretical interest. Given a very large number of particles in the same quantum state, quantum properties (such as wavefunction interference) become visible on a macroscopic scale.[19]

On the other hand, it is interesting to explore the possibility that $\mu = 0$ for bosons. Recall that μ is related to the Lagrange multiplier A, which was deployed as a prefix for the constraint on the total number of particles (or, alternatively, the probability). Setting $\mu = 0 \rightarrow A = \mu/k_B T = 0$ so *the number of particles is not constrained.* As an example, consider a gas of photons in an enclosed reflecting volume; the number of photons is not conserved due to absorption and emission by the walls. In this case, the Bose–Einstein distribution of the form $\mathbb{N}_n/g_n = 1/1 + e^{\varepsilon_n/k_B T}$ is the basis of the *Planck radiation law*, which we will derive at a more opportune point in Section 9.4.2.

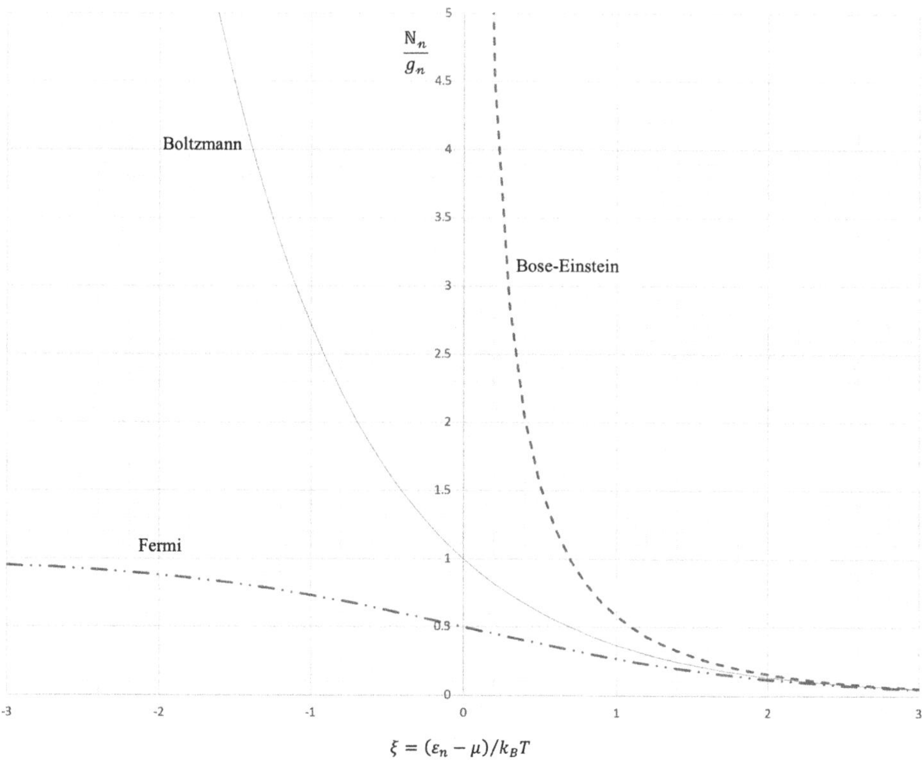

FIGURE 8.9 The three statistical distributions that we have derived plotted using a common parameter.

8.5 COMPARISONS

A convenient way of comparing our three distributions is to plot them using a common parameter:

$$\xi = \frac{\varepsilon_n - \mu}{k_B T} \tag{8.71}$$

In the boson case, having $\mu < 0$ implies that $\xi > 0$, whereas there is no such restriction for fermions. When it comes to the Boltzmann statistics, $\mu = 0$ and $\xi < 0$ corresponds to situations where the total energy of a state is negative. Given those provisos, we can plot all three distributions, as shown in Figure 8.9.

Note that:

- The distributions converge at high values of ξ, consistent with the general principle that quantum effects should blend into conventional classical physics in the macroscopic regime
- The excess of bosons compared to the Boltzmann distribution at lower energies, first alluded to in section 8.4, is now clearly visible. Although to be clear, we must remember that the total number of particles present must be obtained by including the density of states as well. The three distributions are "weighting factors" on their own

NOTES

1 Note that $\boldsymbol{r} \times \boldsymbol{p}$ has the same units as $\hbar$: $\text{Js} = \text{kg m}^2 \text{s}^{-1} = \text{m} \times \text{kg m}^{-1}$.
2 A more sophisticated approach is taken in quantum field theory.
3 Neglecting the small effect of the spin–orbit interaction in an atom. Here, we are dealing with a "gas" of electrons.
4 Here, we are assuming, for simplicity, that the energy of the state is not dependent on the spin component of the electrons. In practice, this means, for example, that there is no magnetic field involved.
5 Enrico Fermi (1901–1954), Nobel Prize in physics, 1938.
6 Paul Adrien Maurice Dirac (1902–1984), Nobel Prize in physics, 1933.
7 Dirac, Paul A. M. *Principles of Quantum Mechanics* (revised 4th ed.). London: Oxford University Press. pp. 210–201, 1967.
8 Recall from Chapter 7 that degenerate states have the same energy values. The two possible spin values of an electron would then constitute degenerate states.
9 Introduction to Solid State Physics, Kittel, Wiley 8th Edition, Figure 3, p. 136.
10 While this looks to be a spurious number, it corresponds to a neat Fermi temperature of 50,000 K.
11 This is distinct from the relativistic effect, which actually increases the mass of electrons moving at close to the speed of light.
12 Although, once again, note that the term *degeneracy* is not being used in this context to indicate states of equal energy.
13 This is why terrestrial fusion reactors have proven so hard to develop: they require temperatures well in excess of those experienced within the core of the star.
14 We have not yet observed a limit to the lifespan of a white dwarf—the earliest ones created are still hanging around.
15 Subrahmanyan Chandrasekhar (1910–1995), a theoretical physicist well-known for his study of compact stars and gravitation, shared the 1983 Nobel Prize for Physics.
16 This, coupled with the similar issue for fermions, explains why the microstate is often defined simply in terms of the number of particles in each energy level. That is fine and correct for identical particles. When we discussed the patterns and arrangements in Chapter 7 leading to the Maxwell–Boltzmann distribution, we assumed that the particles were not quantum identical, so we had to distinguish patterns from states.
17 These are the "partitions" of the partition function…
18 Satyendra Nath Bose was lecturing on radiation theory in Dhaka when he accidentally derived the formula that agreed with experimentation. His intention had been to show the students how conventional theory failed to account for experimental results. His simple mistake in sign leading to the correct Planck radiation law, which convinced Bose that the Maxwell–Boltzmann distribution that he had been using might not be applicable in all circumstances. Having failed to get his ideas in print, he sent a copy to Einstein, who translated the work from English to German and sent it to Zeitschrift fur Physik with the recommendation that it be published. Einstein also sent his companion paper predicting the existence of the condensates, which are now jointly named after the two authors.
19 The 2001 Nobel Prize in Physics was awarded to Cornell, Wieman, and Ketterle "for the achievement of Bose-Einstein condensation in dilute gases of alkali atoms, and for early fundamental studies of the properties of the condensates".

9 The Grand Canonical Ensemble

We developed a reasonably detailed and comprehensive understanding of boson and fermion statistics from the previous chapter. While this is interesting physics in itself, more important for us is an understand of how the symmetry properties of these particles impact on the entropy associated with them. In this chapter, we start by considering situations where the number of particles in a system is not fixed. Next, the theory developed is used to derive quantum statistics again, but this time without recourse to various approximations. Access to the requisite probability functions enables the Gibbs entropy to be calculated for fermions and bosons.

9.1 VARIABLE NUMBERS OF PARTICLES

Up to this point, all the systems we have dealt with have contained a fixed number of particles. However, there are good reasons to consider situations where things are more open. For example, we may come across:

- Systems in which different phases (e.g., solid, liquid, and gas) are co-present and exchanging particles so that the number of particles in any one phase is liable to change
- Systems containing a range of substances that might chemically react with one another, forming new molecules and reducing the number of particles present in any one of the substances (this will be discussed further in Chapter 13)
- Systems where different regions are under different conditions, for example, water flowing down a river and then off a cliff, so that the gravitational potential energy (for example) varies across the system
- Systems where particles are being created and annihilated

Developing the theory needed to deal with such possibilities will also give us another way to think about fermion and boson statistics and the impact of the quantum nature of truly identical particles on the entropies involved.

9.2 THE GRAND CANONICAL ENSEMBLE

In this variation, the ensemble comprises a very large number, $\mathcal{N}$, of identical systems each of which can exchange both energy *and* particles with a reservoir. The reservoirs are assumed to be very much greater in extent than the systems they are connected to. Hence, they brush-off any particle/energy transfers to/from a system without any impact on their own properties. Other parameters, such as the system's volume and temperature are kept constant.

DOI: 10.1201/9781003121053-12

Consequently, across the ensemble, the number of particles in each example system will not be fixed, neither will the energy, as that will depend on the number of particles.

Our aim will be to calculate ensemble averages for the energy and number of particles. They will then represent the observed characteristics of a system that has come into equilibrium with its reservoir.

We will specify the different microscopic states of the system by i and define $\mathcal{N}_i$ as the number of systems which happen to be in microstate i across the ensemble. Hence:

$$\sum_i \mathcal{N}_i = \mathcal{N} \tag{9.1}$$

and the probability of finding this microstate, should you randomly select a system from the ensemble, is:

$$p_i = \frac{\mathcal{N}_i}{\mathcal{N}} \tag{9.2}$$

It follows that:

$$\frac{1}{\mathcal{N}}\sum_i \mathcal{N}_i = \sum_i p_i = 1 \tag{9.3}$$

Every system in microstate i contains $\mathbb{N}_i$ particles and has total energy ε_i (note the different font for the "$\mathbb{N}$", so we don't confuse the number of particles in the system with the number of systems in the ensemble). The ensemble average system energy $\langle E \rangle$ and the ensemble average number of particles in a system $\mathbb{N}$ are:

$$\frac{1}{\mathcal{N}}\sum_i \mathcal{N}_i \varepsilon_i = \langle E \rangle \qquad \frac{1}{\mathcal{N}}\sum_i \mathcal{N}_i \mathbb{N}_i = \langle \mathbb{N} \rangle \tag{9.4}$$

Or equally:

$$\sum_i p_i \varepsilon_i = \langle E \rangle \qquad \sum_i p_i \mathbb{N}_i = \langle \mathbb{N} \rangle \tag{9.5}$$

Each of these relationships can be recast as a constraint:

$$\sum_i p_i - 1 = 0 \qquad \sum_i p_i \varepsilon_i - \langle E \rangle = 0 \qquad \sum_i p_i \mathbb{N}_i - \langle \mathbb{N} \rangle = 0 \tag{9.6}$$

Given the nature of the ensemble, we set about maximizing the Gibbs functional, $\mathcal{S}_G$, subject to these constraints. Hence, we construct:

$$\mathcal{S}_G' = -k_B \sum_i p_i \ln(p_i) + Ak_B\left(\sum_i p_i - 1\right) + Bk_B\left(\sum_i p_i \varepsilon_i - \langle E \rangle\right) + Ck_B\left(\sum_i p_i \mathbb{N}_i - \langle \mathbb{N} \rangle\right) \tag{9.7}$$

where we have used *three* Lagrange multipliers, $Ak_B, Bk_B,$ and Ck_B. As we have chosen to prefix each multiplier with the factor k_B, we can neatly gather terms:

$$\mathcal{S}'_G = -k_B \sum_i p_i \ln(p_i) + k_B \sum_i p_i \left(A + B\varepsilon_i + C\mathbb{N}_i\right) - k_B \left(A + B\langle E \rangle + Ck_B \langle \mathbb{N} \rangle\right) \quad (9.8)$$

so that we can construct the differential form:

$$d\mathcal{S}'_G = -k_B \sum_i dp_i \ln(p_i) + k_B \sum_i dp_i \left(A + B\varepsilon_i + C\mathbb{N}_i\right) \quad (9.9)$$

and hence, find the maximum by setting equal to zero:

$$d\mathcal{S}'_G = -k_B \sum_i \left(\ln(p_i) - A - B\varepsilon_i - C\mathbb{N}_i\right) dp_i = 0 \quad (9.10)$$

For this to be true for every i, we must have (with the help of the multipliers):

$$\ln(p_i) = A + B\varepsilon_i + C\mathbb{N}_i = 0 \quad (9.11)$$

or:

$$p_i = e^{A + B\varepsilon_i + C\mathbb{N}_i} \quad (9.12)$$

We expect that $B = -1/k_B T$, but it is also convenient and conventional to write $C = \mu/k_B T$ giving:

$$p_i = e^{-\varepsilon_i/k_B T + \mu\mathbb{N}_i/k_B T} e^A \quad (9.13)$$

The first Lagrange multiplier, A, is clearly a normalization constant, which is also the partition function, $e^A = 1/Z$. So, we have the *grand canonical probability distribution*:

$$p_i = \frac{1}{Z} e^{(\mu\mathbb{N}_i - \varepsilon_i)/k_B T} \quad (9.14)$$

scaled by the partition function, Z:

$$Z = \sum_{i=1}^{\infty} e^{(\mu\mathbb{N}_i - \varepsilon_i)/k_B T} \quad (9.15)$$

The nature and role of μ is explored more thoroughly in Section 9.2.3.

9.2.1 Ensemble Averages for Particle Number and Energy

Now we have the probability distribution, we can calculate the ensemble average number of particles:

$$\langle \mathbb{N} \rangle = \sum_i p_i \mathbb{N}_i = \frac{1}{Z} \sum_i e^{(\mu\mathbb{N}_i - \varepsilon_i)/k_B T} \mathbb{N}_i \quad (9.16)$$

As:

$$\frac{\partial}{\partial \mu}\left(e^{(\mu \mathbb{N}_i - \varepsilon_i)/k_B T}\right) = \frac{\mathbb{N}_i}{k_B T} e^{(\mu \mathbb{N}_i - \varepsilon_i)/k_B T} \tag{9.17}$$

we can simplify this formula to:

$$\langle \mathbb{N} \rangle = \frac{1}{Z}\sum_i e^{(\mu \mathbb{N}_i - \varepsilon_i)/k_B T}\mathbb{N}_i = \frac{k_B T}{Z}\frac{\partial}{\partial \mu}\left(\sum_i e^{(\mu \mathbb{N}_i - \varepsilon_i)/k_B T}\right) \tag{9.18}$$

Hence:

$$\langle \mathbb{N} \rangle = k_B T \frac{1}{Z}\frac{\partial Z}{\partial \mu} = k_B T \frac{\partial}{\partial \mu}\left(\ln(Z)\right) \tag{9.19}$$

which we will make good use of later.

{Recall from Chapter 7 that we had a similar expression, $\mathcal{S}_G = k_B \frac{\partial}{\partial T}\left(T\ln(Z)\right)$ }

The ensemble average energy is:

$$\langle E \rangle = \sum_i p_i \varepsilon_i = \frac{1}{Z}\sum_i e^{(\mu \mathbb{N}_i - \varepsilon_i)/k_B T}\varepsilon_i \tag{9.20}$$

To make this calculation a little easier, it is helpful to write $\beta = 1/k_B T$ so that:

$$\langle E \rangle = \frac{1}{Z}\sum_i e^{\beta(\mu \mathbb{N}_i - \varepsilon_i)}\varepsilon_i \tag{9.21}$$

We can now see how:

$$\frac{\partial}{\partial \beta}\left(e^{\beta(\mu \mathbb{N}_i - \varepsilon_i)}\right) = \left(\mu \mathbb{N}_i - \varepsilon_i\right)e^{\beta(\mu \mathbb{N}_i - \varepsilon_i)}$$

$$e^{\beta(\mu \mathbb{N}_i - \varepsilon_i)}\varepsilon_i = \mu \mathbb{N}_i e^{\beta(\mu \mathbb{N}_i - \varepsilon_i)} - \frac{\partial}{\partial \beta}\left(e^{\beta(\mu \mathbb{N}_i - \varepsilon_i)}\right) \tag{9.22}$$

which makes:

$$\begin{aligned}\langle E \rangle &= \frac{1}{Z}\sum_i \left\{\mu \mathbb{N}_i e^{\beta(\mu \mathbb{N}_i - \varepsilon_i)} - \frac{\partial}{\partial \beta}\left(e^{\beta(\mu \mathbb{N}_i - \varepsilon_i)}\right)\right\} = \mu \mathbb{N} - \frac{1}{Z}\frac{\partial}{\partial \beta}\left(\sum_i e^{\beta(\mu \mathbb{N}_i - \varepsilon_i)}\right) \\ &= \mu \mathbb{N} - \frac{1}{Z}\frac{\partial Z}{\partial \beta}\end{aligned} \tag{9.23}$$

Hence:

$$\langle E \rangle = \mu \mathbb{N} - \frac{\partial \ln(Z)}{\partial \beta} \tag{9.24}$$

So, we have two expressions:

$$\langle \mathbb{N} \rangle = k_B T \frac{\partial}{\partial \mu}(\ln(Z)) \qquad \langle E \rangle = \mu \langle \mathbb{N} \rangle - \frac{\partial \ln(Z)}{\partial \beta} \tag{9.25}$$

illustrating what a useful universal tool the partition function is.

9.2.2 Fluctuations

The ensemble average quantities have been calculated by maximizing the Gibbs functional to obtain the relevant probability distribution. However, significant fluctuations from these averages could take place.

The general way in which we measure variations from the average is by using the *variance* of a quantity (5.16), $\sigma^2(x) = \langle x^2 \rangle - \langle x \rangle^2$. We then obtain the comparative scale of fluctuations from - $\sigma(x)/x$.

Applying this to particle number gives a very nice result. First, we calculate $\langle \mathbb{N}^2 \rangle$:

$$\langle \mathbb{N}^2 \rangle = \sum_i p_i \mathbb{N}_i^2 = \frac{1}{Z} \sum_i e^{\beta(\mu \mathbb{N}_i - \varepsilon_i)} \mathbb{N}_i^2 \tag{9.26}$$

We already have the result (9.17), which is slightly modified here:

$$\frac{\partial}{\partial \mu}\left(e^{\beta(\mu \mathbb{N}_i - \varepsilon_i)}\right) = \beta \mathbb{N}_i e^{\beta(\mu \mathbb{N}_i - \varepsilon_i)} \tag{9.27}$$

So, we can easily see that:

$$\frac{\partial^2}{\partial \mu^2}\left(e^{\beta(\mu \mathbb{N}_i - \varepsilon_i)}\right) = \beta \frac{\partial}{\partial \mu}\left(\mathbb{N}_i e^{\beta(\mu \mathbb{N}_i - \varepsilon_i)}\right) = \beta^2 \mathbb{N}_i^2 e^{\beta(\mu \mathbb{N}_i - \varepsilon_i)} \tag{9.28}$$

Making:

$$\langle \mathbb{N}^2 \rangle = \frac{1}{\beta^2 Z} \frac{\partial^2 Z}{\partial \mu^2} \tag{9.29}$$

This enables us to construct the variance:

$$\sigma^2(\mathbb{N}) = \langle \mathbb{N}^2 \rangle - \langle \mathbb{N} \rangle^2 = \frac{1}{\beta^2 Z} \frac{\partial^2 Z}{\partial \mu^2} - \left(\frac{1}{\beta Z} \frac{\partial Z}{\partial \mu}\right)^2 = \frac{1}{\beta^2}\left\{\frac{1}{Z} \frac{\partial^2 Z}{\partial \mu^2} - \frac{1}{Z^2}\left(\frac{\partial Z}{\partial \mu}\right)^2\right\} \tag{9.30}$$

As:

$$\frac{\partial^2 \ln(Z)}{\partial \mu^2} = \frac{\partial}{\partial \mu}\left(\frac{1}{Z}\frac{\partial Z}{\partial \mu}\right) = -\frac{1}{Z^2}\left(\frac{\partial Z}{\partial \mu}\right)^2 + \frac{1}{Z}\frac{\partial^2 Z}{\partial \mu^2} \tag{9.31}$$

we have the extremely elegant:

$$\sigma^2(\mathbb{N}) = k_B^2 T^2 \frac{\partial^2 \ln(Z)}{\partial \mu^2} \tag{9.32}$$

To show the scale of typical fluctuations, let's look back to the ideal gas. Assuming the gas to be in contact with a particle reservoir as well as a thermal reservoir, an appropriate *grand canonical partition function* (9.15), can be constructed from that for an ideal gas (consult Appendix PA1):

$$Z = \sum_{i=1}^{\infty} e^{(\mu \mathbb{N}_i - \varepsilon_i)/k_B T} = \sum_{\mathbb{N}=0}^{\infty} e^{\mu \mathbb{N}/k_B T} \frac{1}{\mathbb{N}!}\left(Z_{\text{Ideal}}\right)^{\mathbb{N}} \tag{9.33}$$

where Z_{Ideal} is the single particle ideal gas partition function from Chapter 7 (7.42):

$$Z_{\text{Ideal}} = \int_0^{\infty} g(\varepsilon) e^{-\varepsilon/k_B T} d\varepsilon = \frac{L^3}{\hbar^3}\left(\frac{mk_B T}{2\pi}\right)^{3/2}$$

For what is to come, it helps to stuff some of the constants together into one basket, λ:

$$Z_{\text{Ideal}} = \frac{L^3}{\hbar^3}\left(\frac{mk_B T}{2\pi}\right)^{3/2} = V\left(\frac{mk_B T}{2\pi\hbar^2}\right)^{3/2} = \frac{V}{\lambda^3} \tag{9.34}$$

so that the grand canonical function is now:

$$Z = \sum_{\mathbb{N}=0}^{\infty} e^{\mu \mathbb{N}/k_B T} \frac{1}{\mathbb{N}!}\left(\frac{V}{\lambda^3}\right)^{\mathbb{N}} = \sum_{\mathbb{N}=0}^{\infty} \frac{1}{\mathbb{N}!}\left(\frac{Ve^{\mu/k_B T}}{\lambda^3}\right)^{\mathbb{N}} = \exp\left(\frac{Ve^{\mu/k_B T}}{\lambda^3}\right) \tag{9.35}$$

as the summation in the middle of the line is the exponential function expanded out. Now, we can do some calculations. First:

$$\langle \mathbb{N} \rangle = k_B T \frac{\partial}{\partial \mu}(\ln(Z)) = k_B T \frac{\partial}{\partial \mu}\left(\frac{Ve^{\mu/k_B T}}{\lambda^3}\right) = \frac{Ve^{\mu/k_B T}}{\lambda^3} \tag{9.36}$$

Then:

$$\sigma^2(\mathbb{N}) = k_B^2 T^2 \frac{\partial^2 \ln(Z)}{\partial \mu^2} = k_B T \frac{\partial}{\partial \mu}\left(\frac{Ve^{\mu/k_B T}}{\lambda^3}\right) = \langle \mathbb{N} \rangle \tag{9.37}$$

Finally, we get to the comparative scale of the fluctuations:

$$\frac{\sigma(\mathbb{N})}{\mathbb{N}} = \frac{\sqrt{\langle \mathbb{N} \rangle}}{\langle \mathbb{N} \rangle} = \frac{1}{\sqrt{\langle \mathbb{N} \rangle}} \tag{9.38}$$

This is very significant, as these fluctuations away from the average value die off to zero in the thermodynamic limit. Hence our ensemble averages are realistic estimations of the experimentally obtained values.

9.2.3 Modified First Law

To gain a handle on the meaning of μ in our probability distribution, we're going to calculate the Gibbs entropy and then see how that flows through into the First Law.

Applying the grand canonical probability distribution to the Gibbs functional, will maximize the entropy:

$$\mathcal{S}_G = -k_B \sum_i p_i \ln(p_i) = -k_B \sum_i p_i \ln\left(\frac{e^{(\mu \mathbb{N}_i - \varepsilon_i)/k_B T}}{Z}\right)$$

$$= -k_B \sum_i p_i \left\{\left[\frac{(\mu \mathbb{N}_i - \varepsilon_i)}{k_B T}\right] - \ln(Z)\right\}$$

$$= -k_B \sum_i p_i \left\{\frac{(\mu \mathbb{N}_i - \varepsilon_i)}{k_B T}\right\} + k_B \ln(Z)$$

$$= \frac{1}{T}(\langle E \rangle - \mu \langle \mathbb{N} \rangle) + k_B \ln(Z) \tag{9.39}$$

(using $\sum_i p_i = 1$). Compare this expression with (5.46).

Cross-multiplying gives:

$$T\mathcal{S}_G = \langle E \rangle - \mu \langle \mathbb{N} \rangle + k_B T \ln(Z) \tag{9.40}$$

Now, we can incorporate our results from the ideal gas (Section 9.2.2):

$$T\mathcal{S}_G = \langle E \rangle - \mu \langle \mathbb{N} \rangle + k_B T \ln(Z) = \langle E \rangle - \mu \langle \mathbb{N} \rangle + k_B T \frac{V e^{\mu/k_B T}}{\lambda^3} \tag{9.41}$$

Pinching a result from Chapter 5 (5.37):

$$\mathcal{P} = T \left.\frac{\partial S}{\partial V}\right|_{N,U}$$

and applying it to our case we find, as $U = \langle E \rangle, N = \langle \mathbb{N} \rangle$:

$$\mathcal{P} = k_B T \frac{e^{\mu/k_B T}}{\lambda^3} \tag{9.42}$$

This immediately slots back into (9.41):

$$T\mathcal{S}_G = U - \mu \mathrm{N} + \mathcal{P}V \tag{9.43}$$

Or in a more familiar form:

$$\Delta U = T\Delta\mathcal{S}_G + \mu\Delta\mathrm{N} - \mathcal{P}\Delta V \tag{9.44}$$

which is clearly an appropriate modification to the first law. Although this has been derived in the context of the ideal gas, it is an expression that is generally valid. We see that the term $\mu\Delta\mathrm{N}$ expresses *the change in energy due to adding a particle into the system*, provided V and $\mathcal{S}_G$ are kept constant. However, as it is difficult to add (or subtract) a particle from the total count *without changing the entropy in the process*, many people prefer an alternative way of thinking about μ, which explains why it's often called the *chemical potential*. In this approach, μ is likened to gravitational potential energy. When an object rolls down a hill, we can understand the physics in terms of forces, accelerations, and velocities. Alternatively, we can analyse the situation using gravitational potential energy and kinetic energy. At the top of the hill, the object has a higher gravitational potential energy, which can be released (converted into KE) by moving downwards. Similarly, any region where the density of molecules in a gas (for example) is greater than normal has a higher chemical potential. The natural tendency for molecules to diffuse to other parts of the gas is then explained by their seeking out regions where the chemical potential is smaller.

By the way, and changing the subject for a moment, now that we have (9.42) and (9.36), we can play around:

$$\mathcal{P} = k_B T \frac{e^{\mu/k_B T}}{\lambda^3} \qquad \langle \mathbb{N} \rangle = \frac{V e^{\mu/k_B T}}{\lambda^3} = \frac{\mathcal{P}V}{k_B T} \tag{9.45}$$

or:

$$\mathcal{P}V = \mathrm{N}k_B T \tag{9.46}$$

which brings a smile to our faces.

9.3 RECOVERING FERMION AND BOSON STATISTICS

Using the grand canonical ensemble, we have derived a probability distribution over microstates:

$$p_i = \frac{1}{Z} e^{(\mu \mathrm{N}_i - \varepsilon_i)/k_B T}$$

Interestingly, it is now possible to re-cast the arguments from Sections 8.3.1 and 8.4.1 to obtain the Fermi and Bose–Einstein distributions *without recourse to any approximations*, such as the one we borrowed from Stirling. The process will also shed more light on μ and the nature of these distributions.

In Section 8.2, we discussed the symmetry properties of multi-particle states. If we have just two particles, there are both symmetric and anti-symmetric possibilities:

$$\psi_S = \frac{1}{\sqrt{2}}\left(\varphi_1\psi_2 + \psi_1\varphi_2\right) \qquad \psi_A = \frac{1}{\sqrt{2}}\left(\varphi_1\psi_2 - \psi_1\varphi_2\right) \tag{9.47}$$

both of which are combinations of the *single-particle states*, φ, ψ.

In the special case of non-interacting particles, we can use Schrödinger's equation to calculate a collection of single particle states, which can then be combined to make appropriate multi-particle states for the situation. We are going to use the term *orbitals* to refer to single particle states, to save confusing them with the microstates of the system. It's important to note that *each possible value* of an *internal property* of a particle (such as a spin component) would count as a *separate orbital*. These orbitals can either be empty of particles (vacant) or contain one or more particles (occupied). Their *occupation number* (average number of particles in a state) will be of great interest to us.

Conceptually, its crucial to see that each orbital *can be given its own grand canonical ensemble*. As the particles are non-interacting, we do not expect the occupancy (or otherwise) of any one orbital to have a thermodynamic influence on the others. Hence, their ensembles are independent.

Across one of these ensembles, where every member is the same orbital, the different *microstates* then comprise the different *occupations* of that orbital. Being vacant, having one particle, two particles, etc., would all be different microstates. These microstates can be *indexed* by $\mathbb{N}$ (which is playing the role of i from earlier) the number of particles in the orbital. If we label the orbital by j, then the energy of the microstate is $\mathbb{N}\varepsilon_j$, with ε_j being the characteristic energy of the orbital, and the probability of finding $\mathbb{N}$ particles in the orbital is:

$$p_j(\mathbb{N}) = \frac{e^{\left(\mu\mathbb{N}-\mathbb{N}\varepsilon_j\right)/k_BT}}{Z} = \frac{e^{\mathbb{N}\left(\mu-\varepsilon_j\right)/k_BT}}{Z} \tag{9.48}$$

One crucial task will be evaluating the partition function:

$$Z = \sum_{\mathbb{N}=0}^{\infty} e^{\mathbb{N}\left(\mu-\varepsilon_j\right)/k_BT} \tag{9.49}$$

in various cases.

{In the following, always remember that different j are referring to different *orbitals* and hence different grand canonical *ensembles*, not various *elements* or members within the *same ensemble*…}

9.3.1 Fermions

As fermions obey the Pauli exclusion principle, our life is made easier. That's why we are going to tackle them first.

Each orbital can be vacant or contain at most a single particle. Hence $\mathbb{N} \in \{0,1\}$ and the partition function is a straight sum of two terms:

$$Z_F = \sum_{\mathbb{N}=0}^{\mathbb{N}=1} e^{\mathbb{N}(\mu-\varepsilon_j)/k_B T} = 1 + e^{(\mu-\varepsilon_j)/k_B T} \tag{9.50}$$

We can calculate the average occupancy of the orbital, using (9.19), which we took the trouble to derive earlier, for this purpose:

$$\langle \mathbb{N} \rangle = k_B T \frac{1}{Z} \frac{\partial Z}{\partial \mu}$$

In our specific situation:

$$\langle \mathbb{N}_j \rangle = \frac{k_B T}{1 + e^{(\mu-\varepsilon_j)/k_B T}} \frac{\partial}{\partial \mu} \left(1 + e^{(\mu-\varepsilon_j)/k_B T} \right) = \frac{k_B T}{1 + e^{(\mu-\varepsilon_j)/k_B T}} \times \frac{e^{(\mu-\varepsilon_j)/k_B T}}{k_B T}$$

$$= \frac{e^{(\mu-\varepsilon_j)/k_B T}}{1 + e^{(\mu-\varepsilon_j)/k_B T}} = \frac{1}{e^{-(\mu-\varepsilon_j)/k_B T} \left(1 + e^{(\mu-\varepsilon_j)/k_B T} \right)} \tag{9.51}$$

reducing to:

$$\langle \mathbb{N}_j \rangle = \frac{1}{1 + e^{(\varepsilon_j-\mu)/k_B T}} \tag{9.52}$$

We can recover the earlier version of Fermi-Dirac statistics from Chapter 8 by observing that the average number of particles across the ensemble for orbital j, $\langle \mathbb{N}_j \rangle$, is equivalent to $\mathbb{N}_n / g_n$, from before:

$$\frac{\mathbb{N}_n}{g_n} = \frac{1}{1 + e^{(\varepsilon_n-\mu)/k_B T}} \tag{9.53}$$

9.3.2 Bosons

Calculating the partition function for bosons is trickier, as there is no limit to the number of bosons that can be in any orbital:

$$Z_B = \sum_{\mathbb{N}=0}^{\infty} e^{\mathbb{N}(\mu-\varepsilon_j)/k_B T} = \sum_{\mathbb{N}=0}^{\infty} \left(e^{(\mu-\varepsilon_j)/k_B T} \right)^{\mathbb{N}} \tag{9.54}$$

This is a *geometric progression*—a series with first-term a and multiplying factor r:

$$GP = a + ar + ar^2 + ar^3 \ldots = \sum_{m=1}^{\infty} ar^{m-1} \tag{9.55}$$

Such series can "sum to infinity", i.e., have a defined convergent sum for an infinite number of terms:

$$GP_{\infty} = \frac{a}{1-r} \tag{9.56}$$

provided $-1 < r < 1$.

In this instance, $a = 1$ and $r = e^{(\mu-\varepsilon_j)/k_BT}$ so:

$$Z_B = GP_{\infty} = \frac{a}{1-r} = \frac{1}{1-e^{(\mu-\varepsilon_j)/k_BT}} \tag{9.57}$$

As we did for fermions, we can calculate the average number of particles in the orbital from (9.19):

$$\langle \mathbb{N}_j \rangle = \frac{k_BT}{Z_B}\frac{\partial Z_B}{\partial \mu}$$

$$= k_BT\left(1-e^{(\mu-\varepsilon_j)/k_BT}\right)\frac{\partial}{\partial \mu}\left(\frac{1}{1-e^{(\mu-\varepsilon_j)/k_BT}}\right)$$

$$= k_BT\left(1-e^{(\mu-\varepsilon_j)/k_BT}\right)\frac{e^{(\mu-\varepsilon_j)/k_BT}}{k_BT\left(1-e^{(\mu-\varepsilon_j)/k_BT}\right)^2}$$

$$= \frac{1}{e^{-(\mu-\varepsilon_j)/k_BT}\left(1-e^{(\mu-\varepsilon_j)/k_BT}\right)} \tag{9.58}$$

So, we get:

$$\langle \mathbb{N}_j \rangle = \frac{1}{e^{(\varepsilon_j-\mu)/k_BT}-1} \tag{9.59}$$

or:

$$\frac{\mathbb{N}_n}{g_n} = \frac{1}{e^{(\varepsilon_n-\mu)/k_BT}-1} \tag{9.60}$$

As we have used the sum to infinity for a GP, there is a mathematical constraint to contend with. To make the sum viable, it is necessary that $r < 1$, which in our specific case translates to $r = e^{(\mu-\varepsilon_j)/k_BT} < 1$, *including such occasions when* $\varepsilon_j = 0$. This has an impact on the physics. If $\varepsilon_j = 0$, then $e^{\mu/k_BT} < 1$ so that $\mu < 0$, i.e., *the chemical potential for bosons must be negative.* It can be either positive or negative if we are dealing with fermions.

In this context, identical particles in quantum theory, using the name *chemical potential* for μ is not that appropriate—something like *number potential* or *occupation potential* might be more apt. Whatever we term it, the implication of having a negative value for bosons is an energy *benefit* if we increase the number of particles.

9.4 ON MATTERS OF PROBABILITY

As it is in their nature to obey the exclusion principle, the average occupation for fermions:

$$\langle \mathbb{N}_j \rangle = \frac{1}{1+e^{(\varepsilon_j-\mu)/k_BT}} < 1 \tag{9.61}$$

To the casual glance, this makes it seem like $\langle \mathbb{N}_j \rangle$ has the form of a probability distribution, *which it is not*. Our choice of the symbol $\langle \mathbb{N}_j \rangle$ was partly to offset this impression. It's good to avoid this mistake, as it can cause confusion when compared to the boson case, where:

$$\langle \mathbb{N}_j \rangle = \frac{1}{e^{(\varepsilon_j-\mu)/k_BT} - 1} \geq 1 \tag{9.62}$$

as bosons have no exclusion principle to limit their excesses.

To extract the relevant genuine probabilities, we need to return to the probability function ((9.48) and (9.49)):

$$p_j(\mathbb{N}) = \frac{e^{\mathbb{N}(\mu-\varepsilon_j)/k_BT}}{Z} \qquad Z = \sum_{\mathbb{N}=0}^{\infty} e^{\mathbb{N}(\mu-\varepsilon_j)/k_BT}$$

and apply it to the two different classes of particle:

Fermions

$$p_j(\mathbb{N}) = \frac{e^{\mathbb{N}(\mu-\varepsilon_j)/k_BT}}{Z_F} \quad \text{with} \quad Z_F = \sum_{\mathbb{N}=0}^{\mathbb{N}=1} e^{\mathbb{N}(\mu-\varepsilon_j)/k_BT} = 1 + e^{(\mu-\varepsilon_j)/k_BT} \tag{9.63}$$

Bosons

$$p_j(\mathbb{N}) = \frac{e^{\mathbb{N}(\mu-\varepsilon_j)/k_BT}}{Z_B} \quad \text{with} \quad Z_B = \sum_{\mathbb{N}=0}^{\infty} e^{\mathbb{N}(\mu-\varepsilon_j)/k_BT} = \frac{1}{1-e^{(\mu-\varepsilon_j)/k_BT}} \tag{9.64}$$

It all comes down to the partition function…

Let's examine the situation for fermions in a little more detail.

The probability that orbital j is occupied is found by taking $\mathbb{N} = 1$:

$$p_j(1) = \frac{e^{(\mu-\varepsilon_j)/k_BT}}{1+e^{(\mu-\varepsilon_j)/k_BT}} \tag{9.65}$$

On the other hand, the probability that the orbital is vacant, $\mathbb{N} = 0$ is:

$$p_j(0) = \frac{1}{1+e^{(\mu-\varepsilon_j)/k_BT}} \tag{9.66}$$

Comparing the two results, we see:

$$p_j(1) = e^{(\mu-\varepsilon_j)/k_BT}\, p_j(0) \tag{9.67}$$

Also, its relatively obvious but worth checking that:

$$p_j(1) + p_j(0) = \frac{e^{(\mu-\varepsilon_j)/k_BT}}{1+e^{(\mu-\varepsilon_j)/k_BT}} + \frac{1}{1+e^{(\mu-\varepsilon_j)/k_BT}} = 1 \tag{9.68}$$

or:

$$p_j(1) = 1 - p_j(0) \tag{9.69}$$

Armed with these two probabilities, we can calculate the average occupation of the state "long hand" from:

$$\langle \mathbb{N}_j \rangle = \sum_{\mathbb{N}=0}^{\mathbb{N}=1} p_j(\mathbb{N})\mathbb{N} = 1 \times \left(\frac{e^{(\mu-\varepsilon_j)/k_BT}}{1+e^{(\mu-\varepsilon_j)/k_BT}} \right) + 0 \times \left(\frac{1}{1+e^{(\mu-\varepsilon_j)/k_BT}} \right)$$

$$= \frac{e^{(\mu-\varepsilon_j)/k_BT}}{1+e^{(\mu-\varepsilon_j)/k_BT}} = \frac{1}{1+e^{(\varepsilon_j-\mu)/k_BT}} \tag{9.70}$$

as we are well used to seeing.[1]

Another quite pretty relationship is[2]:

$$\langle \mathbb{N}_j \rangle = \frac{1}{1+e^{(\varepsilon_j-\mu)/k_BT}} = \frac{e^{(\mu-\varepsilon_j)/k_BT}}{e^{(\mu-\varepsilon_j)/k_BT}+1} = p_j(1) \tag{9.71}$$

Now, importantly, this normalised and well-behaved binary probability is *only referring to the selected orbital*. There is a *separate* probability function for *each* of the orbitals, with two values, one for the state being occupied and one for it being vacant. The situation is rather different to the earlier Boltzmann distribution. After all, the Boltzmann canonical ensemble spanned all energy levels, whereas when we derived the Fermi-Dirac distribution here, we set up one ensemble for each orbital.

We're going to need to work a little harder to get an overall fermion probability function *that applies to any state*. More on that shortly (Section 9.4.1).

Turning now to bosons, we have:

$$p_j(\mathbb{N}) = \frac{e^{\mathbb{N}(\mu-\varepsilon_j)/k_BT}}{Z_B} = \left(1-e^{(\mu-\varepsilon_j)/k_BT}\right)\left(e^{(\mu-\varepsilon_j)/k_BT}\right)^{\mathbb{N}} \tag{9.72}$$

If this is behaving itself correctly, $\sum_{\mathbb{N}=0}^{\infty} p_j(\mathbb{N}) = 1$, {again, remember that this sum is normalized for one orbital, j, so this is *not* a probability distribution across orbitals}. To check, we calculate:

$$\sum_{\mathbb{N}=0}^{\infty} p_j(\mathbb{N}) = \left(1-e^{(\mu-\varepsilon_j)/k_BT}\right)\sum_{\mathbb{N}=0}^{\infty}\left(e^{(\mu-\varepsilon_j)/k_BT}\right)^{\mathbb{N}} = \frac{Z_B}{Z_B} = 1 \tag{9.73}$$

Emboldened by this, we try our hand at the average occupation:

$$\langle \mathbb{N}_j \rangle = \sum_{\mathbb{N}=0}^{\mathbb{N}=\infty} p_j(\mathbb{N})\mathbb{N} = \left(1-e^{(\mu-\varepsilon_j)/k_BT}\right)\sum_{\mathbb{N}=0}^{\mathbb{N}=\infty}\left(e^{(\mu-\varepsilon_j)/k_BT}\right)^{\mathbb{N}}\mathbb{N} \tag{9.74}$$

We can clear the undergrowth a bit by writing $x = e^{(\mu-\varepsilon_j)/k_BT}$, so (9.74) now takes the form:

$$(1-x)\sum_0^{\infty} mx^m \tag{9.75}$$

making it easier to spot how to carry out the sum. To deal with $S = \sum_0^{\infty} mx^m = x + 2x^2 + 3x^3 + \ldots$, we start by multiplying by x and then subtracting:

$$S - xS = (1-x)S = \{x + 2x^2 + 3x^3 + \ldots\} - \{x^2 + 2x^3 + 3x^4 + \ldots\} \tag{9.76}$$

Tidying:

$$(1-x)S = x + x^2 + x^3 + \ldots = \frac{x}{1-x} \tag{9.77}$$

as the subtracted series is a GP with $a = r = x$. This gives us $S = x/(1-x)^2$ which we can plug back in:

$$(1-x)\sum_0^{\infty} mx^m = \frac{x}{1-x} \tag{9.78}$$

Re-planting the undergrowth:

$$\langle \mathbb{N}_j \rangle = \sum_{\mathbb{N}=0}^{\mathbb{N}=\infty} p_j(\mathbb{N})\mathbb{N} = \frac{e^{(\mu-\varepsilon_j)/k_BT}}{1-e^{(\mu-\varepsilon_j)/k_BT}} = \frac{1}{e^{(\varepsilon_j-\mu)/k_BT}-1} \tag{9.79}$$

as before.

Having established that this is a properly functioning probability function, applicable to one orbital, we can explore some of its more amusing features. Starting from (9.72):

$$p_j(\mathbb{N}) = \frac{e^{\mathbb{N}(\mu-\varepsilon_j)/k_BT}}{Z_B} = \left(1-e^{(\mu-\varepsilon_j)/k_BT}\right)\left(e^{(\mu-\varepsilon_j)/k_BT}\right)^{\mathbb{N}}$$

it is evident that:

$$p_j(0) = \left(1-e^{(\mu-\varepsilon_j)/k_BT}\right) \qquad p_j(1) = e^{(\mu-\varepsilon_j)/k_BT}p_j(0) \tag{9.80}$$

In fact:

$$p_j(\mathbb{N}) = \left(1-p_j(0)\right)^{\mathbb{N}} p_j(0) \tag{9.81}$$

Our results have consequences for thinking about the average occupation, as:

$$\langle \mathbb{N}_j \rangle = \frac{e^{(\mu-\varepsilon_j)/k_BT}}{1-e^{(\mu-\varepsilon_j)/k_BT}} = \frac{1-p_j(0)}{p_j(0)} \tag{9.82}$$

Inverting:

$$p_j(0) = \frac{1}{1+\langle \mathbb{N}_j \rangle} \tag{9.83}$$

both of which are neat results and should be compared with (9.71):

$$\langle \mathbb{N}_j \rangle = \frac{1}{1+e^{(\varepsilon_j-\mu)/k_BT}} = \frac{e^{(\mu-\varepsilon_j)/k_BT}}{e^{(\mu-\varepsilon_j)/k_BT}+1} = p_j(1)$$

for fermions.

9.4.1 Grand Partition Functions

As we have been at pains to explain, we have used a collection of grand canonical distributions to extract average occupation numbers, and now probabilities, for each orbital. When we set things up, we proposed that the various ensembles were independent, as our particles were not interacting with each other to any significant extent. So, an overall probability distribution can be constructed from a *product of the separate functions*:

Fermions

$$\mathbb{P} = \frac{1}{\mathcal{Z}_{\mathrm{F}}} \prod_j e^{\left(\mu \mathrm{N}_j - \mathrm{N}_j \varepsilon_j\right)/k_B T} \qquad \text{with} \qquad \mathcal{Z}_{\mathrm{F}} = \prod_j \left(1 + e^{\left(\mu - \varepsilon_j\right)/k_B T}\right)$$

$$\mathrm{N}_j \in \{0,1\} \tag{9.84}$$

Bosons

$$\mathbb{P} = \frac{1}{\mathcal{Z}_{\mathrm{B}}} \prod_j e^{\left(\mu \mathrm{N}_j - \mathrm{N}_j \varepsilon_j\right)/k_B T} \qquad \text{with} \qquad \mathcal{Z}_{\mathrm{B}} = \prod_j \left(\frac{1}{1 - e^{\left(\mu - \varepsilon_j\right)/k_B T}}\right)$$

$$\mathrm{N}_j \in \{0,1,2,3\ldots\} \tag{9.85}$$

with $\mathcal{Z}_{\mathrm{F}}$ and $\mathcal{Z}_{\mathrm{B}}$ as the respective *grand partition functions*.

9.4.2 The Planck Radiation Law

Now is a good moment to derive the Planck radiation law, as promised in Section 8.4.2.

A *perfect black body* is a sample of material in thermal equilibrium with its radiation field. While such a perfect thermal source is hard to come across in practice,[3] good experimental and theoretical analogues can be produced using a *radiating cavity*. We imagine (and construct as best we can) a box with reflecting walls warmed to a controlled temperature. One side of the box contains a small cavity, out of which EM radiation can escape for study. Within the box, we have a collection of photons, which are bosons of spin 1. Hence, the situation is tailor-made for our grand partition function and the boson statistics we have been developing.

The photon wave functions within the box are just like those for our particle in a box from Chapter 6. In this case, the wavelengths along the *x*, *y*, and *z* axes will be quantized:

$$\lambda_x = \frac{2L}{n_x}, \lambda_y = \frac{2L}{n_y}, \lambda_z = \frac{2L}{n_z} \tag{9.86}$$

It is more convenient to work in terms of the *wave numbers*:

$$k_x = \frac{n_x}{2L}, k_y = \frac{n_y}{2L}, k_z = \frac{n_z}{2L} \tag{9.87}$$

as for any photon propagating across the box, we can combine the quantized wave numbers, more easily than different wavelength "components":

$$k^2 = k_x^2 + k_y^2 + k_z^2 \tag{9.88}$$

So:

$$k^2 = \frac{1}{4L^2}\left(n_x^2 + n_y^2 + n_z^2\right)$$

$$n^2 = n_x^2 + n_y^2 + n_z^2 = \frac{4L^2}{\lambda^2} \tag{9.89}$$

As before (Section 7.7.1), we can count the number of states by considering a spherical volume in n-space:

$$V = \frac{4}{3}\pi r^3 = \frac{4}{3}\pi\left(n_x^2 + n_y^2 + n_z^2\right)^{3/2} = \frac{4}{3}\pi\left(\frac{4L^2}{\lambda^2}\right)^{3/2} = \frac{4}{3}\pi\left(\frac{8L^3}{\lambda^3}\right) \tag{9.90}$$

We only need the top-right quadrant, so taking 1/8th gives:

$$V' = \frac{4\pi L^3}{3\lambda^3} \tag{9.91}$$

However, photons can occur in one of two *polarizations*, so the total number of modes is doubled:

$$N = \frac{8\pi L^3}{3\lambda^3} \tag{9.92}$$

As $c = f\lambda$, the volume mode density (number of modes per unit volume) in terms of frequency, f is:

$$\rho_V(f) = \frac{8\pi f^3}{3c^3} \tag{9.93}$$

So, finally, we have the volume mode density per frequency range:

$$\frac{d\rho_V(f)}{df} = g_V(f) = \frac{8\pi f^2}{c^3} \tag{9.94}$$

which is effectively the density of states in this case. This will be very useful in a few lines time.

Our next step is to evaluate the appropriate partition function. As the photons are being emitted and absorbed, their number is not conserved, so we need to set $\mu = 0$. To make this more evident, think back to the derivation of the grand canonical distribution, where we set $B = \mu/k_BT$. In turn, B was the Lagrange multiplier we applied to the total number constraint that we put in place when we maximized. If we get rid of μ, and so B, that constraint drops out. This makes the *photon grand partition function*, $\mathcal{Z}_P$ (based on that for bosons (9.85) with $\mu = 0$):

$$\mathcal{Z}_P = \prod_j \left(\frac{1}{1 - e^{-\varepsilon_j/k_BT}}\right) = \prod_j \left(\frac{1}{1 - e^{-hf_j/k_BT}}\right) \tag{9.95}$$

as $\varepsilon_j = hf_j$

Taking logs turns this awkward product into a sum:

$$\ln\left(\mathcal{Z}_P\right) = -\sum_j \ln\left(1 - e^{-hf_j/k_BT}\right) \tag{9.96}$$

and sets up the use of an earlier relationship, which will enter stage left shortly.

The next sleight of hand is to convert this sum into an integral, on the assumption that the gap between quantized photon energies, Δf_j, is small enough to let us morph the count over. The only other thing we need to watch for is the density of states that must be inserted so that the count over states becomes:

$$\ln\left(\mathcal{Z}_{\mathrm{P}}\right) = -V\int_0^{\infty} g_V(f)\ln\left(1-e^{-hf/k_BT}\right)df \tag{9.97}$$

Using (entering stage left) (9.24) with $\mu = 0$:

$$\langle E\rangle = -\frac{\partial \ln(Z)}{\partial \beta}$$

we get:

$$\frac{\langle E\rangle}{V} = -\frac{\partial}{\partial\beta}\left(\int_0^{\infty} g_V(f)\ln\left(1-e^{-hf/k_BT}\right)df\right) = -\frac{\partial}{\partial\beta}\left(\int_0^{\infty} g_V(f)\ln\left(1-e^{-\beta hf}\right)df\right) \tag{9.98}$$

as writing $\beta = 1/k_BT$ makes the calculations slightly easier.

Carrying out the differentiation, which can be taken inside the integral as it is not the subject:

$$\frac{\langle E\rangle}{V} = -\int_0^{\infty} g_V(f)\frac{\partial}{\partial\beta}\left(\ln\left(1-e^{-\beta hf}\right)\right)df = \int_0^{\infty} g_V(f)\frac{hfe^{-\beta hf}}{\left(1-e^{-\beta hf}\right)}df$$

$$= \int_0^{\infty} g_V(f)\frac{hf}{\left(e^{\beta hf}-1\right)}df \tag{9.99}$$

If we express this in terms of the *spectral radiation density*, $u(f,T)$:

$$\langle E\rangle = V\int_0^{\infty} u(f,T)\,df \tag{9.100}$$

we find that the *Planck Radiation Law* takes the form:

$$u(f,T) = g_V(f)\frac{hf}{\left(e^{\beta hf}-1\right)} \tag{9.101}$$

Or, re-inserting the various terms we have hidden for simplicity:

$$u(f,T) = \frac{8\pi hf^3}{c^3}\left(\frac{1}{e^{hf/k_BT}-1}\right) \tag{9.102}$$

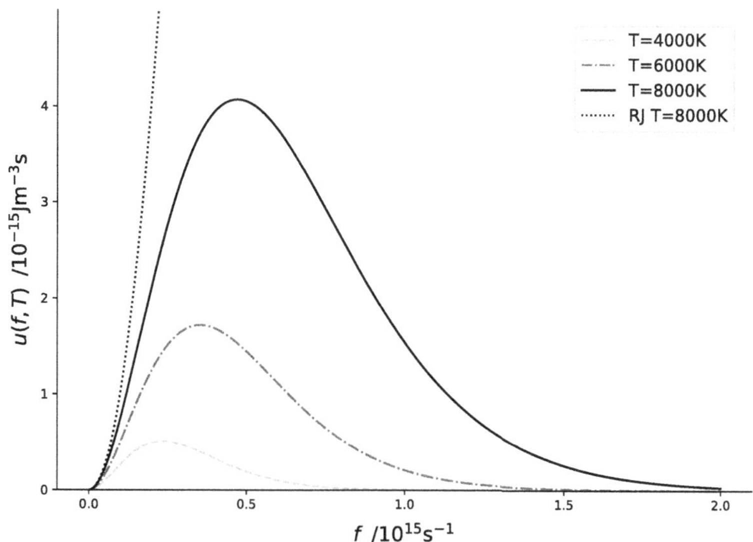

FIGURE 9.1 The black body distribution described by the Planck Radiation Law. The dotted curve is the Raleigh-Jeans law, which was based on classical (non-quantum) thermodynamics and failed to adequately describe the data.

Writing this in terms of wavelength is not as intuitively obvious as you might think, as it needs the relationship[4] $u(f,T)df = u(\lambda,T)d\lambda$:

$$u(\lambda,T) = \frac{8\pi hc}{\lambda^5}\left(\frac{1}{e^{hc/\lambda k_B T} - 1}\right) \tag{9.103}$$

A plot of the frequency curve for various temperatures is shown in Figure 9.1, along with a more "classical" curve which fails to describe the data at high frequencies.[5]

Planck's discovery of this equation, which earned him the Nobel Prize in 1918, followed an entirely different route. At the time, bosons were an unknown feature of the fledgling quantum theory, so Planck deployed a heuristic approach to find the best-fit formula to the experimental data. In searching for a theoretical justification for his result, he was forced to assume the "oscillators" in the material of the black body, which were emitting an absorbing the EM radiation, could only do so in quantized lumps of energy, hf. This started the quantum revolution.

9.4.3 Wien's Law

As it will be needed in Chapter 15, we're going to see how *Wien's distribution law* $\lambda_{\text{Peak}} \propto 1/T$ can be unearthed from the Planck relationship.

To find the peak wavelength, i.e., the wavelength at which most energy is exchanged, we differentiate and set equal to zero, i.e., $\partial u(\lambda T)/\partial\lambda = 0$
As:

$$\frac{\partial}{\partial\lambda}\left(\frac{8\pi hc}{\lambda^5}\left(\frac{1}{e^{hc/\lambda k_B T} - 1}\right)\right) = \frac{-40\pi hc}{\lambda^6}\left(\frac{1}{e^{hc/\lambda k_B T} - 1}\right) - \frac{8\pi hc}{\lambda^5}\left(\frac{-hc}{\lambda^2 k_B T}\right)\left(\frac{e^{hc/\lambda k_B T}}{\left(e^{hc/\lambda k_B T} - 1\right)^2}\right) \tag{9.104}$$

we get:

$$5-\left(\frac{hc}{\lambda k_B T}\right)\left(\frac{e^{hc/\lambda k_B T}}{e^{hc/\lambda k_B T}-1}\right)=0 \tag{9.105}$$

Temporarily setting $\xi=\frac{hc}{k_B T}$, we simplify to:

$$5-\xi\left(\frac{e^{\xi}}{e^{\xi}-1}\right)=5-\xi\left(\frac{1}{1-e^{-\xi}}\right)=0$$

$$\xi=5\left(1-e^{-\xi}\right) \tag{9.106}$$

Solving this innocent-looking equation would mean introducing more mathematical technology than is warranted in a book of this nature, so please accept as taken that the solution is:

$$\xi=\frac{hc}{\lambda k_B T}=5+W_0\left(-5e^{-5}\right)=K\approx 4.965 \tag{9.107}$$

Here W_0 is the so-called *principal branch* of the *Lambert W function*, effectively giving us a constant which we have bundled into *K*. From here, we can conclude:

$$\lambda_{\text{Peak}}=\frac{hc}{Kk_B T}=\frac{W}{T} \tag{9.108}$$

as required, with $W \times 2.9 \times 10^{-3}$ mK.

9.5 ENTROPY

It is high time that we explored the impact of fermion and bosons statistics on the entropy of such particles. After all, as well as being indistinguishable, they have specific symmetry requirements in their multiparticle states, which must impact on the possibilities for arranging them in quantum states.

9.5.1 Fermion Entropy

Having armed ourselves with the fermion grand partition function (9.84):

$$\mathcal{Z}_{\text{Fermion}}=\prod_j\left(1+e^{(\mu-\varepsilon_j)/k_B T}\right)$$

we can use the relationship derived in Chapter 7 (7.36):

$$\mathcal{S}_G=k_B\frac{\partial}{\partial T}\{T\ln(\mathcal{Z})\}=k_B\ln(\mathcal{Z})+\frac{k_B T}{\mathcal{Z}}\frac{\partial \mathcal{Z}}{\partial T}$$

to obtain the appropriate entropy.

Admittedly, working with the grand partition function gets a little messy:

$$\mathcal{S}_{GF} = k_B \ln\left(\prod_j \left(1+e^{(\mu-\varepsilon_j)/k_BT}\right)\right) + \frac{k_BT}{\mathcal{Z}}\frac{\partial}{\partial T}\left(\prod_j \left(1+e^{(\mu-\varepsilon_j)/k_BT}\right)\right)$$

$$= k_B \sum_j \ln\left(1+e^{(\mu-\varepsilon_j)/k_BT}\right) + \frac{k_BT}{\mathcal{Z}} \sum_{j\neq k} -\frac{\mu-\varepsilon_j}{k_BT^2} e^{(\mu-\varepsilon_j)/k_BT} \prod_{k\neq j}\left(1+e^{(\mu-\varepsilon_k)/k_BT}\right)$$

$$= k_B \sum_j \ln\left(1+e^{(\mu-\varepsilon_j)/k_BT}\right) - \left(\frac{1}{T}\right)\left(\frac{1}{\prod_j \left(1+e^{(\mu-\varepsilon_j)/k_BT}\right)}\right) \sum_{j\neq k}\left(\mu-\varepsilon_j\right) e^{(\mu-\varepsilon_n)/k_BT} \prod_{k\neq j}\left(1+e^{(\mu-\varepsilon_k)/k_BT}\right)$$

$$= k_B \sum_j \ln\left(1+e^{(\mu-\varepsilon_j)/k_BT}\right) - \left(\frac{1}{T}\right)\sum_j \left(\mu-\varepsilon_j\right)\frac{e^{(\mu-\varepsilon_j)/k_BT}}{1+e^{(\mu-\varepsilon_j)/k_BT}} \tag{9.109}$$

but finally, after several deep breaths, we get to:

$$\mathcal{S}_{GF} = k_B \sum_j \left\{ \ln\left(1+e^{(\mu-\varepsilon_j)/k_BT}\right) + \left(\frac{1}{k_BT}\right)\frac{\left(\varepsilon_j-\mu\right)}{e^{(\varepsilon_j-\mu)/k_BT}+1} \right\} \tag{9.110}$$

Having laboured to construct this expression, we now need to shift it into a different and interesting form. As this is another slightly tedious and distracting calculation, the details can be found in Appendix PA2, for now we move directly to the result:

$$\mathcal{S}_{GF} = -k_B \sum_j \left\{ \left(1-\langle \mathrm{N}_j\rangle\right)\ln\left(1-\langle \mathrm{N}_j\rangle\right) + \langle \mathrm{N}_j\rangle \ln\left(\langle \mathrm{N}_j\rangle\right) \right\} \tag{9.111}$$

We have already established some useful expressions for fermions ((9.69) and (9.71)):

$$\langle \mathrm{N}_j\rangle = p_j(1) \qquad p_j(0) = 1 - p_j(1)$$

which we use to write:

$$\mathcal{S}_{GF} = -k_B \sum_j \left\{ p_j(0)\ln\left(p_j(0)\right) + p_j(1)\ln\left(p_j(1)\right) \right\} \tag{9.112}$$

The structure of this interesting expression becomes very intelligible with a few thoughts. Fermion statistics are binary; the occupation of each orbital is a simple yes/no situation. So, this relationship follows as in principle *we do not know which is the case*. The entropy includes both options *due to a lack of information on our part*.

If we were calculating the entropy at 0 K, then from Figure 8.4, for example, we would know that all the states up to ε_F were occupied and that all the states after were vacant. The entropy would be:

$$\mathcal{S}_{GF} = -k_B\left\{\left(\sum 1\ln(1)+\sum 0\ln(0)\right)_{\varepsilon_n\le\varepsilon_F}+\left(\sum 0\ln(0)+\sum 1\ln(1)\right)_{\varepsilon_n>\varepsilon_F}\right\}=0 \qquad (9.113)$$

At any other temperature, it is the states in the vicinity of μ that contribute to the entropy, *precisely those where we can't be sure about the occupancy.*

Perhaps a fruitful line to follow would be to consider the relationship between information and entropy. There is certainly information bound up in a quantum state, but by its nature the information is, from a classical perspective, incomplete. Indeed, with multiparticle states, such as the ones considered at the start of this chapter, there is rich information content. We need to consider the entropy within a quantum state, which we will turn to in Chapter 12. Information theory is covered in Chapter 14, and we have further thoughts in Chapter 10.

9.5.2 Boson Entropy

To extract the boson entropy, we again start with 7.36:

$$\mathcal{S}_G = k_B\ln(\mathcal{Z})+\frac{k_BT}{\mathcal{Z}}\frac{\partial\mathcal{Z}}{\partial T}$$

but this time we use (9.85):

$$\mathcal{Z}_B=\prod_j\left(\frac{1}{1-e^{(\mu-\varepsilon_j)/k_BT}}\right)$$

Calculating the first term:

$$\begin{aligned}k_B\ln(\mathcal{Z})&=k_B\ln\left(\prod_j\left(\frac{1}{1-e^{(\mu-\varepsilon_j)/k_BT}}\right)\right)=k_B\sum_j\ln\left(\frac{1}{1-e^{(\mu-\varepsilon_j)/k_BT}}\right)\\&=-k_B\sum_j\ln\left(1-e^{(\mu-\varepsilon_j)/k_BT}\right)\end{aligned} \qquad (9.114)$$

Now we tackle the second term:

$$\frac{k_BT}{\mathcal{Z}}\frac{\partial\mathcal{Z}}{\partial T}=k_BT\prod_j\left(1-e^{(\mu-\varepsilon_j)/k_BT}\right)\frac{\partial}{\partial T}\left(\prod_j\left(\frac{1}{1-e^{(\mu-\varepsilon_j)/k_BT}}\right)\right)$$

$$=k_BT\prod_j\left(1-e^{(\mu-\varepsilon_j)/k_BT}\right)\sum_{k\neq j}\frac{(\mu-\varepsilon_k)e^{(\mu-\varepsilon_k)/k_BT}}{k_BT^2\left(1-e^{(\mu-\varepsilon_k)/k_BT}\right)^2}\left\{\prod_{j\neq k}\left(\frac{1}{1-e^{(\mu-\varepsilon_j)/k_BT}}\right)\right\}$$

$$=k_BT\sum_k\frac{(\mu-\varepsilon_k)e^{(\mu-\varepsilon_k)/k_BT}\left(1-e^{(\mu-\varepsilon_k)/k_BT}\right)}{k_BT^2\left(1-e^{(\mu-\varepsilon_k)/k_BT}\right)^2}$$

$$= \frac{1}{T}\sum_k \frac{(\mu-\varepsilon_k)e^{(\mu-\varepsilon_k)/k_BT}}{1-e^{(\mu-\varepsilon_k)/k_BT}} \tag{9.115}$$

Patching them together gets us to the boson entropy, $\mathcal{S}_{GB}$:

$$\mathcal{S}_{GB} = -k_B\sum_j \ln\left(1-e^{(\mu-\varepsilon_j)/k_BT}\right) - \frac{1}{T}\sum_k \frac{(\mu-\varepsilon_k)e^{(\mu-\varepsilon_k)/k_BT}}{1-e^{(\mu-\varepsilon_k)/k_BT}}$$

$$= -k_B\sum_j\left\{\ln\left(1-e^{(\mu-\varepsilon_j)/k_BT}\right)+\left(\frac{1}{k_BT}\right)\frac{(\varepsilon_j-\mu)}{e^{(\varepsilon_j-\mu)/k_BT}-1}\right\} \tag{9.116}$$

which has a notably similar structure to the fermion expression from earlier (9.110). As before, this can be re-cast in terms of probabilities, but that is another calculation that we have placed in the appendix for later inspection. Moving to the result, we find:

$$\mathcal{S}_{GB} = k_B\sum_j\left(1+\langle \mathrm{N}_j\rangle\right)\ln\left(1+\langle \mathrm{N}_j\rangle\right) - k_B\sum_j\langle \mathrm{N}_j\rangle\ln\left(\langle \mathrm{N}_j\rangle\right) \tag{9.117}$$

(I have split the terms here for comparison and development purposes…)
which compares with the fermion result (9.111), here similarly split in two:

$$\mathcal{S}_{GF} = -k_B\sum_j\left(1-\langle \mathrm{N}_j\rangle\right)\ln\left(1-\langle \mathrm{N}_j\rangle\right) - k_B\sum_j\langle \mathrm{N}_j\rangle\ln\left(\langle \mathrm{N}_j\rangle\right) \tag{9.118}$$

Reassuringly, the entropy defined by these relations is a positive quantity, as expected. However, the positivity of $\mathcal{S}_{GB}$ follows from $\langle \mathrm{N}_j\rangle \geq 0$, and positivity of $\mathcal{S}_{GF}$ from $0 \leq \langle \mathrm{N}_j\rangle \leq 1$.

The boson entropy (9.117) can also be expressed in terms of probability. This is another quite complex calculation which the reader can follow in the appendix if wished. Again, we simply quote the answer here:

$$\mathcal{S}_{GB} = -k_B\sum_j\left(\frac{1}{p_j(0)}\right)\left\{\left(1-p_j(0)\right)\ln\left(1-p_j(0)\right)+p_j(0)\ln\left(p_j(0)\right)\right\} \tag{9.119}$$

This result can be interpreted in a similar way to the equivalent fermion case. After all, the probability that an orbital is occupied is $\left(1-p_j(0)\right)$, which is the total probability for any number of particles in that orbital. The factor of $1/p_j(0)$ at the front acts a weighting factor applied to the occupied/vacant parts of the entropy.

9.5.3 Classical Limits

We can be less concerned about the indistinguishability of the particles if the population density of the orbitals is very low, i.e., $\langle \mathrm{N}_j\rangle \ll 1$. This is sometimes referred to as the *classical limit* in this context, but it is not *fully classical* as we are still treating the particles in a quantum manner, otherwise they would not be constrained to energy levels or orbitals.

In any case, if $\langle \mathbb{N}_j \rangle \ll 1$, then $1-\langle \mathbb{N}_j \rangle \approx 1$ and $1+\langle \mathbb{N}_j \rangle \approx 1$, which has interesting consequences:

$$\mathcal{S}_{GF} = -k_B \sum_j \left(1-\langle \mathbb{N}_j \rangle\right) \ln\left(1-\langle \mathbb{N}_j \rangle\right) - k_B \sum_j \langle \mathbb{N}_j \rangle \ln\left(\langle \mathbb{N}_j \rangle\right)$$

$$\approx -k_B \sum_j \ln(1) - k_B \sum_j \langle \mathbb{N}_j \rangle \ln\left(\langle \mathbb{N}_i \rangle\right) = -k_B \sum_j \langle \mathbb{N}_j \rangle \ln\left(\langle \mathbb{N}_j \rangle\right) \quad (9.120)$$

also:

$$\mathcal{S}_{GB} = k_B \sum_j \left\{ \left(1+\langle \mathbb{N}_j \rangle\right) \ln\left(1+\langle \mathbb{N}_j \rangle\right) - \mathbb{N}_j \ln\left(\langle \mathbb{N}_j \rangle\right) \right\}$$

$$\approx -k_B \sum_j \ln(1) - k_B \sum_j \langle \mathbb{N}_j \rangle \ln\left(\langle \mathbb{N}_j \rangle\right) = -k_B \sum_j \langle \mathbb{N}_j \rangle \ln\left(\langle \mathbb{N}_j \rangle\right) \quad (9.121)$$

i.e., *the two expressions become identical.*

Clearly, the way we have structured (9.117) and (9.118), reveals that the first terms in each results from indistinguishability, expressed by the different symmetry properties of the two classes of particle. The second term is the more "classical" aspect, which is approached in the dilute limit.

NOTES

1 Having used a probability function to derive the occupation distribution reinforces that the latter is not itself a probability.
2 But don't be fooled by this into thinking we have a probability distribution here…
3 Amusingly, the cosmic microwave background radiation, famous as being excellent evidence for a hot big bang model of the early universe, does show a highly satisfactory black body spectrum.
4 Since the energy density of radiation shouldn't depend on whether we label it using f or •, so $u(f,T)df = u(\lambda,T)d\lambda$, so $u(\lambda,T) = u(f,T) \times \left|\frac{df}{d\lambda}\right| = cu(f,T)/\lambda^2$.
5 The failure of classical electromagnetic theory at high frequencies was given the impressive title of *the Ultra-Violet Catastrophe*.

10 Entropy and the Time Evolution of Quantum States

The Second Law of Thermodynamics places strictures on how systems can evolve over time. If we are to understand the fundamental nature of entropy in terms of the microscopic world, its relationship to the Second Law forces us to consider how quantum systems progress with time. It appears at first glance that the unitary time evolution characteristic of the Schrödinger equation conflicts with the Second Law. Resolving this issue will cast a new perspective on the nature of entropy.

10.1 BASIS STATES

In Chapter 6, we derived the amplitude for a particle trapped in a box (6.19)[1]:

$$\Psi_n(x,t)=\sqrt{\frac{2}{L}}\sin\left(\frac{1}{\hbar}\wp_n x\right)e^{-i\frac{1}{\hbar}E_n t}$$

This function follows a general pattern, which is true for any energy *eigenstate*, i.e., that there is a *spatial term* multiplied by a *temporal term*:

$$\Psi_n(x,t)=\varphi_n(x)\phi_n(t) \tag{10.1}$$

The spatial term is always a solution of the *time-independent* Schrödinger equation:

$$-\frac{\hbar^2}{2m}\frac{d^2\varphi_n(x)}{dx^2}+V(x)\varphi_n(x)=E_n\varphi_n(x) \tag{10.2}$$

where $V(x)$ represents the potential energy governing the physics of the system.

The temporal term obeys an adjunct relationship:

$$i\hbar\frac{d\phi_n(t)}{dt}=E_n\phi_n(t) \tag{10.3}$$

which gives the solution:

$$\int\frac{d\phi_n(t)}{\phi_n(t)}=\int\frac{E_n}{i\hbar}dt \tag{10.4}$$

DOI: 10.1201/9781003121053-13

so that:

$$\ln\left(\phi_n(t)\right)+c=-\frac{iE_n}{\hbar}t \tag{10.5}$$

or

$$\phi_n(t)=A_n e^{-i\frac{1}{\hbar}E_n t} \tag{10.6}$$

where the A_n are integration/normalization constants.

Given the mathematical properties of the complex exponential function, $e^{-i\frac{1}{\hbar}E_n t}$, the magnitude of the amplitude does not change with time, but its phase rotates at a rate determined by the energy, E_n. For this reason, energy eigenstates are sometimes referred to as *stationary states*. If a particle in the system is in an appropriate energy eigenstate, then unless the system is perturbed from outside, there the particle will sit forever more.[2] The question of perturbations is developed in Section 10.3.1.

It is frequently the case that the system is not in an energy eigenstate, in which case we construct the relevant state by a sum over possible *basis states*. In many cases the energy eigenstates are a suitable choice, and we have:

$$\Psi(x,t)=\sum_k A_k\varphi_k(x)e^{-iE_k t/\hbar} \tag{10.7}$$

As time passes, the structure of this superposition changes, as the phase of each energy state is rotating at a rate governed by its energy. A varying pattern of interference between the basis states is produced. As there is an organized evolving set of phase differences between the terms in the summation, a superposition of this sort is also known as a *coherent superposition*. A contrasting case would be where each term has some randomly fluctuating phase factor, making the interference wash out due to the shifting, pattern-less, phase relationships. This kind of situation can arise when a system is gently interacting with its surroundings: environmental factors tend to shift in a random manner.

We happen to have picked on energy states as an example of a basis, but any physical observable has a collection of eigenstates which will form a basis. It's then a case of picking a suitable basis for the situation. Other basis sums may not have a simple time evolution such as we have seen in the energy basis, but in all cases, the *time dependent Schrödinger equation* will be governing what happens:

$$-\frac{\hbar^2}{2m}\frac{\partial^2\psi(x,t)}{\partial x^2}+V(x)\,\psi(x,t)=i\hbar\frac{\partial\psi(x,t)}{\partial t} \tag{10.8}$$

where:

$$\Phi(x,t)=\sum_k a_k\phi_k(x,t) \tag{10.9}$$

the $\phi_k(x,t)$ are the chosen eigenstates and a_k the amplitudes. From a physical perspective, *the state remains the same no matter which basis we choose to use in the expansion*. One aspect in the selection of basis will be the fate of the system. If we are going to subject the system to an energy measurement, then it is clearly appropriate to expand the state over energy eigenstates. If we intend to measure some other physical variable, then the most appropriate basis will be the one formed from the eigenstates of that variable. This whole topic is explored in more detail in Chapter 11.

In the Schrödinger equations, the term on the left-hand side:

$$-\frac{\hbar^2}{2m}\frac{\partial^2\psi(x,t)}{\partial x^2}+V(x)\,\psi(x,t) \tag{10.10}$$

is very often repackaged in the form:

$$\left\{-\frac{\hbar^2}{2m}\frac{\partial^2}{\partial x^2}+V(x)\right\}\psi(x,t)=\widehat{H}\psi(x,t) \tag{10.11}$$

by defining $\widehat{H}$ as the *Hamiltonian operator*. This is the quantum version of the classical Hamiltonian, which is a sum of the kinetic and potential energies in the system. Using this terminology, the Schrödinger equations become:

$$\widehat{H}\psi(x,t)=i\hbar\frac{\partial\psi(x,t)}{\partial t}\qquad\qquad \widehat{H}\varphi_n(x)=E_n\varphi_n(x) \tag{10.12}$$

10.1.1 Superposition Amplitudes

If we intend to expand a general state over a basis of eigenstates for a physical variable, then we need to know how to calculate the expansion amplitudes that we need.

With any state constructed as a superposition over eigenstates, $\{\psi_k(x,t)\}$, forming a basis:

$$\Phi(x,t)=\sum_k a_k\psi_k(x,t) \tag{10.13}$$

if it is normalized correctly, then:

$$\int\Phi(x,t)\Phi^*(x,t)dx=1 \tag{10.14}$$

which means:

$$\int\left(\sum_k a_k\psi_k(x,t)\right)\left(\sum_l a_l^*\psi_l^*(x,t)\right)dx=\sum_{k,l}\left\{a_k a_l^*\int\psi_k(x,t)\,\psi_l^*(x,t)dx\right\} \tag{10.15}$$

One of the nice features of a basis set of *eigenfunctions* $\{\psi_k(x,t)\}$ is that being in one of them precludes being in another, thus for all $k\neq l$ the states are *orthogonal*, meaning:

$$\int\psi_k(x,t)\,\psi_l^*(x,t)dx=0 \tag{10.16}$$

whereas for $k = l$ in a normalized set:

$$\int \psi_l(x,t)\,\psi_l^*(x,t)\,dx = 1 \tag{10.17}$$

(While each of the separate functions may be normalized, this does not mean that the combination is; hence, our exploring this argument). A collection of states which are mutually orthogonal and normalized is called an *orthonormal set*. They are ideal for superpositions…

Our superposition is now nicely reduced:

$$\sum_{k,l}\left\{a_k a_l^* \int \psi_k(x,t)\,\psi_l^*(x,t)\,dx\right\} = \sum_k a_k a_k^* = 1 \tag{10.18}$$

which is the normalization condition for the overall state.

This is all the dark magic needed to find the expansion amplitudes, for consider:

$$\int \psi_l^*(x,t)\Phi(x,t)\,dx = \int \psi_l^*(x,t)\sum_k a_k \psi_k(x,t)\,dx$$

$$= \sum_k a_k \int \psi_l^*(x,t)\,\psi_k(x,t)\,dx = a_l \tag{10.19}$$

So, we have a general and useful rule:

$$\Phi(x,t) = \sum_k a_k \psi_k(x,t) \qquad a_k = \int \psi_k^*(x,t)\Phi(x,t)\,dx \tag{10.20}$$

10.2 TIME EVOLUTION OF STATES

To explore the time evolution of a general state using any basis, we start with the approximation:

$$\psi(x,t+\delta t) \approx \psi(x,t) + \frac{\partial \psi(x,t)}{\partial t}\delta t \tag{10.21}$$

which holds for infinitesimal time intervals, δt. Tidying this up is simply a matter of factorizing:

$$\psi(x,t+\delta t) \approx \left(1 + \delta t\frac{\partial}{\partial t}\right)\psi(x,t) \tag{10.22}$$

in which case, the next infinitesimal time-shift results in:

$$\psi(x,t+2\delta t) \approx \left(1 + \delta t\frac{\partial}{\partial t}\right)\psi(x,t+\delta t) \approx \left(1 + \delta t\frac{\partial}{\partial t}\right)\left(1 + \delta t\frac{\partial}{\partial t}\right)\psi(x,t) \tag{10.23}$$

Or, more generally:

$$\psi(x,t+\mathcal{N}\delta t) \approx \left(1 + \delta t\frac{\partial}{\partial t}\right)^{\mathcal{N}}\psi(x,t) \tag{10.24}$$

It now makes sense to break a small but finite time shift, Δt, into a very large number, $\mathcal{N}$, of even smaller moves $\delta t = \Delta t/\mathcal{N}$:

$$\psi(x,t+\Delta t) \approx \left(1+\frac{\Delta t}{\mathcal{N}}\frac{\partial}{\partial t}\right)^{\mathcal{N}} \psi(x,t) \tag{10.25}$$

Citing (10.12) we get:

$$\psi(x,t+\Delta t) \approx \left(1-\frac{i\Delta t}{\mathcal{N}\hbar}\widehat{H}\right)^{\mathcal{N}} \psi(x,t) \tag{10.26}$$

In the limit of allowing $\mathcal{N} \to \infty$, and using the definition of the exponential function[3]:

$$\psi(x,t+\Delta t) = e^{-\frac{i\widehat{H}\Delta t}{\hbar}} \psi(x,t) \tag{10.27}$$

It is conventional to re-brand this exponential as the *time evolution operator*, $\widehat{U}(t)$:

$$\widehat{U}(\Delta t) = e^{-\frac{i\widehat{H}\Delta t}{\hbar}} \qquad \widehat{U}(\Delta t)\,\psi(x,t) = \psi(x,t+\Delta t) \tag{10.28}$$

Importantly, this operator advances a quantum state through time but is *not dependent on any explicit initial time.*

10.2.1 Unitary Time Evolution

The time evolution operator has some interesting and significant features. Firstly, it will also devolve a state backwards from $t \to t-\Delta t$:

$$\widehat{U}(-\Delta t) = e^{+\frac{i\widehat{H}\Delta t}{\hbar}} = \left(e^{-\frac{i\widehat{H}\Delta t}{\hbar}}\right)^{*} = \widehat{U}^{*}(\Delta t) \tag{10.29}$$

using the complex conjugate of the original operator. Taking a system around a little time loop $t \to t+\Delta t \to t$ is equivalent to applying one after the other:

$$\Psi(x,t+\Delta t) = \widehat{U}(\Delta t)\Psi(x,t)$$

$$\Psi(x,t+\Delta t-\Delta t) = \widehat{U}(-\Delta t)\left(\widehat{U}(\Delta t)\Psi(x,t)\right) = \widehat{U}^{*}(\Delta t)\widehat{U}(\Delta t)\Psi(x,t) \tag{10.30}$$

which forces:

$$\widehat{U}^{*}(\Delta t)\widehat{U}(\Delta t) = 1 \tag{10.31}$$

This shows that $\widehat{U}^{*}(\Delta t)$ is the *inverse operator*, $\widehat{U}^{-1}(\Delta t)$, to $\widehat{U}(\Delta t)$:

$$\widehat{U}^{-1}(\Delta t)\widehat{U}(\Delta t)=1 \quad \text{(definition of inverse)} \tag{10.32}$$

and:

$$\widehat{U}^{*}(\Delta t)=\widehat{U}^{-1}(\Delta t) \tag{10.33}$$

Technically, $\widehat{U}$ is a *unitary operator*, a class of operators, $\hat{\mathcal{U}}$, defined by the more general relationship:

$$\int \widehat{\mathcal{U}}^{*}\phi^{*}(x)\widehat{\mathcal{U}}\varphi(x)\,dx=\int \phi^{*}(x)\varphi(x)\,dx \tag{10.34}$$

for any $\{\phi,\varphi\}$.

This rather elegant mathematical property has a significant consequence when it comes to the normalization of states. Given any unitary operator, $\hat{\mathcal{U}}$, such that:

$$\widehat{\mathcal{U}}\psi_k(x,t)=c_k\chi_k(x,t) \tag{10.35}$$

were we to apply this to our initial state (10.13), then:

$$\widehat{\mathcal{U}}\Phi(x,t)=\widehat{\mathcal{U}}\left(\sum_k a_k\psi_k(x,t)\right)=\sum_k a_k\widehat{\mathcal{U}}\psi_k(x,t)=\sum_k b_k\chi_k(x,t) \tag{10.36}$$

To investigate the normalization, we put together the integral:

$$\int\left(\widehat{\mathcal{U}}^{*}\Phi^{*}(x,t)\right)\left(\widehat{\mathcal{U}}\Phi(x,t)\right)dx=\sum_{k,l} b_k^{*}b_l\int \chi_k^{*}(x,t)\chi_l(x,t)\,dx \tag{10.37}$$

Unitarity (10.34) guarantees that the integral contracts:

$$\int \chi_k^{*}(x,t)\chi_l(x,t)\,dx=\int \widehat{\mathcal{U}}^{*}\psi_k^{*}(x,t)\widehat{\mathcal{U}}\psi_l(x,t)\,dx=\int \psi_k(x,t)\psi_l^{*}(x,t)\,dx \tag{10.38}$$

So, we conclude:

$$\sum_k b_k b_k^{*}=1 \tag{10.39}$$

In prosaic terms, the application of any unitary operator, and the time evolution operator is a specific example, preserves the probability sum.[4]

It is also worth remembering that the argument of the start of this section, where we took a state around a time loop, ties the unitarity of the time evolution to its *reversibility*. The reversibility or otherwise of processes will become a key issue in Chapter 12.

10.2.2 Non-unitary Time Evolution

Not every form of quantum time evolution is unitary…

If we have a superposition:

$$\Psi(x,t)=\sum_k a_k\psi_k(x,t) \tag{10.40}$$

over eigenstates for a certain system variable (energy, if you like), then it's difficult to know what to say about the *specific value* of that variable in this instance. What will an experiment reveal?

Each eigenstate ψ_k represents the system having a specific value (*eigenvalue*) of the physical quantity, v_k. If the system was in one of these eigenstates, we would know *with certainty* that the quantity, if measured, would show the eigenvalue.[5] What we should make, ontologically speaking, of a superposition is an ongoing debate in the community,[6] but most hold that in such situations, the *physical quantity in question is not well defined for the system* and that in a real (ontological) sense, *the value does not exist until the moment of measurement.*

If we measure the physical quantity, v, for the superposition $\Psi(x,t)$ then as a matter of verified experimental fact, we always obtain one of the eigenvalues,[7] not some smeared average across the possible values. This is not a matter of dispute. If we do another measurement of the same quantity straight after, we get the same eigenvalue again.[8] Clearly, the act of measurement has in some sense "collapsed" the superposition into one specific instance, but in a random way so we can't predict the v_k that we'll get, only the relative probability, $a_k a_k^* = |a_k|^2$. This is a *quantum probability*, not a *classical probability* used to assist us in complex cases where we can't easily keep track of what is going on, or measure with sufficient precision.

Symbolically:

$$\Psi(x,t) = \sum_k a_k \psi_k(x,t) \Rightarrow \psi_K(x,t) \tag{10.41}$$

with $\Rightarrow$ indicating the process of measurement. As previously said, these aspects are not generally debated. What does create discussion is the ontological standing of what's going on and the physical process[9] that brings it about.

One thing is clear: the collapse of state associated with measurement *is a non-unitary time evolution*. When one instance is picked out from the sum (or *manifest*, depending on your ontological perspective), the superposition has vanished to be replaced by the single appropriate eigenstate. Importantly, *the process is not reversible*. This will be of great significance in Chapter 12. Being non-unitary and irreversible, this is *not a process that is governed by the Schrödinger equation* nor any of the equations used in quantum theory.[10] It is, in a sense, an additional "bolt-on" assumption.

To be clear, once a measurement has collapsed the state in the manner (10.41):

$$\Psi(x,t) = \sum_k a_k \psi_k(x,t) \Rightarrow \psi_K(x,t)$$

the resulting state, $\psi_K(x,t)$, will evolve in time in a unitary manner, until the next measurement takes place.

While these are fascinating and important questions in quantum theory, they are also deeply significant for our discussion of entropy as *unitary time evolution does not change the Gibbs entropy* (5.46):

$$\mathcal{S}_G = -k_B \sum_n p_n \ln(p_n)$$

10.3 TIME EVOLUTION OF THE GIBBS ENTROPY

Any formal exploration of how the Gibbs entropy responds to unitary time evolution is best left until we have the use of *density matrices* in our toolkit. This is done in Chapter 11 and then taken up in Chapter 12. However, we can make a plausible interim case from the following.

Consider an ensemble of systems all of which are in the same energy eigenstate, ψ_K. The probability of picking a system out of the ensemble and measuring energy E_K is 100%. This makes the probability of finding the system in any microstate $p_n = 1$ for $n = K$, $p_n = 0$ for $n \neq K$. Hence, the Gibbs entropy is:

$$\mathcal{S}_G = -k_B\left(1\times\ln(1)+\sum_{n\neq K}0\times\ln(0)\right)=0 \tag{10.42}$$

It is worth noting that in this situation, we have complete knowledge of the systems in the ensemble, at least to the extent that quantum nature allows. Significantly, $\mathcal{S}_G = 0$, which suggests a deep connection between the entropy and our knowledge of the ensemble.

Given the structure of an energy eigenstate:

$$\psi_K = \varphi_K(x)\phi_K(t) = \varphi_K(x)e^{-iE_K t/\hbar} \tag{10.43}$$

its apparent that the phase ticks over with time and nothing much changes. We neither gain or lose information and the Gibbs entropy is hence constant, at zero. Even an energy measurement preserves the eigenstate, so it will have no impact on the ensemble's members.

Alternatively, we may have an ensemble of systems which are all in the same superposition. For ease, let's consider a superposition of energy eigenstates. However, this choice will not undermine the generality of the argument. After all, a state is state, irrespective of what basis we choose to use in expanding it out. Informally, we expect our conclusions to be basis independent; formally, we will be able to establish this in later chapters. So, each member of the ensemble is in the state:

$$\Psi(x,t) = \sum_k a_k\psi_k(x,t) = \sum_k a_k\varphi_K(x)e^{-iE_K t/\hbar} \tag{10.44}$$

Don't be misled by seeing probability amplitudes in this sum. It is certainly true that these amplitudes lead to probabilities via the Born rule, $p_K = |a_K|^2 = a_K^* a_K$, but these are *latent probabilities* that only manifest when a measurement takes place, leading to state collapse. They are *not* probabilities to use in the Gibbs entropy. The Gibbs functional references a probability distribution for microstates in an ensemble. In this case, all the systems are in the same state $\Psi(x,t)$, leading to $\mathcal{S}_G = 0$. Importantly, unitary time evolution does not change the nature of a superposition. After all:

$$\hat{U}(\Delta t)\Psi(x,t) = e^{-\frac{i\hat{H}\Delta t}{\hbar}}\sum_k a_k\varphi_k(x)e^{-iE_k t/\hbar} = \sum_k a_k\varphi_k(x)e^{-iE_k(t+\Delta t)/\hbar}$$

$$= \sum_k a_k e^{-iE_k\Delta t/\hbar}\psi_k(x,t)$$

which is still a superposition. Furthermore, the amplitude to manifest any one of the eigenstates is:

$$a'_K = \int \psi_K^*\Psi(x,t+\Delta t)dx = \int \psi_K^*\sum_l a_l e^{-iE_l\Delta t/\hbar}\psi_l(x,t)dx = a_k e^{-iE_k\Delta t/\hbar}$$

As a result, the probabilities:

$$p_K = |a'_K|^2 = (a'_K)^* a'_K = a_K^* a_K e^{+iE_k\Delta t/\hbar}e^{-iE_k\Delta t/\hbar} = a_K^* a_K$$

do not change. Physically, the state has not altered, and the ensemble has not changed, so the Gibbs entropy must be the same.

If the systems in the ensemble are subjected to measurement, then state collapse will take place separately for each of them. The ensemble becomes a collection of systems spanning a range of energy eigenstates. Potentially, the entropy may have changed as a result. The impact of measurement on entropy is one of the significant topics addressed in Chapter 12. For the moment, we turn to other possibilities.

In practice, a more likely scenario is that our ensemble is comprised of systems that are in eigenstates or superpositions, *but we are not sure which ones*. There is a probability distribution across the ensemble, with p_n being the probability to find an example system in eigenstate/superposition/microstate n out of the collection. This will be a *classical probability distribution*, not one determined by quantum amplitudes. It represents a lack of information on our part, so $\mathcal{S}_G \neq 0$. However, even in this scenario, any unitary time evolution of the systems will not alter the probability distribution. Outside of whichever processes that were part of the ensemble's preparation, the quantum nature of a state does not impact on the probability of finding that state across the ensemble. Unless there is some mechanism that alters our knowledge of the states represented across the ensemble, the entropy will not change. What then of the Second Law? How is this consistent with quantum theory?

10.3.1 Ignorance Is a Sort-of Defence

We need to consider other mechanisms, presumably non-unitary ones, that give rise to changes in the probabilities, p_n. *This would require the system to shift from one possible state to another*. Most of the time, systems wander around the universe according to their own lights and are not generally subject to the poking and prodding of scientists carrying out measurements. Yet, change clearly comes about at macroscopic and microscopic levels, something that is, as yet, not covered in our account of quantum theory.

As mentioned earlier, unitary time evolution is reversible in principle, so we need to be looking for irreversible processes giving rise to probability shifts. Interactions with the environment or other nearby systems would presumably fit the bill.

This would mean an extension to the way in which we have set up the physics of the system. After all, the states derived from a system's Hamiltonian are energy eigenstates that are stationary *precisely because there is no mechanism to trigger transitions between states*. That is the role of an interaction between the system and something outside.

In many cases, we don't have sufficient detailed knowledge of the exact Hamiltonian needed in such complex situations. We have to take a somewhat simplified model, denoted by $\widehat{H}_0$ and add in a small *perturbation term*, $\hat{h}$. As we do not know the exact form of $\hat{h}$, (if we did, we would know the correct Hamiltonian after all), we can't work out exactly what is going to happen as a result of its being there. The perturbation represents something disturbing the system during a (gentle) interaction between the system and its environment, distorting the quantum mechanics of the ordinarily stable system and giving rise to transitions between states. Once the perturbation stops, we expected the system to settle down into its old-style behaviour, albeit in a different configuration.

Typically, the analysis starts by constructing an approximate state as a superposition of energy eigenstates, ψ_n' of $\widehat{H}_0$ – which are not the *true* energy eigenstates of the system, $\widehat{H}_0 + \hat{h}$, while the perturbation is active. Indeed, this is why the perturbation induces transitions.

Now, we are in a different scenario; as the perturbation allows systems to flip between states, *we can envisage that the classical probabilities assigned to the likelihood of finding the system in each state change with time*.

To see how this works, assume that under perturbation, a state of energy ε_n can jump into any state with energy ε_m provided $\varepsilon_n \leq \varepsilon_m \leq \varepsilon_n + \delta\varepsilon$.

The probability per unit time of such a transition, Ξ_{nm}, is given by *Fermi's Golden Rule*[11]:

$$\Xi_{nm} = \frac{2\pi}{\hbar}\left|h_{nm}\right|^2 g_m \tag{10.45}$$

Here, h_{nm} is the amplitude governing the transition under the influence of the perturbation and g_m is the density of states in the region of the final state.

To obtain a 1–1 transition rate from a given initial state with a quantum number n to a given final state of quantum number m, we need to divide by the number of states in the region of state m, i.e., $g_m \delta\varepsilon$ to produce:

$$\xi_{nm} = \frac{2\pi}{\hbar\delta\varepsilon}\left|h_{nm}\right|^2 \tag{10.46}$$

For any amplitude h_{nm}, governing the transition $n \rightarrow m$, the amplitude for the reverse transition, $m \rightarrow n$, is the complex conjugate: $h_{mn} = h^*_{nm}$. This is a basic rule of quantum theory. As $|h_{nm}|^2 = h_{nm}h^*_{nm}$, it follows that $\xi_{nm} = \xi_{mn}$.

Parking this for the moment, consider how the probability of finding the system in any given microstate might evolve over a brief time interval.

We start with an ensemble which at time t contains $\mathbb{N}_n$ systems in state n. During some short time period, Δt, a few of these might make the transition into state m. Equally, some of the examples starting in state m could flip to state n. Hence, we write[12]:

$$\Delta\mathbb{N}_n = \left\{-\mathbb{N}_n \sum_m \xi_{nm} + \sum_m \xi_{mn}\mathbb{N}_m\right\}\Delta t \tag{10.47}$$

Dividing throughout by the total number of systems in our ensemble gives us:

$$\Delta p_n = \left\{-p_n \sum_m \xi_{nm} + \sum_m \xi_{mn} p_m\right\}\Delta t \tag{10.48}$$

Recalling that $\xi_{nm} = \xi_{mn}$, we can simplify:

$$\Delta p_n = \sum_m \xi_{nm}\left(p_m - p_n\right)\Delta t \tag{10.49}$$

or:

$$\frac{dp_n}{dt} = \sum_m \xi_{nm}\left(p_m - p_n\right) \tag{10.50}$$

which is the *Fermi Master Equation*. Significantly, this is *not* time reversible. In simple terms if $t \rightarrow -t$, the left-hand side of the equation changes sign, but the right-hand side doesn't.

Next, we need to change tack once again and return to the Gibbs entropy.

Differentiating the Gibbs entropy with respect to time:

$$\frac{dS_G}{dt} = -k_B \frac{d}{dt}\left\{\sum_{n=1}^{\infty} p_n \ln(p_n)\right\} = -k_B \left\{\sum_{n=1}^{\infty} \frac{dp_n}{dt}\ln(p_n) + p_n \frac{d\ln(p_n)}{dt}\right\}$$

$$= -k_B \left\{\sum_{n=1}^{\infty} \frac{dp_n}{dt}\ln(p_n) + p_n \frac{1}{p_n}\frac{dp_n}{dt}\right\}$$

$$= -k_B \left\{\sum_{n=1}^{\infty} \frac{dp_n}{dt}\ln(p_n) + \frac{dp_n}{dt}\right\}$$

$$= -k_B \sum_{n=1}^{\infty} \frac{dp_n}{dt}\ln(p_n) + \sum_{n=1}^{\infty} \frac{dp_n}{dt}. \tag{10.51}$$

where the second term has been struck out as it is equal to zero. After all, normalization requires that $\sum p_n = 1$ at all times, so the individual rates of change for the separate p_n must sum to zero to maintain this balance. Our result is:

$$\frac{dS_G}{dt} = -k_B \sum_{n=1}^{\infty} \frac{dp_n}{dt}\ln(p_n) \tag{10.52}$$

At this point, we can insert the Fermi master equation (10.50):

$$\frac{dp_n}{dt} = \sum_m \xi_{nm}(p_m - p_n)$$

to produce:

$$\frac{dS_G}{dt} = -k_B \sum_{n=1}^{\infty}\left\{\sum_m \xi_{nm}(p_m - p_n)\right\}\ln(p_n) = -k_B \sum_{m,n} \xi_{nm}(p_m - p_n)\ln(p_n) \tag{10.53}$$

Now consider any pair of states (m,n) out of the sum in (10.53). For transitions between these states, there are two contributions to the rate of change of the respective probabilities:

$$\frac{dp_n}{dt} = \xi_{nm}(p_m - p_n) \qquad \frac{dp_m}{dt} = \xi_{nm}(p_n - p_m) \tag{10.54}$$

So, in the sum of terms in (10.53), we would have:

$$\frac{dp_n}{dt}\ln(p_n) = \xi_{nm}(p_m - p_n)\ln(p_n) \tag{10.55}$$

and:

$$\frac{dp_m}{dt}\ln(p_m) = \xi_{nm}(p_n - p_m)\ln(p_m) \tag{10.56}$$

Merging them gives a contribution to $\frac{dS_G}{dt}$ which looks like:

$$\xi_{nm}(p_m - p_n)(\ln(p_m) - \ln(p_n)) \tag{10.57}$$

Now:

- If $p_m \geq p_n$ the first term is positive (or zero), but so is the second term
- If $p_m \leq p_n$, then the first term is negative (or zero), but so is the second term

Clearly, this paired contribution to the sum *must always be positive or zero.*

The same must be true for *any pair of states* that we choose; hence, $\frac{dS_G}{dt} \geq 0$ and we have established compatibility with the Second Law.

10.3.2 But at What Cost?

The argument of the previous section resolves the apparent clash between the time evolution of quantum systems and the observed macroscopic tendencies enshrined in the Second Law. This has been achieved by introducing a form of coarse graining. In broad terms, coarse graining is always needed whenever we try to model macroscopic properties in terms of microscopic behaviour, as a consequence of our incomplete information. We may lack sufficient resolution in our measuring devices, or the method of measurement itself may introduce some scrambling of the inherent information. The process of coarse graining is not simply a theoretical approximation that we have a measure of control or choice regarding. Some of it lies outside of our influence. However, if entropy is related to information (or lack of it), then any coarse graining must lead to an increase in entropy.

The Fermi Master contains some implicit coarse graining. The basic transition rates:

$$\xi_{nm} = \frac{2\pi}{\hbar\delta\varepsilon}|h_{nm}|^2 \tag{10.58}$$

are proportional to the absolute square of the transition amplitudes, h_{nm}, which irons out their phase information. We have also introduced a basic lack of information by failing to specify the exact nature of the perturbation.

So, the approach of the previous section succeeds due to the coarse graining involved, which once again establishes an "information content" aspect to the fundamental nature of entropy.

Most importantly, introducing the ability of the system to make transitions between states, albeit states which are clearly not eigenstates of the "true" Hamiltonian, brings in the notion of state collapse. As we have discussed, this is not modelled by the basic equations of quantum theory yet is required as a bolt-on assumption to relate calculations to actual measurements.[13] This leads us to the possibility that the unresolved issue of state collapse in the interpretation of quantum theory is *somehow explicitly required for compatibility with the Second Law*. Some physicists consider the reverse, that state collapse is *due* to the Second Law. Here, we are dipping into many complex and unresolved issues in the philosophy of quantum theory.

10.3.3 Another Argument

The discussion in Section 10.3.1 establishes a mechanism by which the Gibbs entropy of a quantum ensemble can change with time. However, perhaps surprisingly, a constant Gibbs entropy is also compatible with the Second Law. The key is to appreciate that the Gibbs functional:

$$\mathcal{S}_G = -k_B \sum_{n=1}^{\infty} p_n \ln\left(p_n\right)$$

is a number that can be calculated for any probability distribution. It is only equivalent to the thermodynamic entropy *if the probability distribution used maximises the value of* $\mathcal{S}_G$.

Now consider two moments in time, t_A and t_B, with t_B being later than t_A, $t_B \geq t_A$.

Let $\{\sigma_n\}$ stand for a probability distribution across a collection of microstates each of which is consistent with the macrostate of the system. The probability distribution is constrained only in the sense that it reproduces the correct values of macroscopic quantities and is normalized. In other words:

$$\sum_n \sigma_n E_n = \langle E\rangle = U \qquad \sum_n \sigma_n = 1$$

The Gibbs functional calculated from this probability distribution, $S_G(\{\sigma_n\})$, is *not necessarily* the maximum possible value, as there are doubtless a variety of possible distributions other than $\{\sigma_n\}$. Somewhere, somewhen, we might find another, $\{\sigma_n'\}$, that gives a larger value, $S_G(\{\sigma_n'\}) > S_G(\{\sigma_n\})$.

Denoting $\{\Sigma_n\}$ as the *specific* probability distribution that *does* maximize the Gibbs functional, we evidently have:

$$S_G\left(\left\{\Sigma_n\right\}\right) \geq S_G\left(\left\{\sigma_n\right\}\right) \tag{10.59}$$

As this maximized Gibbs functional is equal to the thermodynamic entropy:

$$\mathcal{S}_T = S_G\left(\left\{\Sigma_n\right\}\right) \tag{10.60}$$

it follows that the thermodynamic entropy at the time t_B, $\mathcal{S}_T\left(t_B\right)$, must be greater than or equal to the Gibbs functional evaluated for any distribution $\left\{\sigma_n\right\} \neq \left\{\Sigma_n\right\}$ at the same time, t_B:

$$\mathcal{S}_T\left(t_B\right) \geq S_G\left(\left\{\sigma_n\left(t_B\right)\right\}\right). \tag{10.61}$$

If the quantum states undergo simple unitary time evolution between t_A and t_B, $t_A < t_B$, then, according to the argument of Section 10.3, the Gibbs functional does not change[14]:

$$S_G\left(\left\{\sigma_n\left(t_B\right)\right\}\right) = S_G\left(\left\{\sigma_n\left(t_A\right)\right\}\right) \tag{10.62}$$

In which case, the thermodynamic entropy at t_B must *also* be greater than the Gibbs functional for $\{\sigma_n\}$ evaluated at t_A:

$$\mathcal{S}_T(t_B) \geq S_G(\{\sigma_n(t_A)\}) \tag{10.63}$$

If we do happen to pick $\{\sigma_n\} = \{\Sigma_n\}$, then:

$$S_G(\{\Sigma_n(t_A)\}) = \mathcal{S}_T(t_A) \tag{10.64}$$

and so, finally:

$$\mathcal{S}_T(t_B) \geq \mathcal{S}_T(t_A) \tag{10.65}$$

This shows that thermodynamic entropy cannot decrease over time.

This argument establishes the consistency between a constant Gibbs functional/entropy and the Second Law, without recourse to the Fermi Master Equation, and hence some implicit coarse graining, but it does not advance our cause when it comes to exposing some underlying physical interpretation of entropy.

10.3.4 So Does This Help?

In classical statistical thermodynamics, the microstate of a system is denoted by the position, x, and momentum, p, of all the particles within the system. This can be "plotted" as a single point in a multi-dimensional phase space (Section 7.1). The collection of microstates which are all compatible with the observed macrostate forms a region, Γ, of this phase space with a certain "volume", V_Γ. Given that $\{x, p\}$ are continuous quantities for all the particles concerned (even if the system is confined to a limited volume, the span across the volume is still continuous), it is impossible to "count" the number of microstates within the volume. Either a form of coarse graining is required, which partitions the phase space into elemental volumes of our choice, or we call upon quantum theory with its uncertainty principle and declare that the natural "unit" of volume in the phase space is $\hbar^{3\mathcal{N}}$, where $\mathcal{N}$ is the number of particles in the system. The count of microstates is then:

$$\Omega = \frac{V_\Gamma}{\hbar^{3\mathcal{N}}} = \frac{1}{\hbar^{3\mathcal{N}}} \int_\Gamma d\tau \tag{10.66}$$

using the abbreviation:

$$d\tau = \prod_{i=1}^{\mathcal{N}} dx_i dy_i dz_i dp_{x_i} dp_{y_i} dp_{z_i} \tag{10.67}$$

In this sense, the Boltzmann entropy, $\mathcal{S}_B = k_B \ln(\Omega)$ can be considered as the logarithm of the phase space volume of microstates equivalent to a specified macrostate. If we now consider a *probability density* applied to the particles within the phase space volume, $\rho(x, p)$, using x, p to stand for all the particles, we can generalize the Gibbs functional. However, we need to be careful as the probability density, ρ has dimensionality, i.e., it is in units of $\hbar^{-3\mathcal{N}}$. As logs are only defined for dimensionless quantities, we need:

$$\mathcal{S}_G = -k_B \int_{\Gamma} \rho(x,p) \ln\left(\hbar^{3\mathcal{N}} \rho(x,p)\right) d\tau \tag{10.68}$$

the dimensionality of ρ outside the integral being counterbalanced by $d\tau$, which is in units of $\hbar^{3\mathcal{N}}$.

If we are concerned with a microcanonical ensemble, so that each microstate is equally likely,then:

$$\rho(x,p) = \frac{1}{V_\Gamma} \text{ inside } \Gamma \qquad \rho(x,p) = 0 \text{ outside } \Gamma \tag{10.69}$$

That makes:

$$\begin{aligned}\mathcal{S}_G &= +k_B \int_{\Gamma} \frac{1}{V_\Gamma} \ln\left(\hbar^{3\mathcal{N}} V_\Gamma\right) d\tau = \frac{k_B}{V_\Gamma} \int_{\Gamma} \left(\ln\left(V_\Gamma\right) + \ln\left(\hbar^{3\mathcal{N}}\right)\right) d\tau \\ &= k_B \ln\left(V_\Gamma\right) + k_B \ln\left(\hbar^{3\mathcal{N}}\right) - 1 = k_B \ln\left(\hbar^{3\mathcal{N}} \Omega\right) + k_B \ln\left(\hbar^{3\mathcal{N}}\right) - 1 \\ &= k_B \ln(\Omega) + 2k_B \ln\left(\hbar^{3\mathcal{N}}\right) - 1 = k_B \ln(\Omega) + \mathrm{K}\end{aligned} \tag{10.70}$$

If we now define, for compactness, $\mathcal{H}_G$ by:

$$\mathcal{S}_G - K = -k_B \mathcal{H}_G \tag{10.71}$$

It follows that:

$$\Omega = e^{-\mathcal{H}_G} \tag{10.72}$$

a definition we will retain for all ensembles.

Our volume of phase space, in units of $\hbar^{3\mathcal{N}}$, determines the entropy. However, we need to be a little more formal about how we define regions of phase space in the first place. If we gently wave our hands and say[15] "Γ is the region of phase space occupied by reasonably probable microstates that are consistent with the given macrostate", we are certainly open to the criticism of vagueness, but not as much as one might expect. Let's proceed to take a probability density, $\rho(x,p)$, being careful not to specify its exact nature, and define $\Gamma(\mathcal{C})$ as containing states for which $\rho(x,p) \geq \mathcal{C}$, where in turn we have chosen $\mathcal{C}$ so that the total probability to find a system in the region is:

$$\int_{\Gamma(\mathcal{C})} \rho(x,p) d\tau = 1 - \varepsilon \qquad 0 < \varepsilon < 1 \tag{10.73}$$

This, in turn, determines a phase space volume:

$$V_\Gamma(\varepsilon) = \int_{\Gamma(\varepsilon)} d\tau \tag{10.74}$$

and hence:

$$\Omega(\varepsilon) = \frac{V_\Gamma(\varepsilon)}{\hbar^{3\mathcal{N}}} \tag{10.75}$$

It is then possible to show that[16]:

$$\lim_{\mathcal{N}\to\infty}\left\{\frac{\mathcal{H}_G + \ln(\Omega(\varepsilon))}{\mathcal{N}}\right\} = 0 \tag{10.76}$$

Note that the result *does not depend on the choice of* ϵ. Essentially, while changing ϵ does impact on $\Omega(\varepsilon)$, and possibly by a considerable amount, the consequential change in $\ln(\Omega(\varepsilon))$ is somewhat less and in any case counterbalanced by the growth in $\mathcal{N}$. So, $\mathcal{H}_G$ *does* measure the phase space volume of "reasonably probable microstates", *and in a way that does not depend on our choice of what we mean by "reasonably probable" in the thermodynamic limit.*

It is common in the literature to refer to Ω as a phase space volume, when in truth it is the count of states, determined by partitioning the volume into lumps of $\hbar^{3\mathcal{N}}$. We will adopt this phrasing but underline the word volume to indicate this subtle distinction.

Now, let's consider an experiment performed on a specified system. At time $t = 0$, we measure a set of *macroscopic* variables $\chi_0 = \{X_1(0), X_2(0), \ldots, X_k(0)\}$, where the collection is sufficient to uniquely define the thermodynamic state of the system.[17] Such a selection is *not* sufficient to pin down which single *microstate* the system is occupying. In fact, we have simply specified a region of classical phase space, Γ_0, in which there is a high probability that the system will be found, consistent with χ_0. We will say that this region has volume V_{Γ_0} which in our units of $\hbar^{3\mathcal{N}}$ corresponds to Ω_{Γ_0}. Don't forget that we have established that we can be vague regarding "high probability".

If the system is governed by the canonical distribution, then Ω_{Γ_0} will be *the largest phase space volume consistent with* χ_0. If some other probability distribution is in operation, then its associated volume *must fit inside* Ω_{Γ_0}. This is the exact justification for why we use the canonical distribution to describe equilibrium situations.

Echoing a remark first made by Boltzmann,[18] Ω_{Γ_0} effectively *measures our degree of ignorance regarding the true initial microstate of the system*, consistent with our measured macroscopic properties, χ_0. However, this is a rather *subjective* view on things.

Rotating into a more objective standpoint, we could recast by saying that Ω_{Γ_0} *measures our lack of control over the microstate*, when the only variables open to being manipulated are the macroscopic variables denoted by χ_0.

If experiments are to be repeatable, and hence verifiable, *there must be some defined preparation process used to produce systems for each experiment performed*. However, even if we constrain multiple examples of our system to the same set of values for the measured macroscopic variables, χ_0 the collection of systems will vary in their microstates and hence span Ω_{Γ_0}.

So far, of course, we have not carried out any form of experiment, just prepared systems to be manipulated. So, let us allow an adiabatic (reversible, unitary) change to the system. As we have

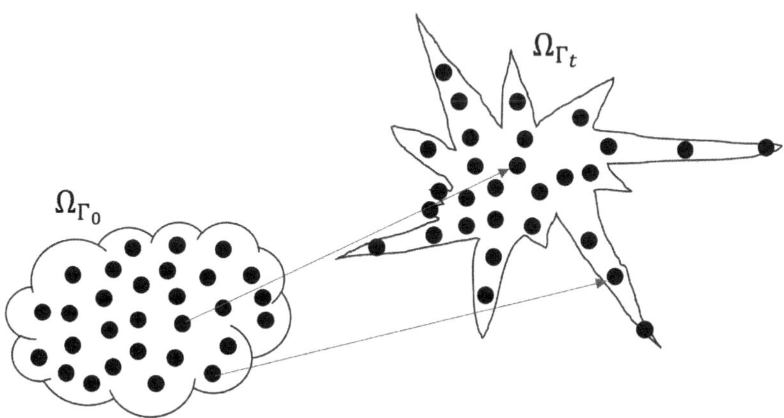

FIGURE 10.1 As a system evolves, each microstate in the initial phase space volume, Ω_{Γ_0}, will uniquely map into a state in the final phase space volume, Ω_{Γ_t}. Our experimental procedures are not able to pin down a single microstate in Ω_{Γ_t}, so the resulting state in Ω_{Γ_t} will be randomly distributed across repeated experiments. However, the regions Ω_{Γ_0} and Ω_{Γ_t} will be the same volume as the Gibbs entropy is constant during the experiment.

seen, the Gibbs functional will remain constant, hence $\mathcal{H}_G$ is constant and so $\Omega_{\Gamma_t} = \Omega_{\Gamma_0}$. For each (unknown) microstate in Ω_{Γ_0} there is a unique state in Ω_{Γ_t} and as we repeat the experiment, the final microstate will be a random selection from those inside Ω_{Γ_t} (Figure 10.1). Note that we should not expect that Ω_{Γ_t} is a simple clone of Ω_{Γ_0}. It is more likely that the evolved structure will gain complexity and be somewhat "fractal-like" in nature, hence the somewhat spiky volume shown in this diagram.

If we pause to measure the macroscopic parameters as the adiabatic change comes to a halt, we will have a collection $\mathcal{X}_t = \{X_1(t), X_2(t), \ldots, X_k(t)\}$. In principle, this new collection $\mathcal{X}_t$ is consistent with a phase space <u>volume</u> Ω'_t which spans all possible microstates compatible with $\mathcal{X}_t$, *even if they could not have been reached from one of the initial microstates in* Ω_{Γ_0}.

If we spoke to another physicist and simply passed on the information that the system was in the macrostate defined by $\mathcal{X}_t$, without telling her anything about an initial state, she would calculate the entropy, consistent with the Clausius entropy extracted from the collection of macroscopic variable values $\mathcal{X}_t$, as:

$$\mathcal{S}_B(t) = \mathcal{S}_T(t) = k_B \ln(\Omega'_t) \tag{10.77}$$

It is evident that the phase space <u>volume</u> Ω_{Γ_t} is a sub-<u>volume</u> within Ω'_t, or our purported experiment would not be repeatable (Figure 10.2). In other words, $\Omega_{\Gamma_t} \leq \Omega'_t$, but as $\Omega_{\Gamma_t} = \Omega_{\Gamma_0}$ then:

$$\mathcal{S}_B(0) = \mathcal{S}_T(0) = k_B \ln(\Omega_0) \leq k_B \ln(\Omega'_t) = \mathcal{S}_T(t) \tag{10.78}$$

and we have the Second Law once again.

Now, we have a justifiable physical interpretation of why the Second Law comes about.

As phase space volume is conserved during the system's unitary evolution, if any experiment is to be repeatable, then *the total phase space volume compatible with the final state,* Ω'_t, *cannot*

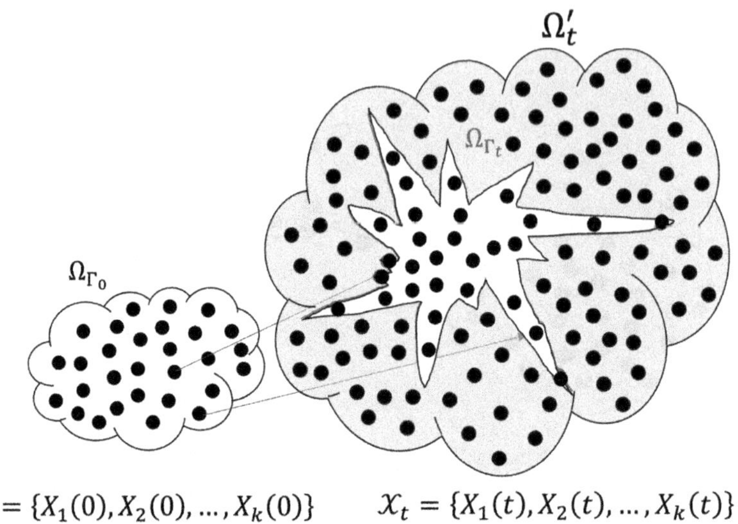

FIGURE 10.2 The macrostate defined by the measurements $\mathcal{X}_t$, spans a phase space volume Ω'_t. This volume is greater than Ω_{Γ_t}, which the volume of states accessible from Ω_{Γ_0}, which is itself defined by the collection of measurement values $\mathcal{X}_0$.

be less than the phase space volume that describes our ability to prepare the initial state, Ω_{Γ_0} in a reproducible way.

In a nutshell, the thermodynamic entropy of the final state of a system is not a time-evolution of the Gibbs entropy for the initial state, for that quantity remains constant. Rather, *the thermodynamic entropy of the final state is a re-evaluated Gibbs entropy using a phase space volume that cannot be smaller than that which constrains the initial state.*

10.4 ANTHROPOMORPHIC ENTROPY

One theme throughout these discussions has been a drive to extract an objective definition of entropy, grounded in microphysics that is consistent with the classical entropy from Section 1. This is an important aim for two reasons.

Firstly, while the classical entropy based on thermodynamic variables is precise and unambiguous, it is rather nebulous when it comes to physical content. We know what it *is*; we just don't know what it *means*.

Secondly, it is evident that *all* macroscopic physical parameters and theories must to some extent be underpinned by microphysics, quantum theory in fact.

While any approach based on coarse graining can, with some justification, be regarded as objective, as any impact from our subjective choices of how we go about the coarse graining becomes vanishingly small in the thermodynamic limit, this is still conceptually somewhat unsatisfactory. Is there not *something* that all coarse-graining approaches converge on?

The argument of Section 10.3.4 is more appealing as it dispenses with coarse graining in any form. However, there is another point that needs to be made: *there is something of a subjective element to what we mean by the classical entropy in any case!*

Jaynes[19] argues that any thermodynamic description of a system is constructed from a selection of thermodynamic variables (temperature, pressure, volume, etc.) which is *context dependent* (governed by the sorts of experiments that you want to do) and not necessarily an exhaustive selection of the variables that the system possesses. We choose a sub-set of the system's variables and either ignore or control the remainder. Our calculations of entropy then depend on the sub-set that we select. So, can we be sure that the entropy is the same *independent of the collection of variables that we have elected to use?*

Indeed, is it possible to define an entropy that is *not* choice dependent in this manner?

Jaynes cites a comment that he attributes to Wigner: entropy is an *anthropomorphic concept.*

In taking up the argument, Jaynes writes[20]:

> One might reply that in each part of the experiments cited, we have used only part of the degrees of freedom of the system, and there is a 'true' entropy which is a function of all these parameters simultaneously...There is no end to this search for the ultimate 'true' entropy until we have reached the point where we control the location of each atom independently. But just at that point the notion of entropy collapses and we are no longer talking about thermodynamics.
>
> From this we see that entropy is an anthropomorphic concept, not only in the well-known statistical sense that it measures the extent of human ignorance as to the microstate. Even at the purely phenomenological level, entropy is an anthropomorphic concept. For it is a properly, not of the physical system, but of the experiments you or I choose to perform on it. [Our emphasis].

NOTES

1 In what follows, we are working in one dimension for simplicity. This will not alter any conclusions and the argument can easily be developed for three-dimensional dependencies.

2 This raises an interesting question with regard to energy levels in atoms. It is well known that electrons in higher energy levels will tend to fall into lower energy states and radiate photons. If the energy levels are eigenstates and hence stationary, the electrons should not do that. The solution to this comes from the perturbations that arise in the quantum structure of the electrical field binding the electrons to the nucleus.

3 The exponential function has many equivalent definitions, but the one employed here is $e^x = \lim_{n\to\infty}(1+x/n)^n$. Taking the exponential of an operator is interpreted via this expansion.

4 However, there is an on-going debate regarding role of singularities found at the core of black holes. In principle, any quantum system that runs across a singularity is pulled out of existence, hence its time evolution, is distinctly non-unitary. It then becomes a question of what happens to the *information* that system represents. This is further discussed in Chapter 15.

5 Glossing over minor issues of degeneracy…

6 At least among those interested in such matters. Many simply "shut up and calculate".

7 At least to within experimental uncertainty.

8 The "immediate" proviso being a shorthand for the system not interacting or otherwise evolving away from the eigenstate between measurements.

9 If there is one. Some physicists hold that all that is changing is the state of our knowledge.

10 Some physicalists, notably Roger Penrose, suggest that the Schrödinger equations need to be modified to allow non-unitary evolution and stochastic state collapse. Penrose's specific suggestion, which we will return to later, is that gravity is a key factor in this.

11 We state this, without proof. The interested reader will be able to find an appropriate argument in any book on quantum mechanics that covers perturbation theory.

12 Assuming that Δt is so short that multiple sequential transitions do not take place.

13 The Many Worlds interpretation of quantum theory suggests that there is no state collapse. The wavefunction describes an expanding series of parallel worlds which steadily decohere from each other. Arguably, this is also a form of coarse graining as phase information gets scrambled along the way.

14 This would be true even if the probability distribution changed, provided the new distribution reproduced the correct average values of physical quantities when compared to their experimentally measured values.

15 Here we follow the argument of E. T Jaynes in the paper Jaynes E T 1965 *Gibbs vs Boltzmann Entropies, AJP* 33 391–398.

16 First proven by Shannon: C. E. Shannon, *Bell Syst. Tech. J.* 27, 379, 623 (1948); reprinted in C. E. Shannon and W. Weaver, *The Mathematical Theory of Communication* (University of Illinois Press, Urbana, IL, 1949).

17 Clearly these variables could be some collection such as pressure, volume, temperature, etc., but here we don't want to be tied down to exactly what they might be.
18 As referenced by Jaynes in the paper cited in endnote 15.
19 Same paper as endnote 15.
20 Same paper as endnote 15.

11 The Density Matrix

The quantum state contains all accessible information about a system. However, for various reasons, the information is less complete than we might expect given the classical states we are used to. This suggests the possibility that the state itself contains a measure of entropy (lack of information), which could potentially arise in two ways. Firstly, quantum states contain an inherent uncertainty with respect to the measurement outcomes of conjugate variables. Secondly, due to limitations in the preparation process, we may not be sure which state a system is occupying. The best we can do in this situation is to assign each potential state a classical probability and construct a specialist mathematical representation to deal with the situation. This is known as the density matrix. It turns out to be a rather handy tool in general and crucial for the exploration of quantum entropy. So, in this chapter, we are going to extend our technical understanding of quantum theory and the representation of states. This will arm us sufficiently to venture into the physics of the measurement problem and the entropy of quantum systems, which we will tackle in Chapter 12.

11.1 OBSERVABLES

Like any scientific theory, it is vitally important that the predictions of quantum theory can be checked by some experimental process. This hinges on our ability to extract values for *observables*, i.e. measurable physical quantities, from the quantum state, which is a role performed by specific mathematical operators.[1] Classically, we understand the properties and dynamics of particles using observables such as momentum, energy, and position. Each of these has a corresponding operator within quantum theory.

Consider our normalized wave function for a particle in a box (6.16):

$$\Psi_n(x,t)=\sqrt{\frac{2}{L}}e^{-i\frac{1}{\hbar}E_n t}\sin\left(\frac{1}{\hbar}\wp_n x\right)$$

(We have tweaked the symbol for momentum, as in this chapter, there is a danger of it being confused with probability.)

We can see the energy, E_n, sitting inside the exponential. To formally extract it, we need some operator, $\hat{O}$, cunningly constructed so that:

$$\hat{O}\Psi_n(x,t)=E_n\Psi_n(x,t) \tag{11.1}$$

Note that in this equation, the energy has been "pulled out", but the wave function is otherwise unharmed in the process. In situations where this is the case, $\Psi_n(x,t)$ is an *eigenstate* of the operator,

DOI: 10.1201/9781003121053-14

and E_n an *eigenvalue*. We will deal with situations where the wave function is not an eigenstate shortly.

It takes a little mathematical imagination to see that the operator we need is:

$$\hat{E} = i\hbar\frac{\partial}{\partial t} \tag{11.2}$$

so that:

$$\begin{aligned}\hat{E}\Psi_n(x,t) &= i\hbar\frac{\partial\Psi_n(x,t)}{\partial t} = i\hbar\sqrt{\frac{2}{L}}\sin\left(\frac{1}{\hbar}\wp_n x\right)\frac{\partial e^{-i\frac{1}{\hbar}E_n t}}{\partial t} \\ &= i\hbar\sqrt{\frac{2}{L}}\sin\left(\frac{1}{\hbar}\wp_n x\right)\times\left(-\frac{i}{\hbar}E_n\right)e^{-i\frac{1}{\hbar}E_n t} \\ &= E_n\Psi_n(x,t)\end{aligned} \tag{11.3}$$

as requested. Measuring the energy of a particle with wave function Ψ_n would reveal the value[2] E_n. Furthermore, after the measurement, the particle would retain its original wave function, until such time as it interacted with something else.

11.1.1 Great Expectations

In a more general situation, the particle may not be in an energy eigenstate (stationary state), and its wave function would be something slightly meatier; a superposition such as:

$$\Phi = \sum_n a_n\sqrt{\frac{2}{L}}e^{-i\frac{1}{\hbar}E_n t}\sin\left(\frac{1}{\hbar}\wp_n x\right) \tag{11.4}$$

Here a mixing coefficient, such as a_n, represents the complex amplitude to find the particle in eigenstate n, yielding, via the Born rule, $\text{prob}(\text{state } n) = a_n^* a_n$. There is an additional requirement:

$$\sum_n a_n^* a_n = 1 \tag{11.5}$$

which guarantees that our sum spans all the relevant possibilities.

Deploying our energy operator produces:

$$\begin{aligned}\hat{E}\Phi &= \hat{E}\left(\sum_n a_n\sqrt{\frac{2}{L}}e^{-i\frac{1}{\hbar}E_n t}\sin\left(\frac{1}{\hbar}\wp_n x\right)\right) \\ &= \sum_n a_n\sqrt{\frac{2}{L}}\sin\left(\frac{1}{\hbar}\wp_n x\right)i\hbar\frac{\partial e^{-i\frac{1}{\hbar}E_n t}}{\partial t} \\ &= \sum_n a_n E_n\sqrt{\frac{2}{L}}\sin\left(\frac{1}{\hbar}\wp_n x\right)e^{-i\frac{1}{\hbar}E_n t} = \sum_n a_n E_n\Psi_n(x,t)\end{aligned} \tag{11.6}$$

Physically, if a system's state is a superposition, a measurement will trigger the state to collapse into one of the eigenstates within the sum, a process that happens randomly governed by the probabilities $a_n^* a_n$. Hence, any single measurement will result in a value from the set $\{E_n\}$ but we can't predict ahead of time which one.

After a run of measurements, using an ensemble of identical systems in the same superposition, there will be a spectrum of results with the average value:

$$\langle E \rangle = \sum_n \frac{\mathbb{N}_n E_n}{\mathbb{N}} \tag{11.7}$$

where $\mathbb{N}_n$ is the number of occasions that the value E_n has popped up and $\mathbb{N}$ the total number of measurements. Clearly, as $\mathbb{N} \to \infty$, so $\mathbb{N}_n/\mathbb{N} \to p_n$, the probability that the state n is manifest. Hence, we can modify (11.7) to be:

$$\langle E \rangle = \sum_n p_n E_n \tag{11.8}$$

If we are to compare theory to experiment, this average must be calculable from the state. First, we should note that the quantum probability $p_n = a_n^* a_n$. Initially, this seems a rather cryptic clue, but if we form the construction:

$$\Phi^* \hat{E} \Phi = \left(\sum_k a_k^* \sqrt{\frac{2}{L}} e^{+i\frac{1}{\hbar}E_k t} \sin\left(\frac{1}{\hbar} \wp_k x \right) \right) \hat{E} \left(\sum_l a_l \sqrt{\frac{2}{L}} e^{-i\frac{1}{\hbar}E_l t} \sin\left(\frac{1}{\hbar} \wp_l x \right) \right)$$

$$= \frac{2}{L} \sum_k \sum_l a_k^* a_l E_l e^{+i\frac{1}{\hbar}(E_k - E_l)t} \sin\left(\frac{1}{\hbar} \wp_k x \right) \sin\left(\frac{1}{\hbar} \wp_l x \right) \tag{11.9}$$

and then integrate:

$$\langle \hat{E} \rangle = \frac{2}{L} \int_0^L \sum_k \sum_l a_k^* a_l E_l e^{+i\frac{1}{\hbar}(E_k - E_l)t} \sin\left(\frac{1}{\hbar} \wp_k x \right) \sin\left(\frac{1}{\hbar} \wp_l x \right) dx$$

$$= \frac{2}{L} \sum_k \sum_l a_k^* a_l E_l e^{+i\frac{1}{\hbar}(E_k - E_l)t} \int_0^L \sin\left(\frac{1}{\hbar} \wp_k x \right) \sin\left(\frac{1}{\hbar} \wp_l x \right) dx \tag{11.10}$$

Inserting our quantization relationship, $\wp_n = \dfrac{\hbar n \pi}{L}$:

$$\langle \hat{E} \rangle = \frac{2}{L} \sum_k \sum_l a_k^* a_l E_l e^{+i\frac{1}{\hbar}(E_k - E_l)t} \int_0^L \sin\left(\frac{k\pi}{L} x \right) \sin\left(\frac{l\pi}{L} x \right) dx$$

$$= \frac{1}{L} \sum_k \sum_l a_k^* a_l E_l e^{+i\frac{1}{\hbar}(E_k - E_l)t} \int_0^L \left\{ \cos\left(\frac{(k-l)}{L} \pi x \right) - \cos\left(\frac{(k+l)}{L} \pi x \right) \right\} dx$$

$$=\frac{1}{L}\sum_k\sum_l a_k^* a_l E_l e^{+i\frac{1}{\hbar}(E_k-E_l)t}\left[\frac{L}{(k-l)\pi}\sin\left(\frac{(k-l)}{L}\pi x\right)-\frac{L}{(k+l)\pi}\sin\left(\frac{(k+l)}{L}\pi x\right)\right]_0^L$$

$$=\frac{1}{L}\sum_k\sum_l a_k^* a_l E_l e^{+i\frac{1}{\hbar}(E_k-E_l)t}\left\{\frac{L}{(k-l)\pi}\sin\left(\pi(k-l)\right)-\frac{L}{(k+l)\pi}\sin\left(\pi(k+l)\right)\right\} \quad (11.11)$$

If $k \neq l$, then the two sine terms are simple multiples of π, and hence always zero. If $k = l$ then matters alter somewhat:

$$\langle\hat{E}\rangle=\frac{1}{L}\sum_k a_k^* a_k E_k\int_0^L\left\{1-\cos\left(\frac{2k}{L}\pi x\right)\right\}dx$$

$$=\frac{1}{L}\sum_k a_k^* a_k E_k\left[x-\frac{L}{2k\pi}\sin\left(\frac{2k}{L}\pi x\right)\right]_0^L$$

$$=\frac{1}{L}\sum_k a_k^* a_k E_k\left\{L-\frac{L}{2k\pi}\sin(2k\pi)\right\}$$

$$=\sum_k a_k^* a_k E_k \quad (11.12)$$

Which is the result we wished for. As a general principle, the *expectation value* of an operator, $\hat{O}$, applied to a state ψ:

$$\langle\hat{O}\rangle=\int\psi^*(x)\hat{O}\psi(x)dx \quad (11.13)$$

is equal to the average value achieved via a series of measurements on an ensemble of systems in the same state.

11.1.2 Momentum, in Passing

A suitable operator for momentum along the x-axis is $\hat{\wp}_x=-i\hbar\frac{\partial}{\partial x}$, which we can apply to our wave function for the particle in the box:

$$\hat{\wp}_n\psi_n(x,t)=-i\hbar\frac{\partial}{\partial x}\left(\sqrt{\frac{2}{L}}e^{-i\frac{1}{\hbar}E_n t}\sin\left(\frac{1}{\hbar}\wp_n x\right)\right)$$
$$=-i\hbar\sqrt{\frac{2}{L}}e^{-i\frac{1}{\hbar}E_n t}\frac{\wp_n}{\hbar}\cos\left(\frac{1}{\hbar}\wp_n x\right)$$

$$=-i\wp_n\sqrt{\frac{2}{L}}e^{-i\frac{1}{\hbar}E_n t}\cos\left(\frac{1}{\hbar}\wp_n x\right)\neq\wp_n\Psi_n(x,t) \quad (11.14)$$

Clearly, our state is *not* an eigenstate of this operator, which at first glance seems a puzzle. After all, the wave function contains a singular value of momentum, $\wp_n$. However, when we set this state up, we deliberately included both *directions* of momentum, $\pm\wp_n$. We should now expect that the expectation value,[3] composed of both $+\wp_n$ and $-\wp_n$ with equal probability, should be zero. Accordingly, we calculate:

$$\langle\hat{\wp}\rangle = \int_0^l \left(\sqrt{\frac{2}{L}}e^{+i\frac{1}{\hbar}E_nt}\sin\left(\frac{1}{\hbar}\wp_n x\right)\right)\hat{\wp}\left(\sqrt{\frac{2}{L}}e^{-i\frac{1}{\hbar}E_nt}\sin\left(\frac{1}{\hbar}\wp_n x\right)\right)dx$$

$$= \frac{2\wp_n}{\hbar L}\int_0^L \sin\left(\frac{1}{\hbar}\wp_n x\right)\cos\left(\frac{1}{\hbar}\wp_n x\right)dx$$

$$= \frac{2\wp_n}{\hbar L}\left[-\frac{\hbar}{2\wp_n}\cos^2\left(\frac{1}{\hbar}\wp_n x\right)\right]_0^L = -\frac{1}{L}\left\{\cos^2\left(\frac{1}{\hbar}\wp_n L\right)-1\right\} \quad (11.15)$$

Now, we deploy our quantization:

$$\wp_n = \frac{\hbar n\pi}{L} \quad (6.18)$$

so, the expectation value becomes:

$$\langle\hat{\wp}\rangle = -\frac{1}{L}\left\{\cos^2\left(\frac{1}{\hbar}\frac{\hbar n\pi}{L}L\right)-1\right\} = -\frac{1}{L}\left\{\cos^2(n\pi)-1\right\} = 0 \quad (11.16)$$

11.1.3 Position Operator

Coming up with an operator to represent the position of a particle is easy, but some of the consequences are slightly harder to deal with. An immediate clue to the operator comes from a consideration of its expectation value, which ought to be the average position of the particle. Given that the complex square of the wave function represents the probability to find a particle in a region, this immediately suggests that the average position ought to be:

$$\langle x\rangle = \int \Psi_n^*(x,t)\,x\Psi_n(x,t)\,dx$$

which in the case of our particle in the box is:

$$\langle x\rangle = \frac{2}{L}\int_0^L e^{+i\frac{1}{\hbar}E_nt}\sin\left(\frac{1}{\hbar}\wp_n x\right)xe^{-i\frac{1}{\hbar}E_nt}\sin\left(\frac{1}{\hbar}\wp_n x\right)dx$$

$$= \frac{2}{L}\int_0^L x\sin^2\left(\frac{1}{\hbar}\wp_n x\right)dx = \frac{2}{L}\int_0^L x\sin^2\left(\frac{n\pi}{L}x\right)dx$$

$$= \frac{1}{L}\int_0^L x\left(1-\cos\left(\frac{2n\pi}{L}x\right)\right)dx$$

$$= \frac{1}{L}\int_0^L xdx - \int_0^L x\cos\left(\frac{2n\pi}{L}x\right)dx$$

$$= \frac{1}{2L}L^2 - \frac{L^2}{4\pi^2 n^2}\left(2n\pi\sin(2n\pi) + \cos(2n\pi) - 1\right) = \frac{L}{2}$$

as we would guess, even given the work to get to this point.

So, the position operator, $\hat{x}$, is simply[4] $x\times$, i.e. multiply the wave function by x.

11.1.4 Operator Properties

The operators that we have considered so far are part of a general class which serves to extract information about observables from wave functions. This is a very different way of working to that in classical physics and requires a re-assessment of these physical properties at the conceptual level. Momentum, for example, with the operator $\widehat{\wp}_x = -i\hbar\,\partial/\partial x$, should be considered as the rate of change of the wave function's phase with distance. Energy, $\widehat{E} = i\hbar\,\partial/\partial t$, is the rate of change of phase with time. The somewhat more obscure quantum property, *spin*, can be related to the rate at which the wave function changes with angle if you choose to rotate the system about some axis.

While each specific observable has its own properties, there is one general rule that each operator must adhere to. As we know, the wave function is expressed as a complex number, but the measured value of an observable must be real. Hence, *the eigenvalue of any operator that represents an observable must be real.* This places a general constraint on the mathematical properties of any operator purporting to represent a physical variable in quantum theory.

If all the o_i are real-valued and eigenvalues of an operator $\widehat{O}$ to boot, then the expectation value (11.13) must be real as well:

$$\left\langle \hat{o}^* \right\rangle = \left[\int \psi^*(x)\hat{o}\psi(x)dx\right]^* = \int \psi(x)\hat{o}^*\psi^*(x)dx = \int \psi^*(x)\hat{o}\psi(x)dx = \left\langle \hat{o} \right\rangle \quad (11.17)$$

Operators with real eigenvalues are *Hermitian* and subject to the defining constraint:

$$\int \psi^*(x)\hat{o}\psi(x)dx = \int \psi(x)\hat{o}^*\psi^*(x)dx \quad (11.18)$$

11.1.5 Compound Operators

Clearly, physical variables can be combined to obtain other quantities of interest. The energy of a particle, for example, would be:

$$\frac{\wp^2}{2m} + V \quad (11.19)$$

while moving through a potential energy, V. It would be a major limitation if quantum theory did not allow us to combine observables in a similar fashion. Taking our cue from the classical equation above, the Hamiltonian operator for a particle's energy is:

$$\widehat{H} = \frac{1}{2m}\widehat{\wp}\times\widehat{\wp} + V\times\hat{I} = -\frac{\hbar^2}{2m}\frac{\partial^2}{\partial x^2} + V\times\hat{I}$$

where we have used $\hat{I}$ as an *identity operator* (i.e., multiply by one).

In broad terms, any classical function of observables $f(x,y)$ can be obtained in quantum mechanics by the same function of the appropriate operators $f(\hat{x},\hat{y})$. However, there is one extremely important proviso. Not all operators *commute* with each other. Take position and momentum for example. Acting on a wave function with both operators will produce different results depending on the order:

$$\widehat{\wp}_x x\psi(x) = -i\hbar\frac{\partial}{\partial x}\big(x\psi(x)\big) = -i\hbar\psi(x) - i\hbar x\frac{\partial\psi}{\partial x}$$

$$x\widehat{\wp}_x\psi(x) = -i\hbar x\frac{\partial\psi}{\partial x}$$

Two operators which can be applied in any order are said to *commute*, otherwise they are *non-commutative*. Interestingly, and very usefully, commuting operators share the same eigenstates. Let operator $\widehat{A}$ have an eigenstate ϕ with eigenvalue a, then for another operator, $\widehat{B}$, that commutes with $\widehat{A}$:

$$\widehat{B}\left(\widehat{A}\phi\right) = \widehat{B}a\phi = a\widehat{B}\phi$$

Also, by the commutation property:

$$\widehat{B}\left(\widehat{A}\phi\right) = \widehat{A}\left(\widehat{B}\phi\right) = a\left(\widehat{B}\phi\right)$$

so that $\widehat{B}\phi$ must also be an eigenstate of $\widehat{A}$. For this to be true, $\widehat{B}\phi = b\phi$ making ϕ an eigenstate of $\widehat{B}$ as well.

It is no coincidence that x and $\widehat{\wp}_x$, which are the observables linked by the uncertainty principle, do not commute. As they do not share the same eigenstates, one set can be expanded in terms of the other—that is what leads to the uncertainty relationship. Specifically, any pair of non-commuting operators (conjugate variables) will be linked by an uncertainty relationship.

11.1.6 Energy-Time Inequality

Along with the famous uncertainty relationship connecting position and momentum (6.30)

$$\Delta\wp\Delta x \geq \hbar/2$$

there is another often-quoted result linking energy and time:

$$\Delta E\,\delta t \geq \hbar/2 \tag{11.20}$$

This relationship is not, technically, an uncertainty. Uncertainty relationships come about when you are dealing with conjugate variables. In which case, *there is no basis corresponding to eigenstates of both variables*. For $\Delta E\delta t \geq \hbar/2$ to be an uncertainty relationship of this form, there would have to be an energy operator, $\widehat{E}$, which is fine, but also a time operator, $\hat{t}$, which is most certainly not fine. In this inequality, δt is a *time duration, not an uncertainty relating to a time measurement,* Δt. There is no measurement of a quantum system that tells you what date and time it is. We have used δt rather than Δt in (11.20) to emphasise this distinction.

The inequality, (11.20), relates the uncertainty in successive energy measurements, ΔE, to the time interval between the measurements, δt. If the system is in an energy eigenstate, $\Delta E = 0$

(successive energy measurements do not disturb the state) then $\delta t \rightarrow \infty$, i.e., we would have to wait for an indefinite time before we could be sure that the system had evolved. In other words, the system is *not evolving* in this case (it's stationary).

While it would be better to refer to (11.20) as an energy-time inequality, as we have here, we suspect that battle has been lost long ago, and it will always be known as an energy-time uncertainty relationship.

11.2 DISCRETE SITUATIONS

Up to now, we have been dealing with observables that have a continuous spectrum. In our box, the energy is quantized to discrete values, but that does not stop energy from being continuous in other contexts (a free particle, for example). Now, we must turn to cases where the physical variable we are interested in has a small set of discrete values, *by its nature*.

As a classic example, consider the spin components of a spin ½ particle, such as an electron. If we measure the z-axis spin component of such a particle, we get either $+1/2\hbar$ or $-1/2\hbar$. As there are only these two possible values, the spin state of the particle cannot be as expressed as a wave function.[5] Also, the operator we need to extract the spin component observable cannot be written in functional terms. Instead, we need to expand our mathematical vocabulary to encompass *matrices* and *vectors*. We need a mathematical detour.

11.2.1 Enter the Matrix

A matrix is an array of numbers or algebraic functions, or similar:

$$A = \begin{pmatrix} a_{11} & a_{12} & a_{13} \\ a_{21} & a_{22} & a_{23} \\ a_{31} & a_{32} & a_{33} \end{pmatrix} \tag{11.21}$$

In a more compact form, the matrix is an ordered collection of *elements* $\{a_{RC}\}$ where R is the *row index* and C the *column index*.

Given another matrix:

$$B = \begin{pmatrix} b_{11} & b_{12} & b_{13} \\ b_{21} & b_{22} & b_{23} \\ b_{31} & b_{32} & b_{33} \end{pmatrix} \tag{11.22}$$

their product is defined:

$$\begin{aligned} A \times B &= \begin{pmatrix} a_{11} & a_{12} & a_{13} \\ a_{21} & a_{22} & a_{23} \\ a_{31} & a_{32} & a_{33} \end{pmatrix} \begin{pmatrix} b_{11} & b_{12} & b_{13} \\ b_{21} & b_{22} & b_{23} \\ b_{31} & b_{32} & b_{33} \end{pmatrix} \\ &= \begin{pmatrix} a_{11}b_{11} + a_{12}b_{21} + a_{13}b_{31} & a_{11}b_{12} + a_{12}b_{22} + a_{13}b_{32} & a_{11}b_{13} + a_{12}b_{23} + a_{13}b_{33} \\ a_{21}b_{11} + a_{22}b_{21} + a_{23}b_{31} & a_{21}b_{12} + a_{22}b_{22} + a_{23}b_{32} & a_{21}b_{13} + a_{22}b_{23} + a_{23}b_{33} \\ a_{31}b_{11} + a_{32}b_{21} + a_{33}b_{31} & a_{31}b_{12} + a_{32}b_{22} + a_{33}b_{32} & a_{31}b_{13} + a_{32}b_{23} + a_{33}b_{33} \end{pmatrix} \end{aligned} \tag{11.23}$$

Extracting the wood from the trees, the term located at the intersection of row R and column C has the form:

$$[A \times B]_{RC} = \sum_j a_{Rj} b_{jC} \tag{11.24}$$

In this summation, the index j is counting across the columns of the first matrix and down the rows of the second. So, for this rule to work, the number of columns in the first matrix must equal the number of rows in the second. This means that some matrices can't be multiplied together. If the first matrix has $[\#\text{rows}, \#\text{columns}] = [P, Q]$, then the second must have $[\#\text{rows}, \#\text{columns}] = [Q, R]$. Consequently, the product has $[\#\text{rows}, \#\text{columns}] = [P, R]$.

A matrix multiplication of this form is known as an *inner product*. The *outer product* will be encountered later.

Amusingly, and not always conveniently, matrix multiplication is *non-commutative*, $A \times B \neq B \times A$, although there are some special cases (e.g. $A^2 = A \times A$). As an example, consider:

$$A = \begin{pmatrix} 1 & 2 \\ 3 & 4 \end{pmatrix} \qquad B = \begin{pmatrix} 5 & 6 \\ 7 & 8 \end{pmatrix} \tag{11.25}$$

Then:

$$A \times B = \begin{pmatrix} 1 & 2 \\ 3 & 4 \end{pmatrix}\begin{pmatrix} 5 & 6 \\ 7 & 8 \end{pmatrix} = \begin{pmatrix} 19 & 22 \\ 43 & 50 \end{pmatrix} \quad B \times A = \begin{pmatrix} 5 & 6 \\ 7 & 8 \end{pmatrix}\begin{pmatrix} 1 & 2 \\ 3 & 4 \end{pmatrix} = \begin{pmatrix} 23 & 34 \\ 31 & 46 \end{pmatrix} \tag{11.26}$$

which are clearly very different.

11.2.2 Vectors

Matrix multiplication also applies to vectors, although, given the way the rule works, we need both column and row vector forms. For example, with a vector $\boldsymbol{v} = \{v_i\}$ we can multiply the column form from the left:

$$A \times \boldsymbol{v} = \begin{pmatrix} a_{11} & a_{12} & a_{13} \\ a_{21} & a_{22} & a_{23} \\ a_{31} & a_{32} & a_{33} \end{pmatrix}\begin{pmatrix} v_1 \\ v_2 \\ v_3 \end{pmatrix} = \begin{pmatrix} a_{11}v_1 + a_{12}v_2 + a_{13}v_3 \\ a_{21}v_1 + a_{22}v_2 + a_{23}v_3 \\ a_{31}v_1 + a_{32}v_2 + a_{33}v_3 \end{pmatrix}$$

$$[A \times \boldsymbol{v}]_R = \sum_i a_{Ri} v_i \tag{11.27}$$

Equally, we can multiply the row form from the right:

$$\boldsymbol{v} \times A = \begin{pmatrix} v_1 & v_2 & v_3 \end{pmatrix}\begin{pmatrix} a_{11} & a_{12} & a_{13} \\ a_{21} & a_{22} & a_{23} \\ a_{31} & a_{32} & a_{33} \end{pmatrix}$$
$$= \begin{pmatrix} v_1a_{11} + v_2a_{21} + v_3a_{31} & v_1a_{12} + v_2a_{22} + v_3a_{32} & v_1a_{13} + v_2a_{23} + v_3a_{33} \end{pmatrix}$$

$$[v \times A]_C = \sum_i v_i a_{iC} \tag{11.28}$$

Finally, two vectors can be multiplied together, in the correct order:

$$v \times u = \begin{pmatrix} v_1 & v_2 & v_3 \end{pmatrix} \begin{pmatrix} u_1 \\ u_2 \\ v_3 \end{pmatrix} = v_1 u_1 + v_2 u_2 + v_3 v_3$$

$$v \times u = \sum_k v_k u_k \tag{11.29}$$

Technically, the process of converting a column form into a row form is taking the *transpose*. This can also be done for a matrix, so that:

$$\begin{pmatrix} a_{11} & a_{12} & a_{13} \\ a_{21} & a_{22} & a_{23} \\ a_{31} & a_{32} & a_{33} \end{pmatrix}^T = \begin{pmatrix} a_{11} & a_{21} & a_{31} \\ a_{12} & a_{22} & a_{32} \\ a_{13} & a_{23} & a_{33} \end{pmatrix} \tag{11.30}$$

If you take the complex conjugate at the same time, you create the *adjoint matrix*:

$$\left(\begin{pmatrix} a_{11} & a_{12} & a_{13} \\ a_{21} & a_{22} & a_{23} \\ a_{31} & a_{32} & a_{33} \end{pmatrix}^T \right)^* = \begin{pmatrix} a_{11} & a_{12} & a_{13} \\ a_{21} & a_{22} & a_{23} \\ a_{31} & a_{32} & a_{33} \end{pmatrix}^\dagger = \begin{pmatrix} a_{11}^* & a_{21}^* & a_{31}^* \\ a_{12}^* & a_{22}^* & a_{32}^* \\ a_{13}^* & a_{23}^* & a_{33}^* \end{pmatrix} \tag{11.31}$$

All such manipulations can be very useful when dealing with matrices and vectors in quantum theory. Here endeth the mathematical diversion.

11.2.3 Spin Representations

As previously mentioned, measuring the z-axis spin component of a spin 1/2 particle, produces either $+1/2\hbar$ or $-1/2\hbar$. We can use column vectors to represent the eigenstates:

$$\left| +\frac{1}{2}\hbar \right\rangle = \begin{pmatrix} 1 \\ 0 \end{pmatrix} \quad \left| -\frac{1}{2}\hbar \right\rangle = \begin{pmatrix} 0 \\ 1 \end{pmatrix} \tag{11.32}$$

Introducing the notation $| \ \rangle$, which is used by professional quantum mechanics to indicate the state of a system, often with some identifying symbol in between the bracket forms. Dirac[6] invented this mathematical terminology, which is very useful for writing formulas that are equally valid in continuous or discrete cases.

In this format, a general spin state would be:

$$|\Psi\rangle = a\left| +\frac{1}{2}\hbar \right\rangle + b\left| -\frac{1}{2}\hbar \right\rangle = \begin{pmatrix} a \\ b \end{pmatrix} \tag{11.33}$$

with a being the amplitude to detect $+1/2\hbar$ when a z-axis spin component is measured. Correspondingly, b is the amplitude to detect $-1/2\hbar$. This is just like the superposition (11.4).

The operator corresponding to a z-axis spin component measurement is the matrix:

$$\hat{S}_Z = \frac{1}{2}\hbar\begin{pmatrix} 1 & 0 \\ 0 & -1 \end{pmatrix} \tag{11.34}$$

which we can easily confirm using a simple manipulation:

$$\hat{S}_Z\left|+\frac{1}{2}\hbar\right\rangle = \frac{1}{2}\hbar\begin{pmatrix} 1 & 0 \\ 0 & -1 \end{pmatrix}\begin{pmatrix} 1 \\ 0 \end{pmatrix} = \frac{1}{2}\hbar\begin{pmatrix} 1 \\ 0 \end{pmatrix} = \frac{1}{2}\hbar\left|+\frac{1}{2}\hbar\right\rangle \tag{11.35}$$

Equally:

$$\hat{S}_Z\left|-\frac{1}{2}\hbar\right\rangle = \frac{1}{2}\hbar\begin{pmatrix} 1 & 0 \\ 0 & -1 \end{pmatrix}\begin{pmatrix} 0 \\ 1 \end{pmatrix} = -\frac{1}{2}\hbar\begin{pmatrix} 0 \\ 1 \end{pmatrix} = -\frac{1}{2}\hbar\left|+\frac{1}{2}\hbar\right\rangle \tag{11.36}$$

To calculate the expectation value of a spin component measurement, we need one further piece of mathematical machinery. The vector equivalent of the complex conjugate wave function is the *conjugate transpose row vector*:

$$\langle\Psi| = \left(\begin{pmatrix} a \\ b \end{pmatrix}^T\right)^* = \begin{pmatrix} a^* & b^* \end{pmatrix} \tag{11.37}$$

making the expectation value:

$$\begin{aligned}\left\langle\hat{S}_Z\right\rangle &= \left\langle\Psi\left|\hat{S}_Z\right|\Psi\right\rangle = \begin{pmatrix} a^* & b^* \end{pmatrix}\frac{\hbar}{2}\begin{pmatrix} 1 & 0 \\ 0 & -1 \end{pmatrix}\begin{pmatrix} a \\ b \end{pmatrix} = \frac{\hbar}{2}\begin{pmatrix} a^* & b^* \end{pmatrix}\begin{pmatrix} a \\ -b \end{pmatrix} \\ &= \frac{\hbar}{2}\left(a^*a - b^*b\right)\end{aligned} \tag{11.38}$$

which is the discrete version of our earlier continuous definition (11.13):

$$\left\langle\hat{O}\right\rangle = \int\psi^*(x)\hat{O}\psi(x)\,dx$$

11.2.4 Operators as Matrices

Let's now imagine that we have an ensemble of systems with each member in the *same* quantum state $|\psi_i\rangle$, which is one of a collection of $\mathcal{N}$ possible quantum states $\{|\psi_n\rangle\}$.

For discrete observables, each state could be represented as a column vector:

$$|\psi_1\rangle = \begin{pmatrix} 1 \\ 0 \\ 0 \\ \dots \\ 0 \\ 0 \end{pmatrix} \quad |\psi_2\rangle = \begin{pmatrix} 0 \\ 1 \\ 0 \\ \dots \\ 0 \\ 0 \end{pmatrix} \quad |\psi_i\rangle = \begin{pmatrix} 0 \\ 0 \\ 0 \\ \dots \\ 1 \\ 0 \end{pmatrix} \quad |\psi_n\rangle = \begin{pmatrix} 0 \\ 0 \\ 0 \\ \dots \\ 0 \\ 1 \end{pmatrix} \tag{11.39}$$

with their conjugate transpose row vector equivalents:

$$\langle\psi_i| = (1 \quad 0 \quad 0 \quad \dots \quad 0 \quad 0) \qquad \langle\psi_i| = (0 \quad 0 \quad 0 \quad \dots \quad 1 \quad 0) \text{ etc.} \tag{11.40}$$

An operator, $\hat{O}$, acting on one of its eigenstates, $|\psi_i\rangle$, produces an eigenvalue, $\hat{O}|\psi_i\rangle = o_i|\psi_i\rangle$, the left-hand side of which renders in component form as (using (11.27)):

$$\left[\hat{O}|\psi_i\rangle\right]_R = \sum_k \hat{O}_{Rk}\left[|\psi_i\rangle\right]_k \tag{11.41}$$

As we have noted, the state vector has only one element, in the row corresponding to the quantum number, i. In component form the vector is then:

$$\left[\psi_i\right]_R = \delta_{Ri} \tag{11.42}$$

using a useful shorthand in the *Kronecker delta* $\delta_{ij} = 1$ if $i = j$, $\delta_{ij} = 0$ if $i \neq j$. Despite having two indices, δ_{Ri} is *not* a matrix—the subscript i is the quantum state, while R is the row in the vector. Hence δ_{Ri} is a *conditional number*.

This means that (11.41) can be written:

$$\left[\hat{O}|\psi_i\rangle\right]_R = \sum_k \hat{O}_{Rk}\delta_{ki} \tag{11.43}$$

As the right-hand side of our equation, $o_i|\psi_i\rangle$, takes the form $o_R\delta_{Ri}$, we have:

$$\sum_k \hat{O}_{Rk}\delta_{ki} = o_R\delta_{Ri} \tag{11.44}$$

The effect of the Kronecker delta, δ_{ki}, on the summation is to zero out each term unless $k = i$ (remember, i is not changing—that is fixed by the quantum state we are dealing with). So, with this in mind:

$$\hat{O}_{Ri} = o_R\delta_{Ri} \tag{11.55}$$

On the other side, δ_{ri} removes all row entries, other than the row with index i. We end up with:

$$\hat{O}_{RR} = o_R \tag{11.56}$$

The elements within the matrix representing our operator (which has eigenstates $|\psi_i\rangle$) *are all zero, aside from those on the diagonal, which take the eigenvalues.* This is exactly what the z-axis spin operator looked like earlier (11.34):

$$\hat{S}_Z = \frac{1}{2}\hbar\begin{pmatrix} 1 & 0 \\ 0 & -1 \end{pmatrix}$$

In conclusion, a matrix operator takes the form:

$$\left[\hat{O}\right]_{RC} = o_R \delta_{RC} \tag{11.45}$$

Note that this is a specific *representation* of the operator. We are assuming that the states we are using (which have a direct impact on the matrix elements in the operator) are the eigenstates. This is why the matrix has this simple diagonal form. The matrix can be constructed using other basis sets which are not eigenstates, in which case it will have off-diagonal terms in it.

11.2.5 More Expectations

In Section 11.2.3, we calculated the expectation value of the z-component spin operator (11.38), $\langle \hat{S}_Z \rangle = \langle \psi | \hat{S}_Z | \psi \rangle$. Now we can consider the more general case.

Suppose we have a discrete observable mathematically rendered by the operator matrix $[\hat{O}]_{RC}$ in its eigenstate representation. Each eigenstate is a vector $|\psi_i\rangle_R = \delta_{Ri}$, so a superposition of states would be:

$$|\Phi\rangle = \sum_i a_i |\psi_i\rangle \qquad \left[|\Phi\rangle\right]_R = \sum_i a_i \delta_{Ri} \qquad \left[|\Phi^*\rangle\right]_C = \sum_j a_j^* \delta_{Cj} \tag{11.46}$$

Firstly, using (11.27) we write:

$$\left[\hat{O}|\psi_i\rangle\right]_R = \sum_j \left[\hat{O}\right]_{Rj} \sum_i a_i \delta_{ji} = \sum_j \sum_k \sum_i o_k \delta_{Rk} \delta_{jk} a_i \delta_{ji} \tag{11.47}$$

We let each sum run, one by one, allowing the deltas to work their magic:

$$\left[\hat{O}|\psi_i\rangle\right]_R = \sum_j \sum_k o_k \delta_{Rk} \delta_{jk} a_j = \sum_j o_j \delta_{Rj} a_j \tag{11.48}$$

Next, we multiply by the transpose column from the left, using the appropriate rule, (11.29):

$$\left\langle \psi_i \middle| \hat{O} \middle| \psi_i \right\rangle = \sum_t \left[|\psi_i^*\rangle\right]_t \sum_j o_j \delta_{tj} a_j = \sum_t \sum_m \sum_j a_m^* \delta_{tm} o_j \delta_{tj} a_j$$

$$= \sum_t \sum_m a_m^* \delta_{tm} o_t a_t = \sum_t a_t a_t^* o_t \tag{11.49}$$

giving us the expectation value, which is a relief.

11.2.6 Change of Basis

Any quantum state can be written as a superposition of eigenstates of some appropriate observable. If we are working with observable O and its eigenstates, $\{|\psi_n\rangle\}$, then our state would be of the form:

$$|\Psi\rangle = \sum_i a_i |\psi_i\rangle \tag{11.50}$$

In this basis, the matrix version of the operator, $\hat{O}$, representing the observable will be diagonal and so of the form (11.45) $\left[\hat{O}\right]_{RC} = o_R \delta_{RC}$.

Another physicist might be working with an identical ensemble of systems, in the same state $|\Psi\rangle$' but chooses to measure O' instead and is hence working with the eigenstates $\{|\varphi_n\rangle\}$, where $\hat{O}'|\varphi_n\rangle = o_n'|\varphi_n\rangle$. They would expand the state differently:

$$|\Psi\rangle = \sum_i b_i |\varphi_i\rangle \tag{11.51}$$

but this is the *same state*, just in a different *representation*. The operator $\hat{O}'$ would be diagonal using these eigenstates, $\{|\varphi_n\rangle\}$, as a basis, but the operator $\hat{O}$ would not be. Just as the state is the same, no matter what basis set of eigenstates we choose to expand it over, the operator is the same; it's the matrix representation that has changed. Think of it this way: there is a distinction between *physics* and *mathematics*. In terms of physics, any equivalent mathematical form of the operator $\hat{O}$ can be used to represent the same observable. It is a question of convenience depending on the whim of the physicist and a utility factor related to the problem they are dealing with.

The mathematical process by which bases, or representations, can be morphed from one into another was the work of Dirac. The basic idea is to construct a matrix which carries us from one basis into another:

$$|\psi_i\rangle = \mathbb{U}|\varphi_i\rangle \tag{11.52}$$

Once again, we must be very careful to distinguish indices we are using to *name* states, from those we need for rows and columns. With that in mind, we write:

$$\left[|\psi_i\rangle\right]_R = \sum_k \mathbb{U}_{Rk}\left[|\varphi_i\rangle\right]_k \qquad \left[\langle\psi_i|\right]_C = \sum_j \mathbb{U}_{jC}^*\left[\langle\varphi_i|\right]_j \tag{11.53}$$

{When we are dealing with column vectors, which have a single column, our indices are going down the rows. With row vectors, we step across columns}.

If $\mathbb{U}$ is to function correctly, it must start from a collection of states that are orthonormal, so that $\langle \varphi_i | \varphi_j \rangle = \delta_{ij}$ and transform them into a set which are similarly well-behaved: $\langle \psi_i | \psi_j \rangle = \delta_{ij}$. To test this, we build $\langle \psi_i | \psi_i \rangle$:

$$\langle \psi_i | \psi_i \rangle = \sum_F \left[\langle \varphi_i | \right]_F \left[| \psi_i \rangle \right]_F = \sum_F \left(\sum_k \mathbb{U}_{Fk} \left[| \varphi_i \rangle \right]_k \right) \left(\sum_j \mathbb{U}^*_{jF} \left[\langle \varphi_i | \right]_j \right)$$

$$= \sum_F \sum_k \sum_j \mathbb{U}^*_{jF} \mathbb{U}_{Fk} \left[\langle \varphi_i | \right]_j \left[| \varphi_i \rangle \right]_k \tag{11.54}$$

If we make the suggestion:

$$\sum_F \mathbb{U}^*_{jF} \mathbb{U}_{Fk} = \delta_{jk} \tag{11.55}$$

our calculation becomes:

$$\langle \psi_i | \psi_i \rangle = \sum_k \sum_j \delta_{jk} \left[\langle \varphi_i | \right]_j \left[| \varphi_i \rangle \right]_k = \sum_k \left[\langle \varphi_i | \right]_k \left[| \varphi_i \rangle \right]_k = 1 \tag{11.56}$$

Our sneaky suggestion translates into the matrix, $\mathbb{U}$, being *unitary*, i.e., $\mathbb{U}^\dagger \mathbb{U} = \mathbb{U}\mathbb{U}^\dagger = 1$. Switching from one basis to another is a *unitary transformation*, which is an important mathematical and physical point.

Now let's consider what happens to an operator during a basis switch. In Dirac notation, the matrix elements of an operator, $\hat{O}$ are:

$$\left[\hat{O}' \right]_{RC} = \left\langle \psi_R \middle| \hat{O}' \middle| \psi_C \right\rangle \tag{11.57}$$

As the states in the basis $\{ | \psi_n \rangle \}$ are not eigenstates for $\hat{O}'$, this matrix would not be diagonal. But let us see what happens if we use $\mathbb{U}$ to switch bases:

$$\left[| \psi_C \rangle \right]_r = \sum_k \mathbb{U}_{rk} \left[| \varphi_C \rangle \right]_k \qquad \left[\langle \psi_R | \right]_c = \sum_j \mathbb{U}^*_{jc} \left[\langle \varphi_R | \right]_j \tag{11.58}$$

Making:

$$\left[\hat{O}' \right]_{RC} = \left(\sum_j \mathbb{U}^*_{jc} \left[\langle \varphi_R | \right]_j \right) \hat{O}' \left(\sum_k \mathbb{U}_{rk} \left[| \varphi_C \rangle \right]_k \right) \tag{11.59}$$

We can now allow the operator to act on its eigenstate to the right:

$$\left[\hat{O}'\right]_{RC} = \left(\sum_j \mathbb{U}^*_{jc}\left[\langle\varphi_R|\right]_j\right)\left(\sum_k \mathbb{U}_{rk} o'_C \left[|\varphi_C\rangle\right]_k\right)$$

$$= \sum_j \sum_k o'_c \mathbb{U}^*_{jc} \mathbb{U}_{rk}\left[\langle\varphi_R|\right]_j\left[|\varphi_C\rangle\right]_k \tag{11.60}$$

so that $[\hat{O}']_{RR} = o'_R$. In other words, we have diagonalized the matrix. Therefore, we have a very useful set of transformations:

$$|\psi_i\rangle = \mathbb{U}|\varphi_i\rangle \qquad \langle\psi_i| = \langle\varphi_i|\mathbb{U}^\dagger \qquad \hat{O}_\psi = \mathbb{U}^\dagger \hat{O}_\varphi \mathbb{U} \tag{11.61}$$

11.2.7 Hermiticity Again...

In Section 11.1.4, we explained that the operator representing an observable must be Hermitian, so that eigenvalues and expectation values come out to be real numbers.

The same must be true of operators rendered as matrices. The definition is slightly different to account for a matrix representation, but in spirit it works out the same.

Suppose:

$$\hat{O}|\varphi_i\rangle = o_i|\varphi_i\rangle = (a+ib)|\varphi_i\rangle \tag{11.62}$$

then:

$$\left(\left(\hat{O}|\varphi_i\rangle\right)^*\right)^T = \langle\varphi_i|\hat{O}^\dagger = (a-ib)\langle\varphi_i| \tag{11.63}$$

Meaning that:

$$\left\langle\varphi_i\middle|\hat{O}^\dagger\middle|\varphi_i\right\rangle = (a-ib) \tag{11.64}$$

We already know:

$$\left\langle\varphi_i\middle|\hat{O}\middle|\varphi_i\right\rangle = (a+ib) \tag{11.65}$$

So, requiring that $\hat{O}^\dagger = \hat{O}$, forces:

$$(a-ib) = \left\langle\varphi_i\middle|\hat{O}^\dagger\middle|\varphi_i\right\rangle = \left\langle\varphi_i\middle|\hat{O}\middle|\varphi_i\right\rangle = (a+ib) \tag{11.66}$$

Hence, $b = 0$ and the expectation value is real. In which case, a matrix is Hermitian if $\hat{O}^\dagger = \hat{O}$.

11.2.8 The Trace

There is one final piece of mathematical technology that will prove useful: taking the *trace*, Tr, of a matrix, i.e., adding the elements lying along the diagonal.[7]

Such terms must have $R = C$, so for a matrix $[A]_{RC} = a_{RC}$:

$$\mathrm{Tr}[A] = \sum_K a_{KK} \tag{11.67}$$

The only criterion is that the matrix must be a square array, otherwise there is no defined diagonal. For a product matrix, the trace takes the form:

$$\mathrm{Tr}[A \times B] = \sum_K \sum_j a_{Kj} b_{jK} \tag{11.68}$$

The trace has some useful properties:

T1. The trace is *multiplicative:*

$$\mathrm{Tr}[cA] = c\mathrm{Tr}[A]$$

If you multiply each element in a matrix by the same constant, c, then the sum over the diagonals must also have been multiplied by c.

T2. The trace is *additive*:

$$\mathrm{Tr}[A + B] = \mathrm{Tr}[A] + \mathrm{Tr}[B]$$

If each term on the diagonal is the sum of two others, then the sum of diagonals must be the sum of the two separate sets of terms added together.

T3. The trace is *linear*:

$$\mathrm{Tr}[aA + bB] = a\mathrm{Tr}[A] + b\mathrm{Tr}[B]$$

which follows from T1 and T2.

T4. The trace is *symmetric*:

$$\mathrm{Tr}[A \times B] = \mathrm{Tr}[B \times A]$$

The argument runs:

$$\mathrm{Tr}[A \times B] = \sum_K \sum_j a_{Kj} b_{jK}, \mathrm{Tr}[B \times A] = \sum_M \sum_i b_{Mi} a_{iM} = \sum_M \sum_i a_{iM} b_{Mi} = \sum_i \sum_M a_{iM} b_{Mi}$$

Note that this is true *even if the matrix multiplication itself is not commutative*, i.e.,

$$\mathrm{Tr}[A \times B] = \mathrm{Tr}[B \times A] \quad \text{even for } A \times B \neq B \times A$$

T5. The trace is *cyclic*:

$$\mathrm{Tr}[A \times B \times C] = \mathrm{Tr}[C \times A \times B] = \mathrm{Tr}[B \times C \times A] \neq \mathrm{Tr}[A \times C \times B]$$

This follows from the previous property. After all:

$$\mathrm{Tr}\big[A \times [B \times C]\big] = \mathrm{Tr}\big[[B \times C] \times A\big] = \mathrm{Tr}[B \times C \times A]$$

T6. The trace is unaffected by a unitary transformation:

$$\mathrm{Tr}\big[A_2\big] = \mathrm{Tr}\big[\mathbb{U}^{\dagger} A_1 \mathbb{U}\big] = \mathrm{Tr}\big[\mathbb{U}\mathbb{U}^{\dagger} A_1\big] = \mathrm{Tr}\big[A_1\big]$$

where we have deployed the cyclical property of the trace and the definition of unitarity.

There is one further property of the trace (**T7**) that will crop up in Section 11.5.3 where it is more easily introduced in a specific context.

11.3 TYPES OF STATE

The work that we have done so far has been predicated on the assumption that we have complete knowledge of a system's wavefunction. This is tantamount to possessing an ensemble of systems all of which are in the same (known) state. In which case, we have a *pure state*.

However, this does not mean that we have complete knowledge of measurement outcomes. If our pure state happens to be an eigenstate of the observable in question, we can predict with certainty that a measurement will yield the specific eigenvalue. However, an eigenstate for one observable does not have to be an eigenstate for another. It could be that $|\Phi\rangle$ is an eigenstate for observable O, but as we are intending to measure O' instead, we need to expand $|\Phi\rangle$ as a series over the eigenstates of O', i.e., we construct a superposition. Archetypically, this is the case for position and momentum, which is why they are linked by an uncertainty principle. In fact, *there is no quantum state that is simultaneously an eigenstate for all relevant observables*, which is one of the reasons that a quantum state cannot provide as complete a description as a classical state.

Prime face, superpositions look like the classical situation where we model the complex motion of a system as a blended sum over its normal modes of oscillation. However, the classical behaviour observed in such cases is a weighted average over the separate oscillatory motions. When a system in a quantum superposition interacts with a measuring device, the state collapses, and one of the eigenvalues is revealed, not an average over all of them. In fact, *a single measurement can't distinguish between an eigenstate and a superposition.* That can only be shown by a series of repeated measurements on identically prepared systems. The eigenstate will always give the same value, whereas a superposition generates different results, with an average equal to the expectation value. Hence the comment earlier: knowing the state with certainty does not yield certainty in measurement outcomes. Indeed, one could argue that the whole idea of a quantum state is a concept best applied to a collection of systems and loses meaning if applied to the individual. After all as a single measurement can't separate eigenstates from superpositions, any experimental determination of a state requires an ensemble of systems.

It does not matter if we are presented with a single eigenstate or a superposition; if we know what we are dealing with, the state is pure. The defining characteristic is not the nature of the state, but rather our knowledge regarding the state. An ensemble of systems, each of which is known to be in the same superposition/eigenstate would be a pure state ensemble.

If we don't have full information about the nature of the state, then probabilities need to be employed, as we will see in the next section. However, these are classical probabilities, not related to quantum amplitudes. The quantum probabilities that arise when we are dealing with measurements made on a superposition are *inherent in the nature of the state*, and not a feature of our knowledge.

11.3.1 Mixed States

Given an ensemble of systems, we may know that each system is in one of a collection of states $\{|\Psi_n\rangle\}$ *without knowing for sure which one*. As we don't have full knowledge of the state, this cannot be pure. Instead, we have a *mixed state*. In such cases, when we pluck any system from the collection, we will find it to be in state $|\Psi_n\rangle$ with the classical probability p_n. We are *not* collapsing a superposition here, simply expressing our ignorance of the state of the system, a situation that in practice is very likely to occur.[8] This is explored further in Section 11.5.2.

To hammer the point home, consider this scenario. You are conducting an experiment on an ensemble of systems. Members of the technical department know that it is possible to produce systems in state $|1\rangle$ (with eigenvalue $o_1 = 1$) 1/3 of the time and in state $|2\rangle$ (with eigenvalue $o_2 = 2$) 2/3 of the time, but that no absolute control is possible. In other words, they are preparing a mixed state ensemble.

Unfortunately, someone makes an error and reports to you that all members of the ensemble are in state:

$$|\chi\rangle = \frac{1}{3}|1\rangle + \frac{2}{3}|2\rangle \tag{11.69}$$

and so is the ensemble is pure. Your suspicions are immediately aroused, as this is not a normalized state. However, putting this down to a mathematical slip you convert the state appropriately:

$$|\chi'\rangle = \frac{9}{5}\left(\frac{1}{3}|1\rangle + \frac{2}{3}|2\rangle\right) \tag{11.70}$$

and predict an expectation value of:

$$\langle\hat{O}\rangle_P = \frac{9}{5}\left(\frac{1}{9}o_1 + \frac{4}{9}o_2\right) = \frac{1}{5}(1+8) = \frac{9}{5} \tag{11.71}$$

Unfortunately, when measuring O over a sufficiently large number of repeated experiments, you obtain the average value $5/3$.

Knowing that the ensemble is actually mixed, we can easily see why you got this result.

$$\langle\hat{O}\rangle_M = \frac{1}{3}o_1 + \frac{2}{3}o_2 = \frac{1}{3} + \frac{4}{3} = \frac{5}{3} \tag{11.72}$$

If we were dealing with two possible states, $\{|\psi_1\rangle, |\psi_2\rangle\}$, then we can (roughly) say:

- It's a *superposition* if the system is in $|\psi_1\rangle \& |\psi_2\rangle$ *at the same time*
- It's a *mixed state* it the system could be in *either* $|\psi_1\rangle$ or $|\psi_2\rangle$, *but we don't know which*

As it stands, we have no mathematical way of describing a mixed state. We need the density matrix.

11.4 DENSITY MATRICES

We already know how to calculate the expectation value of an operator:

$$\langle\hat{O}\rangle = \int \psi^*(x)\hat{O}\psi(x)dx \tag{11.73}$$

but now consider the following expression, where we have added extra wave function terms of a different variable, $\underline{\psi(x')}$:

$$I = \iint \psi^*(x)\,\underline{\psi(x')\,\psi^*(x')}\hat{O}\psi(x)\underline{dx'}dx \tag{11.74}$$

the additions having been underlined as an aid to the eye.

If the wave function has been normalized correctly,[9] then the underlined terms integrate out as:

$$\underline{\int \psi(x')\psi^*(x')dx'} = 1 \tag{11.75}$$

So, adding them in has made no difference:

$$I = \iint \psi^*(x)\,\underline{\psi(x')\,\psi^*(x')}\hat{O}\psi(x)dx'dx = \int \psi^*(x)\hat{O}\psi(x)dx = \hat{O} \tag{11.76}$$

Admittedly, all we seem to be doing here is complicating things to no real effect, but now consider another sleight of hand, as a visual and conceptual aid. If we write $D(x,x') = \psi^*(x)\,\psi(x')$ and $\mathbb{O}(x',x) = \psi^*(x')\hat{O}\psi(x)$, our integral becomes:

$$I = \underline{\iint D(x,x')\mathbb{O}(x',x)dx'dx} \tag{11.77}$$

which *looks like a continuous version of the trace of a product matrix* (11.68)

$$\mathrm{Tr}[A\times B] = \sum_R\sum_j a_{Rj}b_{jR}$$

if we interpret the variables x, x' as playing the role of row and column indices, respectively:

$$I = \mathrm{Tr}[D\times\mathbb{O}] = \underline{\iint D(x,x')\mathbb{O}(x',x)dx'dx} \tag{11.78}$$

Following on, the continuous inner product must be:

$$[D\times\mathbb{O}](r,c) = \int D(r,x')\mathbb{O}(x',c)dx' \tag{11.79}$$

Here x' has been chosen as the dummy variable for consistency, but it's important to understand that it is a "free choice" as it is being integrated out. We could equally well have:

$$[D\times\mathbb{O}](r,c)=\int D(r,z)\mathbb{O}(z,c)\,dz \tag{11.80}$$

The combination $\mathbb{O}(x',x)=\psi^*(x')\hat{O}\psi(x)$ is called the *matrix element* of the operator $\hat{O}$ and our new construction:

$$D(x,x')=\psi^*(x)\,\psi(x') \tag{11.81}$$

is the *density matrix*.[10]

So, we have a new way to calculate expectation values:

$$\langle\hat{O}\rangle=\mathrm{Tr}[D\times\mathbb{O}] \tag{11.82}$$

Using the density matrix to calculate expectation values encapsulates within a single description all possible measurement scenarios, factoring in the influence of any lack of knowledge we may have in each circumstance. Put that way, it's clear that this mathematical object will be crucial in our quantum exploration of entropy.

11.4.1 A Discrete Density Matrix

The density matrix for a situation like our spin case, would be:

$$D=|\Psi\rangle\langle\Psi|=\begin{pmatrix}a\\b\end{pmatrix}\otimes\begin{pmatrix}a^* & b^*\end{pmatrix}=\begin{pmatrix}a\begin{pmatrix}a^* & b^*\end{pmatrix}\\b\begin{pmatrix}a^* & b^*\end{pmatrix}\end{pmatrix}=\begin{pmatrix}aa^* & ab^*\\ba^* & bb^*\end{pmatrix} \tag{11.83}$$

using the *outer or tensor product*[11], not the traditional (inner) product that we have employed thus far[12]:

{Note that the density matrix from the previous section:

$$D(x,x')=\psi^*(x)\,\psi(x') \tag{11.81}$$

was built with the continuous outer product.[13]}

Now we can try out our new version of the expectation value (11.82):

$$\langle\hat{O}\rangle=\mathrm{Tr}[D\times\mathrm{O}]$$

using our z-component spin operator as an example. Starting from:

$$D\times\hat{S}_Z=\begin{pmatrix}aa^* & ab^*\\ba^* & bb^*\end{pmatrix}\frac{\hbar}{2}\begin{pmatrix}1 & 0\\0 & -1\end{pmatrix}=\frac{\hbar}{2}\begin{pmatrix}aa^* & -ab^*\\ba^* & -bb^*\end{pmatrix} \tag{11.84}$$

Making the expectation value:

$$\left\langle \hat{S}_Z \right\rangle = \mathrm{Tr}\left[D \times \hat{S}_Z \right] = \mathrm{Tr}\left[\frac{\hbar}{2} \begin{pmatrix} aa^* & -ab^* \\ ba^* & -bb^* \end{pmatrix} \right] = \frac{\hbar}{2}\left(aa^* - bb^* \right) \tag{11.85}$$

confirming the result from before (11.38).

11.4.2 Density Matrices in Various Scenarios

Let's now take an ensemble of systems with each member in the *same* state $|\psi_i\rangle$, from a collection of $\mathcal{N}$ possible quantum states, $\{|\Psi_n\rangle\}$, $1 \le n \le \mathcal{N}$.

For discrete observables, states are column vectors:

$$\left|\psi_1\right\rangle = \begin{pmatrix} 1 \\ 0 \\ 0 \\ \cdots \\ 0 \\ 0 \end{pmatrix} \qquad \left|\psi_2\right\rangle = \begin{pmatrix} 0 \\ 1 \\ 0 \\ \cdots \\ 0 \\ 0 \end{pmatrix} \qquad \left|\psi_i\right\rangle = \begin{pmatrix} 0 \\ 0 \\ 0 \\ \cdots \\ 1 \\ 0 \end{pmatrix} \qquad \left|\psi_n\right\rangle = \begin{pmatrix} 0 \\ 0 \\ 0 \\ \cdots \\ 0 \\ 1 \end{pmatrix} \tag{11.86}$$

with their conjugate transpose row vector versions:

$$\left\langle \psi_1 \right| = \begin{pmatrix} 1 & 0 & 0 & \dots & 0 & 0 \end{pmatrix} \qquad \left\langle \psi_i \right| = \begin{pmatrix} 0 & 0 & 0 & \dots & 1 & 0 \end{pmatrix} \tag{11.87}$$

etc.

We can form density matrices for the states, along the lines:

$$D_1 = \begin{pmatrix} 1 \\ 0 \\ 0 \\ \cdots \\ 0 \\ 0 \end{pmatrix} \begin{pmatrix} 1 & 0 & 0 & \dots & 0 & 0 \end{pmatrix} = \begin{pmatrix} 1 & 0 & 0 & \dots & 0 & 0 \\ 0 & 0 & 0 & \dots & 0 & 0 \\ 0 & 0 & 0 & \dots & 0 & 0 \\ \cdots & \cdots & \cdots & \cdots & \cdots & \cdots \\ 0 & 0 & 0 & \dots & 0 & 0 \\ 0 & 0 & 0 & \dots & 0 & 0 \end{pmatrix} \tag{11.88}$$

$$D_2 = \begin{pmatrix} 0 \\ 1 \\ 0 \\ \cdots \\ 0 \\ 0 \end{pmatrix} \begin{pmatrix} 0 & 1 & 0 & \dots & 0 & 0 \end{pmatrix} = \begin{pmatrix} 0 & 0 & 0 & \dots & 0 & 0 \\ 0 & 1 & 0 & \dots & 0 & 0 \\ 0 & 0 & 0 & \dots & 0 & 0 \\ \cdots & \cdots & \cdots & \cdots & \cdots & \cdots \\ 0 & 0 & 0 & \dots & 0 & 0 \\ 0 & 0 & 0 & \dots & 0 & 0 \end{pmatrix} \tag{11.89}$$

Using component notation, for example with $|\psi_2\rangle$

$$[D_2]_{rc} = \delta_{2r}\delta_{2c} = 0,\ r \neq c$$

$$[D_2]_{rc} = \delta_{2r}\delta_{2c} = 1,\ r = c = 2 \tag{11.90}$$

meaning the only non-zero component of the matrix, is at row 2 column 2.

The density matrix for arbitrary $|\psi_i\rangle$ is then:

$$[D_i]_{rc} = \delta_{ir}\delta_{ic} \tag{11.91}$$

Specifically, the density matrix containing a single state $|\psi_i\rangle$ has a lone component, equal to one, located on the diagonal at row i. It will be an $\mathcal{N} \times \mathcal{N}$ matrix, and so can be traced.

This can be generalized to the continuous case; in which case we would write[14]:

$$D_i(r,c) = \psi_i^*(r)\psi_i(c)\delta(r-c) \tag{11.92}$$

Here $\delta(r-c)$ is the *Dirac delta function*,[15] which plays a similar role to the Kronecker delta that we introduced earlier.

11.4.3 Density Matrices of Superpositions

As we are well used to seeing, a superposition is a state of the form (in Dirac's notation for a change…):

$$|\Phi\rangle = \sum_i a_i |\psi_i\rangle \tag{11.93}$$

leading to a density matrix which look like:

$$D_\phi = \left(\sum_i a_i |\psi_i\rangle\right)\left(\sum_j a_j^* \langle\psi_j|\right) = \sum_i\sum_j a_j^* a_i |\psi_i\rangle\langle\psi_j| \tag{11.94}$$

Our example from earlier (11.83):

$$D = |\Psi\rangle\langle\Psi| = \begin{pmatrix} a \\ b \end{pmatrix} \otimes \begin{pmatrix} a^* & b^* \end{pmatrix} = \begin{pmatrix} aa^* & ab^* \\ ba^* & bb^* \end{pmatrix}$$

is precisely of this form. Note that the density matrix of a superposition has off-diagonal terms, which are known in the trade as *coherences*. The elements running down the diagonal are *populations*. If we write the amplitudes in the superposition as complex numbers with explicit phases:

$$a_n = |a_n| e^{i\phi_n}$$

then the population terms:

$$a_n^* a_n = |a_n| e^{-i\phi_n} |a_n| e^{i\phi_n} = |a_n|^2$$

will be phase independent, while the coherences:

$$a_n^* a_m = |a_n| e^{-i\phi_n} |a_m| e^{i\phi_m} = |a_n||a_m| e^{i(\phi_m - i\phi_n)}$$

give information about the relative phases of the different component states in the superposition. This is where the terms get their name. If these phases also contain a time dependence, as they might, given the nature of the time evolution operator, then the net phases in the coherences will shift with time. When we get to density matrices for situations that are not pure states, the coherences will generally be a sum of several complex number terms, and inevitably the phase relationships will be all over the place. For this reason, as we will see later, these situations are called *incoherent superpositions*. A state such as (11.93) is a *coherent superposition*.

One active area of study is the impact of environmental interactions on coherence terms. Generally, they can be shown to average out to zero (a process known as *decoherence*), so turn the density matrix into a summation over pure states, e.g.:

$$\begin{pmatrix} aa^* & ab^* \\ ba^* & bb^* \end{pmatrix} \overset{decoherence}{\rightarrow} \begin{pmatrix} aa^* & 0 \\ 0 & bb^* \end{pmatrix} \tag{11.95}$$

Arguably, the density matrix is a more flexible representation of a state than the wave function.

11.4.4 Properties of Pure State Density Matrices

The density matrices appropriate to pure states have some amusing properties, which we can verify in both continuous and element form.[16]

D1. They are *idempotent*, i.e., $D^2 = D$

In element form:

$$D^2 = \begin{pmatrix} aa^* & ab^* \\ ba^* & bb^* \end{pmatrix} \begin{pmatrix} aa^* & ab^* \\ ba^* & bb^* \end{pmatrix}$$

$$= \begin{pmatrix} aa^*aa^* + ab^*ba^* & aa^*ab^* + ab^*bb^* \\ ba^*aa^* + bb^*ba^* & ba^*ab^* + bb^*bb^* \end{pmatrix}$$

$$= \begin{pmatrix} aa^*\left(aa^* + b^*b\right) & ab^*\left(aa^* + bb^*\right) \\ ba^*\left(aa^* + bb^*\right) & bb^*\left(aa^* + b^*b\right) \end{pmatrix}$$

given the normalization, condition $aa^* + b^*b = 1$, we get:

$$D^2 = \begin{pmatrix} aa^* & ab^* \\ ba^* & bb^* \end{pmatrix} = D$$

or, more generally:

$$D^2 = \left(|\psi\rangle\langle\psi|\right)\left(|\psi\rangle\langle\psi|\right) = |\psi\rangle\langle\psi|\psi\rangle\langle\psi| = |\psi\rangle\langle\psi| = D$$

seeing as $\langle\psi|\psi\rangle = 1$.

For continuous situations:

$$D^2[r,c] = \int D(r,x')D(x',c)dx' = \int \psi^*(r)\,\psi(x')\,\psi^*(x')\,\psi(c)dx'$$

$$= \psi^*(r)\,\psi(c) = D[r,c]$$

D2. They are *Hermitian*, i.e., $D[r,c] = D^*[c,r]$

In element form:

$$(D^*)^T = \left(\begin{pmatrix} aa^* & ab^* \\ ba^* & bb^* \end{pmatrix}^*\right)^T = \left(\begin{pmatrix} a^*a & a^*b \\ b^*a & b^*b \end{pmatrix}\right)^T = \begin{pmatrix} a^*a & b^*a \\ a^*b & b^*b \end{pmatrix} = D$$

For continuous situations:

$$D^*[r,c] = \left(\psi^*(r)\right)^*\left(\psi(c)\right)^* = \psi(r)\,\psi^*(c) = \psi^*(c)\,\psi(r) = D[c,r]$$

It's interesting that being Hermitian makes the density matrix a candidate to represent an observable … more of which in the next section.

D3. They are *normalized*, i.e., $\mathrm{Tr}[D] = 1$

In element form:

$$\mathrm{Tr}[D] = \mathrm{Tr}\left[\begin{pmatrix} a^*a & b^*a \\ a^*b & b^*b \end{pmatrix}\right] = a^*a + b^*b = 1$$

For continuous situations:

$$D(r,c) = \psi^*(r)\,\psi(c)$$

hence:

$$\mathrm{Tr}[D] = \int \psi^*(r)\,\psi(r)dr = 1$$

Note that the combination of properties 1 and 3 implies that a pure state density matrix will have $\mathrm{Tr}\left[D^2\right]=1$. This turns out to be a defining characteristic allowing us to spot when we have a pure state in play.

11.4.5 The Density Matrix as an Observable

We know from Section 11.2.6 that an eigenstate of one observable may well be a coherent superposition in the basis formed by eigenstates of another observable. Furthermore, there is always a unitary transformation that takes us from one basis to another. In density matrix form, a superposition has off-diagonal elements, whereas an eigenstate does not. Hence, *we can always diagonalize a density matrix with the right transformation*. A diagonalized density matrix would look like:

$$D=\begin{pmatrix} p_1 & 0 & \dots & 0 \\ 0 & p_2 & \dots & 0 \\ \dots & \dots & \dots & \dots \\ 0 & 0 & 0 & p_{\mathbb{N}} \end{pmatrix} \tag{11.96}$$

It is tempting to see this as an example of a diagonalized operator from Section 11.2.4, with eigenvalues running down the diagonal. As the density matrix is Hermitian (D2), which is a necessary condition for an operator to represent an observable, it is worth exploring the notion that the density matrix is an observable of some form. In which case there would have to be a collection of eigenstates $\{|\chi_n\rangle\}$ and eigenvalues λ_n where $\lambda_n^2=p_n$: $D|\chi_n\rangle=\lambda_n|\chi_n\rangle$. If we have these eigenstates, then the density matrix can always be written as a sum over the separate density matrices built from these states:

$$D=\sum_n p_n\left|\chi_n\right\rangle\left\langle\chi_n\right| \tag{11.97}$$

As we have repeatedly emphasized, a coherent superposition takes the form:

$$\left|\Psi\right\rangle=\sum_i a_i\left|\psi_i\right\rangle \tag{11.98}$$

and according to the Born rule, the probability of $\left|\Psi\right\rangle\rightarrow\left|\psi_I\right\rangle$ is $\left|a_I\right|^2$. Another way of seeing this is to use the Dirac notation construction:

$$\left\langle\psi_I\middle|\Psi\right\rangle=\left\langle\psi_I\right|\left(\sum_i a_i\left|\psi_i\right\rangle\right)=\sum_i a_i\left\langle\psi_I\middle|\psi_i\right\rangle=\sum_i a_i\delta_{Ii}=a_I \tag{11.99}$$

making the probability $\left|\left\langle\psi_I\middle|\Psi\right\rangle\right|^2$. There is no new physics here, just a modified way of expressing the result. For a quick density matrix demonstration, consider the situation where:

$$\left|\Psi\right\rangle=\begin{pmatrix} a_1 \\ a_2 \end{pmatrix} \qquad \left|\psi_I\right\rangle=\begin{pmatrix} 1 \\ 0 \end{pmatrix} \tag{11.100}$$

with:

$$D_{\Psi} = |\Psi\rangle\langle\Psi| = \begin{pmatrix} a_1^* & a_2^* \end{pmatrix}\begin{pmatrix} a_1 \\ a_2 \end{pmatrix} = \begin{pmatrix} a_1^* a_1 & a_2^* a_1 \\ a_1^* a_2 & a_2^* a_2 \end{pmatrix}$$

$$D_{\psi_I} = |\psi_I\rangle\langle\psi_I| = \begin{pmatrix} 1 & 0 \end{pmatrix}\begin{pmatrix} 1 \\ 0 \end{pmatrix} = \begin{pmatrix} 1 & 0 \\ 0 & 0 \end{pmatrix} \tag{11.101}$$

If we take the product $D_{\Psi} \times D_{\psi_I}$, we get:

$$D_{\Psi} \times D_{\psi_I} = \begin{pmatrix} a_1^* a_1 & a_2^* a_1 \\ a_1^* a_2 & a_2^* a_2 \end{pmatrix}\begin{pmatrix} 1 & 0 \\ 0 & 0 \end{pmatrix} = \begin{pmatrix} a_1^* a_1 & 0 \\ a_1^* a_2 & 0 \end{pmatrix} \tag{11.102}$$

Making:

$$\mathrm{Tr}\left[D_{\Psi} \times D_{\psi_I} \right] = a_1^* a_1 = \left|\langle \psi_I | \Psi \rangle\right|^2 \tag{11.103}$$

This can be interpreted as *the expectation value of the operator D_{ψ_I}, yielding the probability of that state*. Consequently, we can argue that the eigenstates of the density matrix are the pure states represented within the matrix and that the eigenvalues are the probability of finding that state in the ensemble. The density matrix can then be thought to represent the observable action of "choosing" or "selecting" a system out of an ensemble. This line is followed up in Chapter 12 when we consider projection operators (Section 12.4.5)

11.5 THE MIXED STATE

We have now reached the point where we have sufficient tools at our disposal to circle back to the original question: how to represent a mixed state in a mathematical formalism.

To be specific, we envisage an ensemble of $\mathbb{N}$ systems, each of which has a collection of quantum states described by the set of wave functions $\{\psi_i\}$. For each individual system, we do not know the state it has selected, only that the probability of finding it in state $|\psi_i\rangle$ is p_i where:

$$\frac{\mathbb{N}_i}{\mathbb{N}} \xrightarrow{\mathbb{N}\to\infty} p_i \tag{11.104}$$

It follows that each separate quantum state has a density matrix $D_i = |\psi_i\rangle\langle\psi_i|$.

Our aim now is to construct a density matrix, $\mathcal{D}$, appropriate to the ensemble of systems distributed across the quantum states.

We can see that this is a worthy aim, as well as gaining a clue as to how to set about doing it, by considering a series of measurements made on the ensemble. A system is selected at random, and a measurement made of a chosen observable, receiving a result, o_i, according to:

$$\hat{O}|\psi_i\rangle = o_i|\psi_i\rangle \tag{11.105}$$

If we repeat the measurements $\mathcal{N}$ times, then the average value is:

$$\langle o_i \rangle = \sum_i \frac{\mathrm{N}_i}{\mathcal{N}} o_i \tag{11.106}$$

Eventually, of course, with a large enough ensemble and sufficient repeats, we can write:

$$\langle o_i \rangle = \langle \hat{o} \rangle = \sum_i p_i o_i \tag{11.107}$$

If we have a mixed ensemble density matrix, $\mathcal{D}$, then we could also try:

$$\langle \hat{o} \rangle = \mathrm{Tr}[\mathcal{D} \times \mathrm{O}] \tag{11.108}$$

So, the key is to find $\mathcal{D}$, so that these two calculations match. One possibility suggests itself:

$$\mathcal{D} = \sum_i p_i D_i = \sum_i p_i |\psi_i\rangle\langle\psi_i| \tag{11.109}$$

i.e., a sum over the separate density matrices, weighted by the probability of each state occurring in the ensemble. Alternatively, think of this as *the average of the pure state density matrices*. As this is an *incoherent sum*, we are not suggesting that any interference is taking place. The relative phases of the different probabilities are random, hence cannot be accessed experimentally. As noted earlier (Section 11.4.3), the coherence terms will be complex number sums with scrambled phases between them. This is categorically not a coherent superposition.[17]

Writing this matrix in terms of rows and columns is a little tricky, as we need to keep their symbols separate from the index over states. We already know how to write the individual density matrices:

$$[D_i]_{rc} = \delta_{ir}\delta_{ic} \tag{11.110}$$

which makes:

$$[\mathcal{D}]_{rc} = \sum_i p_i \delta_{ir}\delta_{ic} \tag{11.111}$$

Multiplying into the operator:

$$[\mathcal{D} \times \mathrm{O}]_{RC} = \sum_k \mathcal{D}_{RK}\mathrm{O}_{KC} = \sum_K \left(\sum_i p_i \delta_{iR}\delta_{iK} \right)\left(\sum_j o_j \delta_{Kj}\delta_{Cj} \right)$$

$$= \sum_K \sum_i \sum_j p_i \delta_{iR}\delta_{iK} o_j \delta_{Kj}\delta_{Cj} \tag{11.112}$$

Now we let the sums play out with the Kronecker deltas taking huge bites as they go:

$$[\mathcal{D}\times \mathrm{O}]_{RC} = \sum_K\sum_i p_i\delta_{iR}\delta_{iK}o_C\delta_{KC} = \sum_K p_K\delta_{KR}o_C\delta_{KC} \tag{11.113}$$

Finally we take the trace, by letting $R = C$:

$$\mathrm{Tr}[\mathcal{D}\times \mathrm{O}] = \sum_K p_K\delta_{KR}o_R\delta_{KR} = \sum_K p_K o_K \tag{11.114}$$

precisely as we predicted calculating directly from the ensemble. Consequently, and very importantly, *we can represent an ensemble of systems, when we do not know the specific quantum state for each system by the mixed state density matrix*:

$$\mathcal{D} = \sum_i p_i|\psi_i\rangle\langle\psi_i| \tag{11.115}$$

While this is known as a mixed state, a more appropriate term might be a *mixed ensemble*, echoing the idea that the density matrix is describing the ensemble, not the systems with in.

Although we will not deal explicitly with such situations, it is also worth noting that the collection of states $\{|\psi_n\rangle\}$ in the ensemble does not have to be a basis. If there is even less control or knowledge of how the ensemble systems are prepared, then it could be a collection of states $\{|\psi_1\rangle, |\varphi_5\rangle, |\chi_{12}\rangle\cdots\}$ that are unrelated to each other, and so not orthonormal. The density matrix would still effectively represent the statistical nature of the ensemble, provided we had the probability distribution across the systems.

11.5.1 Properties of Mixed State Density Matrices

First, we check to see if $\mathcal{D}$ is idempotent:

$$\mathcal{D}^2 = \sum_j\sum_k p_j p_k|\psi_j\rangle\langle\psi_j|\psi_k\rangle\langle\psi_k| \tag{11.116}$$

If the basis states are correctly set up, they will be normalized, $\langle\psi_k|\psi_k\rangle = 1$, and orthogonal[18] $\langle\psi_k|\psi_j\rangle = 0$, $k \neq j$. In other words, $\langle\psi_k|\psi_j\rangle = \delta_{kj}$. In which case, our sum contracts down to:

$$\mathcal{D}^2 = \sum_j p_j^2|\psi_j\rangle\langle\psi_j| \neq \mathcal{D} \tag{11.117}$$

It follows that, unlike the pure state, $\mathrm{Tr}[\mathcal{D}^2] \neq 1$, which is why we said earlier that it is a defining characteristic of purity. Of course, it remains the case that $\mathrm{Tr}[\mathcal{D}] = 1$, being a sum over the probabilities and we assuming that all possible states are represented in the ensemble.

Also, $\mathrm{Tr}[\mathcal{D}^2] \leq 1$, as $p_j^2 \leq p_j$. In some quarters, $\mathrm{Tr}[\mathcal{D}^2]$ is known as the *purity of the state*; the closer this being to 1, the more nearly pure the state, or the ensemble, is.

11.5.2 Where Do Mixed States Come From?

Broadly, there are three related situations where we may have to deal with a mixed state.

We have already alluded to one as a motivation for constructing the density matrix. Specifically, the case where we have an ensemble of identical systems each of which can be found in one of the states $\{|\psi_n\rangle\}$ but ahead of any test on the system, we have no further information.

Consider the light being emitted from an incandescent bulb. Without diverting into too much detail, photons of light can exist in two polarization states: $|R\rangle$-*right circular polarization* or $|L\rangle$-*left circular polarization.* Coherent superpositions of such states, e.g.:

$$|\Phi\rangle = \alpha|R\rangle + \beta|L\rangle \qquad \alpha^2 + \beta^2 = 1 \tag{11.118}$$

are also possible. The specific superpositions:

$$|V\rangle = \frac{1}{\sqrt{2}}\left(|R\rangle + |L\rangle\right) \qquad |H\rangle = \frac{1}{\sqrt{2}}\left(|R\rangle - |L\rangle\right) \tag{11.119}$$

physically correspond to *vertical linear polarization* and *horizontal linear polarization* respectively. Passing a collection of photons in state $|V\rangle$ through a polarizing filter which only allows right circular photons to pass will cut the beam intensity in half. From a measurement theory perspective, the filter collapses the superposition randomly with 50% probability, $\left(1/\sqrt{2}\right)^2$, so that after the filter a photon is in state $|R\rangle$ with certainty (those that collapse into $|L\rangle$ being absorbed). However, we must not conclude that the photons started with 50% in state $|R\rangle$ and 50% in state $|L\rangle$. Passing them through a vertical linear filter would hardly reduce the intensity,[19] whereas using a horizontal filter would cut the beam out entirely. All of which helps to emphasize, once again, how the coherent superposition is a different kind of state to those found in classical physics. However, the unpolarized light from the filament of our bulb is a different situation. This is an ensemble of photons, 50% being in $|R\rangle$ and 50% in $|L\rangle$. The physics producing the photons does not specify the polarization. This is a mixed state. The best we can do is the density matrix:

$$D = 0.5\begin{pmatrix} 1 & 0 \\ 0 & 0 \end{pmatrix} + 0.5\begin{pmatrix} 0 & 0 \\ 0 & 1 \end{pmatrix} = \begin{pmatrix} 0.5 & 0 \\ 0 & 0.5 \end{pmatrix} \tag{11.120}$$

Another way in which preparation can lead to a mixed state is via the process of measurement. Say we start with an ensemble of systems in a pure state $|\psi_i\rangle$ with an observable O. Next, the systems are all subjected to a measurement of a different observable O'. From this perspective, each system starts in a coherent superposition of the eigenstates of $\widehat{O}'$, $\{|\phi_k\rangle\}$:

$$|\psi_i\rangle = \sum_k a_k |\phi_k\rangle \tag{11.121}$$

After the measurement, we have a new ensemble with the same number of systems as before, but now each one is in one of the $\{|\phi_k\rangle\}$ with probability a_k^2. The team carrying out the measurements would know, from the results, which system was in which state, but if we pass the ensemble on to another team, without telling them the identity of each system, the best they could do would be the density matrix:

$$[D]_{rc} = \sum_k a_k^2 \delta_{rk}\delta_{ck} \tag{11.122}$$

From their perspective, this is a mixed state. Interestingly, in the photon case the probabilities along the diagonal of the density matrix are "classical type" expressing our ignorance at the system level. In the measurement ensemble we have just discussed, it's a matter of perspective. The first experimental team knows that the probabilities are quantum and determined by the relevant amplitudes. The second team lacks this information, so for them the probabilities are "classical".

Our third case brings us to *entanglement.*

As with coherent superpositions, entangled states are a quantifiably new aspect of reality that we did not encounter until we ventured into the quantum realm. Formally, two systems are entangled if the wave function needed to describe both systems at the same time cannot be factored into a product of separate functions:

$$\Psi(x_1, x_2, t) \neq \psi(x_1, t)\varphi(x_2, t) \tag{11.123}$$

From a quantum perspective, the two systems have lost a degree of autonomy: what happens to one now has an influence on the other. In a typical experiment, some decay process results in two photons one of which is in state $|R\rangle$ and the other $|L\rangle$. Typically, the combined quantum state looks like:

$$|\Phi\rangle = \frac{1}{\sqrt{2}}\left(|R\rangle|L\rangle - |L\rangle|R\rangle\right) \tag{11.124}$$

Here the order of states is crucial. The first term has photon 1 in state $|R\rangle$ and photon 2 in state $|L\rangle$. The second term has the attribution reversed. Note that $|\Phi\rangle$ cannot be factorized into a product where everything to do with photon 1 is separate from the information regarding photon 2. Note also that the total spin of the combined state is zero.

Characteristically, the two photons are travelling colinearly but in opposite directions. Photon 2, moving to the right, now reaches an experimental station, conventionally referred to by its operator's name, *Alice*. Here a measurement of polarization is made, revealing photon 2 to be either $|L\rangle$ or $|R\rangle$ with 50:50 probability. This collapses the state appropriately. However, as the state is not factored into separate components for the two photons, this has an immediate impact on photon 1. It must now be in the opposite state, i.e., $|R\rangle$ or $|L\rangle$ with 50:50 probability. As yet, this photon has not reached *Bob*, the operator responsible for measurements on the left. Experiments testing this arrange for Bob to be so far from Alice, information regarding Alice's results, travelling no faster than the speed of light, would not have had a chance to reach Bob before he measures photon 2. As far as Bob is concerned the photons are coming to him in a mixed state, even though they were prepared in a pure state. In planning for his results, Bob sets up an appropriate density matrix:

$$D_{\text{Bob}} = \begin{pmatrix} 1/2 & 0 \\ 0 & 1/2 \end{pmatrix} \tag{11.125}$$

It now becomes a case of seeing how this density matrix can be extracted from the full information as possessed by Alice.

11.5.3 Composite Systems and the Partial Trace

Sometimes we wish to consider an overall system that can be cleanly and clearly partitioned into two (or more) sub-systems. Let's say that system 1 has states $\{|\psi_n\rangle\}$ with system 2 possessing states $\{|\varphi_k\rangle\}$. We also assume that the overall state can be characterized by product states, such as:

$$|\Phi\rangle = |\psi_n\rangle|\varphi_k\rangle \tag{11.126}$$

as the two sub-systems are not entangled.

To construct a density matrix for this scenario we must employ the tensor/outer product. Given two matrices:

$$A = \begin{pmatrix} a_{11} & a_{12} \\ a_{21} & a_{22} \end{pmatrix} \qquad B = \begin{pmatrix} b_{11} & b_{12} \\ b_{21} & b_{22} \end{pmatrix} \tag{11.127}$$

their tensor product is:

$$A \otimes B = \begin{pmatrix} a_{11}\begin{pmatrix} b_{11} & b_{12} \\ b_{21} & b_{22} \end{pmatrix} & a_{12}\begin{pmatrix} b_{11} & b_{12} \\ b_{21} & b_{22} \end{pmatrix} \\ a_{21}\begin{pmatrix} b_{11} & b_{12} \\ b_{21} & b_{22} \end{pmatrix} & a_{22}\begin{pmatrix} b_{11} & b_{12} \\ b_{21} & b_{22} \end{pmatrix} \end{pmatrix} = \begin{pmatrix} a_{11}b_{11} & a_{11}b_{12} & a_{12}b_{11} & a_{12}b_{12} \\ a_{11}b_{21} & a_{11}b_{22} & a_{12}b_{21} & a_{12}b_{22} \\ a_{21}b_{11} & a_{21}b_{12} & a_{22}b_{11} & a_{22}b_{12} \\ a_{21}b_{21} & a_{21}b_{22} & a_{22}b_{21} & a_{22}b_{22} \end{pmatrix} \tag{11.128}$$

This is quite tricky to write in pure component terms, which probably explains why it is so rarely done:

$$[A \otimes B]_{p(r-1)+v,q(s-1)+w} = a_{rs}b_{vw} \tag{11.129}$$

expressing the size of matrix B as $p \times q$, or 2×2 in this case. Note that we must introduce a comma between the row and column indexes in the product to prevent ambiguity.

So, the term $a_{22}b_{12}$, for example, is found at $[A \otimes B]_{2(2-1)+1,2(2-1)+2} = [A \otimes B]_{3,4}$

For our factorizable composite system:

$$D_{12} = D_1 \otimes D_2 \tag{11.130}$$

It can be shown quite readily that:

T7

$$\mathrm{Tr}[D_{12}] = \mathrm{Tr}[D_1 \otimes D_2] = \mathrm{Tr}[D_1]\mathrm{Tr}[D_2] \tag{11.131}$$

which is the extra trace property alluded to in Section 11.2.8.

Using 11.122 with $D_1 = A, D_2 = B$, we see:

$$\mathrm{Tr}[D_1]\mathrm{Tr}[D_2] = (a_{11} + a_{22})(b_{11} + b_{22}) = a_{11}b_{11} + a_{11}b_{22} + a_{22}b_{11} + a_{22}b_{22} = \mathrm{Tr}[D_{12}]$$

Now consider a situation similar to Bob's. Someone has experimented with system 2, but neglects to pass that information on to Bob. As the two systems are not entangled, in this case, this is of little consequence—the results that Bob finds will be independent of whatever happened to system 1 when Alice experimented on it. However, it does help us develop our techniques for the more critical entanglement case.

In essence, we wish to reduce the full information contained in D_{12} by wiping out anything to do with system 2. We can do this by tracing over D_2, as follows:

$$\mathrm{Tr}_2\left[D_{12}\right] = \mathrm{Tr}_2\left[\begin{pmatrix} a_{11}\begin{pmatrix} b_{11} & b_{12} \\ b_{21} & b_{22} \end{pmatrix} & a_{12}\begin{pmatrix} b_{11} & b_{12} \\ b_{21} & b_{22} \end{pmatrix} \\ a_{21}\begin{pmatrix} b_{11} & b_{12} \\ b_{21} & b_{22} \end{pmatrix} & a_{22}\begin{pmatrix} b_{11} & b_{12} \\ b_{21} & b_{22} \end{pmatrix} \end{pmatrix}\right]$$

$$= \begin{pmatrix} a_{11}\mathrm{Tr}\begin{pmatrix} b_{11} & b_{12} \\ b_{21} & b_{22} \end{pmatrix} & a_{12}\mathrm{Tr}\begin{pmatrix} b_{11} & b_{12} \\ b_{21} & b_{22} \end{pmatrix} \\ a_{21}\mathrm{Tr}\begin{pmatrix} b_{11} & b_{12} \\ b_{21} & b_{22} \end{pmatrix} & a_{22}\mathrm{Tr}\begin{pmatrix} b_{11} & b_{12} \\ b_{21} & b_{22} \end{pmatrix} \end{pmatrix}$$

$$= \begin{pmatrix} a_{11}\left(b_{11}+b_{22}\right) & a_{12}\left(b_{11}+b_{22}\right) \\ a_{21}\left(b_{11}+b_{22}\right) & a_{22}\left(b_{11}+b_{22}\right) \end{pmatrix}$$

$$= \left(b_{11}+b_{22}\right)\begin{pmatrix} a_{11} & a_{12} \\ a_{21} & a_{22} \end{pmatrix} = D_1 \tag{11.132}$$

Here we have introduced the notation $\mathrm{Tr}_2[D_{12}]$ to indicate the *partial trace* of the matrix removing factors to do with system 2. The last step follows as the trace of a density matrix, in this case $(b_{11}+b_{22})$, is always equal to 1. In this simple example, the partial trace has produced an appropriate *reduced matrix*, which is D_1.

As the trace is a sum over diagonal terms, we can write it in Dirac form as:

$$\mathrm{Tr}\left[D\right] = \sum_k \langle k|D|k\rangle \tag{11.133}$$

Then the partial trace is:

$$\mathrm{Tr}_2\left[D_{12}\right] = \sum_k \langle k|_2\, D_{12}\,|k\rangle_2 \tag{11.134}$$

or equally:

$$\mathrm{Tr}_1\left[D_{12}\right] = \sum_k \langle k|_1 D_{12}\,|k\rangle_1 \tag{11.135}$$

where we have explicitly indicated which states apply to which systems. It also follows that:

PT1
$$\mathrm{Tr}\left[D_{12}\right] = \mathrm{Tr}_2\left[\mathrm{Tr}_1\left[D_{12}\right]\right] = \mathrm{Tr}_1\left[\mathrm{Tr}_2\left[D_{12}\right]\right] \tag{11.136}$$

Now we need to see if the same technique will work for our entangled systems, where the density matrix cannot be factored, as neither can the states.

11.5.4 Partial Trace Over Entangled States

The density matrix for our entangled state (11.124) is built in the normal way:

$$D_{12} = |\Phi\rangle\langle\Phi| = \frac{1}{2}\left(|R\rangle_1 |L\rangle_2 - |L\rangle_1 |R\rangle_2\right)\left(\langle R|_1 \langle L|_2 - \langle L|_1 \langle R|_2\right)$$

$$= \frac{1}{2}\left(|R\rangle_1 |L\rangle_2 \langle R|_1 \langle L|_2 - |R\rangle_1 |L\rangle_2 \langle L|_1 \langle R|_2 - |L\rangle_1 |R\rangle_2 \langle R|_1 \langle L|_2 + |L\rangle_1 |R\rangle_2 \langle L|_1 \langle R|_2\right) \tag{11.137}$$

In each case, we are taking the outer product of vectors. So, for example, if we set things up as:

$$|R\rangle_1 |R\rangle_2 = \begin{pmatrix} 1 \\ 0 \\ 0 \\ 0 \end{pmatrix} \quad |R\rangle_1 |L\rangle_2 = \begin{pmatrix} 0 \\ 1 \\ 0 \\ 0 \end{pmatrix} \quad |L\rangle_1 |R\rangle_2 = \begin{pmatrix} 0 \\ 0 \\ 1 \\ 0 \end{pmatrix} \quad |L\rangle_1 |L\rangle_2 = \begin{pmatrix} 0 \\ 0 \\ 0 \\ 1 \end{pmatrix} \tag{11.138}$$

then:

$$|R\rangle_1 |L\rangle_2 \langle R|_1 \langle L|_2 = \begin{pmatrix} 0 \\ 1 \\ 0 \\ 0 \end{pmatrix}\begin{pmatrix} 0 & 1 & 0 & 0 \end{pmatrix} = \begin{pmatrix} 0 & 0 & 0 & 0 \\ 0 & 1 & 0 & 0 \\ 0 & 0 & 0 & 0 \\ 0 & 0 & 0 & 0 \end{pmatrix} \tag{11.139}$$

Working on from there, we end up with:

$$D_{12} = \frac{1}{2}\begin{pmatrix} 0 & 0 & 0 & 0 \\ 0 & 1 & -1 & 0 \\ 0 & -1 & 1 & 0 \\ 0 & 0 & 0 & 0 \end{pmatrix} \tag{11.140}$$

{You may wish to convince yourself that this is a pure state by calculating $\mathrm{Tr}\left[D^2\right]\ldots$}

To convert this into the density matrix as seen by Bob, we need to wipe out, erase or "forget" the information related to Alice and hence photon 2. Hence, we need to take the partial trace over photon 2's states. It is illuminating to see how this calculation is done in practice, but as it is rather fiddly and involves a lot of matrix manipulation, we have placed the full details in Appendix PA1.6. The result is as desired:

$$D_1 = \mathrm{Tr}_2\left[D_{12}\right] = \mathrm{Tr}_2\left[\frac{1}{2}\begin{pmatrix} 0 & 0 & 0 & 0 \\ 0 & 1 & -1 & 0 \\ 0 & -1 & 1 & 0 \\ 0 & 0 & 0 & 0 \end{pmatrix}\right] = \begin{pmatrix} 1/2 & 0 \\ 0 & 1/2 \end{pmatrix} \tag{11.141}$$

Note that this density matrix represents a mixed state, which we can confirm as $D_1^2 \neq D_1$. In essence, we have asked how subsystem 1 looks if we "forget" that subsystem 2 exists. As a result, *we generate a mixed state for 1*. Given that we started in a pure state, it means that *some information about 1 got lost in the process of forgetting 2*. This is characteristic of an entanglement: *the individual autonomy of the sub-systems has been lost.*

11.5.5 Partial Trace, Eigenstates, and Eigenvalues

With a composite system, D_{12}, that cannot be factored into separate states, the partial traces effectively define density matrices for the separate systems:

$$D_1 = \mathrm{Tr}_2\left[D_{12}\right] \qquad D_2 = \mathrm{Tr}_1\left[D_{12}\right] \tag{11.142}$$

Surprisingly, and very significantly, these density matrices turn out to have *the same eigenvalues*, provided the composite system is in a pure state.

We will demonstrate this using continuous variables, for a change...

Starting with a composite system in the pure state $D_{12} = \Psi^*(x_1, x_2)\Psi(x_1', x_2')$, we can form the density matrices for the separate systems by wiping information in the partial trace:

$$D_1(r,c) = \int \Psi^*(r,\beta)\Psi(c,\beta)\,d\beta \tag{11.143}$$

$$D_2(R,C) = \int \Psi^*(\alpha,R)\Psi(\alpha,C)\,d\alpha \tag{11.144}$$

Parking D_2 for the moment, we define an eigenstate of D_1 to be $\phi(x_1)$ with eigenvalue λ. This definition is crystalized by the relationship:

$$\left[D_1\phi\right]_r = \int D_1(r,c)\phi(c)\,dc = \lambda\phi(r) \tag{11.145}$$

which expands out to be:

$$\iint \Psi^*(r,\beta)\Psi(c,\beta)\phi(c)\,d\beta dc = \lambda\phi(r) \tag{11.146}$$

As the density matrix is Hermitian, it is also true that:

$$\iint \Psi^*(c,\beta)\Psi(r,\beta)\phi^*(c)\,d\beta dc = \lambda\phi^*(r) \tag{11.147}$$

with the eigenvalue being real.

Now, by enlightened guesswork, we propose an eigenstate for D_2, $\chi(x_2)$:

$$\chi(C) = \int \Psi^*(\alpha',C)\phi^*(\alpha')\,d\alpha' \tag{11.148}$$

To check that this is indeed all that we hope for, we act on the state with D_2:

$$\left[D_2\phi\right]_R = \int D_2(R,C)\chi(C)dC = \iint \Psi^*(\alpha,R)\Psi(\alpha,C)\chi(C)d\alpha dC$$

$$= \iiint \Psi^*(\alpha,R)\underline{\Psi(\alpha,C)\Psi^*(\alpha',C)}\phi^*(\alpha')\underline{d\alpha'}d\alpha\underline{dC} \tag{11.149}$$

If we take note of the underlined terms, we see that they are exactly of the form shown in (11.147), hence this nasty looking triple integral collapses down to be:

$$\left[D_2\phi\right]_R = \lambda\int \Psi^*(\alpha,R)\phi^*(\alpha)d\alpha = \lambda\chi(R) \tag{11.150}$$

demonstrating that this density matrix has the same eigenvalue. The use of Ψ^* in the constructed eigenstate ensures that this is only true if the original bi-partite system is in a pure state. It is worth noting that $\text{Tr}\left[D_1\right] = \text{Tr}\left[D_2\right]$, which follows immediately from their having the same eigenvalues. This will play out quite significantly Section 12.3.7.

11.5.6 Partially Mixed States

Say, for example, we know that 75% of the members of an ensemble of electrons are in state $\left|+\tfrac{1}{2}\hbar\right\rangle$ and that the other 25% are in the coherent superposition $\left|\Psi\right\rangle = \tfrac{1}{\sqrt{2}}\left\{\left|+\tfrac{1}{2}\hbar\right\rangle - \left|-\tfrac{1}{2}\hbar\right\rangle\right\}$. In which case, interference effects within the ensemble would be reduced, due to the randomized phasing between the two sub-ensembles, but not eliminated as they would be present within the sub-ensemble in the superposition. This would be a *partially mixed state*.

11.5.7 A Spotter's Guide to States

As a means of summarizing, it is worth considering some simple examples of the type of state/ensemble that we have come across in this chapter. You might wish to consider for yourself how to categorize the following density matrices, before reading on:

$$\underset{\text{A}}{\begin{pmatrix} 1/2 & 1/2 \\ 1/2 & 1/2 \end{pmatrix}} \qquad \underset{\text{B}}{\begin{pmatrix} 3/4 & 0 \\ 0 & 1/4 \end{pmatrix}} \qquad \underset{\text{C}}{\begin{pmatrix} 1 & 0 \\ 0 & 0 \end{pmatrix}} \qquad \underset{\text{D}}{\begin{pmatrix} 7/8 & -1/8 \\ -1/8 & 1/8 \end{pmatrix}}$$

A: This is a *pure state*. The simplest way to tell is that $D^2 = D$. The presence of coherence terms (off diagonals) tells us that the density matrix contains superpositions

B: This is a *mixed state*. There are no coherence terms, the matrix is in its diagonal representation, so the states are not superpositions. However, $D^2 \neq D$

C: This is a *pure state*. Clearly $D^2 = D$, but also there is a single eigenstate of the density matrix present in the ensemble

D: This is a *partially mixed state*. We can tell it is mixed as $D^2 \neq D$. The presence of coherence terms tells us that there is a superposition involved as well. In fact, this is the state mentioned in the previous section

11.6 THERMAL STATES

Barring any quantum issues to do with identical particles, we expect an ensemble of systems in contact with a thermal reservoir to follow Maxwell–Boltzmann statistics. We would then have access to (5.48):

$$F = -k_B T \ln(Z)$$

from which to extract anything of thermodynamic interest.

From our new perspective, this ensemble will be in a mixed state, described by an appropriate density matrix. Furthermore, the probabilities at work will be given by the Boltzmann factor.

At first glance, we would construct a density matrix on the following lines:

$$D' = \sum_j e^{-\hat{H}/k_B T} |j\rangle\langle j| \tag{11.151}$$

where $\widehat{H}$ is the Hamiltonian operator for the systems in the ensemble. However, this is not quite right as we need to ensure that the probability distribution is normalized; a role generally carried out by the partition function. In this case, an appropriate factor is:

$$Z = \mathrm{Tr}\left[\sum_j e^{-\hat{H}/k_B T} |j\rangle\langle j|\right] \tag{11.152}$$

making the density matrix:

$$D_T = \frac{\sum_j e^{-\hat{H}/k_B T} |j\rangle\langle j|}{\mathrm{Tr}\left[\sum_j e^{-\hat{H}/k_B T} |j\rangle\langle j|\right]} \tag{11.153}$$

If we knew all the energy eigenstates, $\{|\omega_n\rangle\}$ of the system, then our expansion would run over those, and the density matrix would become:

$$D = \frac{\sum_n e^{-\varepsilon_n/k_B T} |\omega_n\rangle\langle \omega_n|}{\mathrm{Tr}\left[\sum_n e^{-\varepsilon_n/k_B T} |\omega_n\rangle\langle \omega_n|\right]} \tag{11.154}$$

but in practice, that is a tall order.

NOTES

1 We have already encountered the unitary time evolution operator, which advances states through time, but one of the crucial issues in quantum theory is that time is *not* an observable.
2 To within experimental error of course.
3 Insert own Monty Python Spanish Inquisition joke at this point (JA). No, please don't (SJH).
4 At least, in the position representation it is. Also, there are issues when it comes to defining the eigenstates…
5 i.e., a continuous functional expression like 6.20.
6 Paul Adrien Maurice Dirac, 1902–1984, Nobel Prize in Physics, 1933, for the formulation of a relativistic quantum mechanical wave equation.
7 Clearly, with a square matrix there are two diagonals. By convention, we are always referring to the diagonal that runs top left of the square array to bottom right. That is *the* diagonal.
8 We don't generally have sufficient control of our preparation processes to pin down a single state all the time.
9 A matter of professional pride and courtesy…
10 The name derives from the role that this object takes in quantum theory, which is like that of a probability density in classical statistics. It is absolutely nothing to do with mass/volume! To emphasise that, the symbol *D* is being used rather than the more conventional ρ.
11 Mathematically, there are subtle differences between outer and tensor products, but in our contexts they amount to the same result.
12 Note that the order of terms here is different to what we had before. When we first defined the density matrix for the continuous case, we wrote $D(x,x') = \psi^*(x)\psi(x')$ with the conjugated wave function first. In the matrix form, we have $D = |\psi\rangle\langle\psi|$, which puts the conjugated version last. However, in the matrix version, due to the rules of matrix multiplication, we must take the conjugate transpose, to make sure we get a row vector. The calculations all come out the same.
13 Strictly of course a state and its conjugate transpose, in vector terms.
14 In this expression, r and c are different values of the same continuous observable. This terminology is being used to show how the continuous case relates to rows and columns in discrete matrices.
15 In a strict mathematical sense, it is not a function, but the name has stuck…
16 We will shortly discover that some of these properties are violated for density matrices in the mixed state scenario.
17 For one thing, the outer product of the state and its conjugate prevents that…
18 The orthogonality condition is interesting, as it means that one state can't make a transition into another. This is precisely the issue we engaged with earlier discussing the evolution of the Gibbs entropy and the Fermi master equation.
19 Filters are basically lumps of glass, so there is always some absorption in the material.

12 Quantum Entropy

From a classical perspective, a quantum state is a surprising mixture of information and ignorance. If a system is sitting in an eigenstate of an observable, we can predict with confidence that a measurement result will yield the appropriate eigenvalue. With a superposition, certainty is not granted to us. Unlike the closest possible classical case, where a superposition represents a genuine blending of classical states, the quantum superposition is a strange overlapping of states simultaneously occupied by the system. On measurement, this ontologically curious animal collapses into one of the overlapping states, a process that is not well understood to this day. The collapse is random, a-causal, but governed by quantum amplitudes. Heading into a measurement, we do not know what we are going to get out. Given the evident link between entropy and information, the probabilities and uncertainties baked into a quantum state may well be a fertile ground for another form of entropy. The density matrix gives us a flexible tool for studying the properties of quantum ensembles of any kind. In this chapter, we will see how von Neumann helped develop and use the density matrix to study the problem of measurement and the information/entropy content of quantum states.

12.1 THE PROBLEM OF MEASUREMENT

> We have reached a point at which it becomes necessary to resort to the thermodynamic method of analysis, because it alone makes it possible for us to come to a correct understanding of the difference between 1 [state collapse] and 2 [unitary time evolution], a distinction into which reversibility questions obviously enter.
>
> *J von Neumann*[1] [our additions]

The origin of the density matrix dates to independent work by two important and highly talented physicists. Lev Landau[2] was motivated to overcome the shortcomings of the wave function when it came to representing quantum systems in all cases. John von Neumann[3] was seeking to develop the theory of quantum statistical mechanics, which he believed would illuminate the measurement problem.[4] As we have seen, in classical statistical thermodynamics, knowing the probability distribution and the partition function gets us a long way and essentially allows everything of interest to be calculated. The density matrix plays a similar role in quantum statistical mechanics.

In Chapter 10, we distinguished between the unitary time evolution of a state, something von Neumann termed process **2**, and the non-unitary state collapse, process **1**. In his groundbreaking book[5] *Mathematical Foundations of Quantum Mechanics*, von Neumann carried out a detailed analysis of measurement and kick-started quantum statistical mechanics. To see how he arrived at a new gauge of entropy, we need to start with a further discussion of processes **1** and **2**.

DOI: 10.1201/9781003121053-15

12.1.1 Time Evolution of the Density Matrix

From Chapter 10, we know that von Neumann's process **2** is governed by the time evolution operator (10.28):

$$\widehat{U}(\Delta t) = e^{-\frac{i\widehat{H}\Delta t}{\hbar}}$$

so that:

$$\psi(x,t+\Delta t) = \widehat{U}(\Delta t)\,\psi(x,t) \qquad \left|\psi(x,t+\Delta t)\right\rangle = \widehat{U}(\Delta t)\left|\psi(x,t)\right\rangle \tag{12.1}$$

Equally:

$$\psi^*(x,t+\Delta t) = \widehat{U}^*(\Delta t)\,\psi^*(x,t) \qquad \left\langle\psi(x,t+\Delta t)\right| = \left\langle\psi(x,t)\right|\widehat{U}^\dagger(\Delta t) \tag{12.2}$$

Evidently, for a pure state density matrix $D = |\psi\rangle\langle\psi|$:

$$\begin{aligned} D(t+\Delta t) &= \left|\psi(x,t+\Delta t)\right\rangle\left\langle\psi(x,t+\Delta t)\right| \\ &= \widehat{U}(\Delta t)\left|\psi(x,t)\right\rangle\left\langle\psi(x,t)\right|\widehat{U}^\dagger(\Delta t) \\ &= \widehat{U}(\Delta t)D(t)\widehat{U}^\dagger(\Delta t) \end{aligned} \tag{12.3}$$

There are some key points regarding process **2** that we can conclude from this:

- It is completely deterministic or "casual"
- It is reversible in a thermodynamic sense as well as practically. We can see this from the following argument

$$\begin{aligned} \widehat{U}(-\Delta t)D(t+\Delta t)\widehat{U}^\dagger(-\Delta t) &= \widehat{U}^\dagger(\Delta t)D(t+\Delta t)\widehat{U}(\Delta t) \\ &= \widehat{U}^\dagger(\Delta t)\widehat{U}(\Delta t)D(t)\widehat{U}^\dagger(\Delta t)\widehat{U}(\Delta t) = D(t) \end{aligned} \tag{12.4}$$

as[6] $\widehat{U}^\dagger = \widehat{U}^{-1}$.

- Process **2** turns pure states into pure states and mixed states into mixed states

This latter point is especially important in contrast to process **1**. We saw in section 11.5.2 that the collapse of a state, or process **1** in von Neumann's terminology, turns an ensemble of pure states into a mixed state. This fundamentally alters the information that we have available to us. He makes the comment[7]: "the development of a state according to **1** is *statistical*, while according to **2** it is *causal*" [our emphasis].

Process **1** is irreversible. No matter how many times **1** happens, it can't turn a mixed state back into a pure state. We can do that by subtracting out of the ensemble systems in a given state after the measurement has taken place, but that is an active intervention on our part. We can also convert mixed states into pure states by applying some unitary transformation to the systems. For a practical example, consider an electron. As they are spin ½ particles, electrons can be in one of two states for their z-axis spin components,

$$\left|+\frac{1}{2}\hbar\right\rangle = \begin{pmatrix}1\\0\end{pmatrix} \qquad \left|-\frac{1}{2}\hbar\right\rangle = \begin{pmatrix}0\\1\end{pmatrix} \tag{12.5}$$

For our initial state, $|\Psi\rangle$, the electron could be "spin up" along some other axis at a selected angle. Restoring them to this angle could then be achieved via some magnetic steering. Again, however, these are active steps that we must take.

For von Neumann, the irreversibility of **1** was an essential clue and an indication that entropy must be a factor in understanding what was happening during state collapse.

12.2 CYCLING A QUANTUM IDEAL GAS

As a means of developing his thinking, von Neumann constructed an interesting thought experiment, presented here in a slightly simplified form.

The starting point is a collection of $\mathcal{N}$ particles, each of which is in the superposition, $|\Psi\rangle = \alpha|\varphi_1\rangle + \beta|\varphi_2\rangle$ where $\{|\varphi_1\rangle, |\varphi_2\rangle\}$ are the single-particle eigenstates of an observable O. To isolate them from each other and any external influence, each particle, i, is placed in an impenetrable box, K_i. The boxes are small but large enough that their positions and momenta can be treated classically. Importantly, the state of the particle in the box has no effect on the thermodynamics of the box itself.

Each box is placed in a much larger box K which has volume V and is coupled to a thermal reservoir of temperature T. In this way, the boxes can be treated as a classical ideal gas, with each box playing the role of a "molecule".

These rather arcane steps ensure that we are dealing with an ensemble of non-interacting systems which are identical other than their internal quantum state. The ensemble is described by a density matrix, $\mathbb{D} = |\Psi\rangle\rangle\langle\Psi|$.

von Neumann describes the setup in the following terms[8]:

> In the terminology of classical statistical mechanics, we are dealing with a Gibbs ensemble; i.e., the application of statistics and thermodynamics will be made not on the (interacting) components of a single, very complicated mechanical system with many (only imperfectly known) degrees of freedom – but on an ensemble of very many (identical) mechanical systems, each of which may have an arbitrarily large number of degrees of freedom, and each of which is entirely separated from the others, and does not interact with any of them.
>
> Quote 12.1

The experiment now proceeds through a series of steps.

Step 0: Each box is opened in turn, and a measurement made of O. This causes state collapse (Process **1**) for each particle into either $|\varphi_1\rangle$, with probability α^2, or $|\varphi_2\rangle$, with probability β^2. We now have a mixed-state ensemble inside K with $\mathcal{N}_1 = \alpha^2\mathcal{N} = p_1\mathcal{N}$, $\mathcal{N}_2 = \beta^2\mathcal{N} = p_2\mathcal{N}$, assuming a sufficiently large number of particles/boxes (Figure 12.1). The density matrix is now:

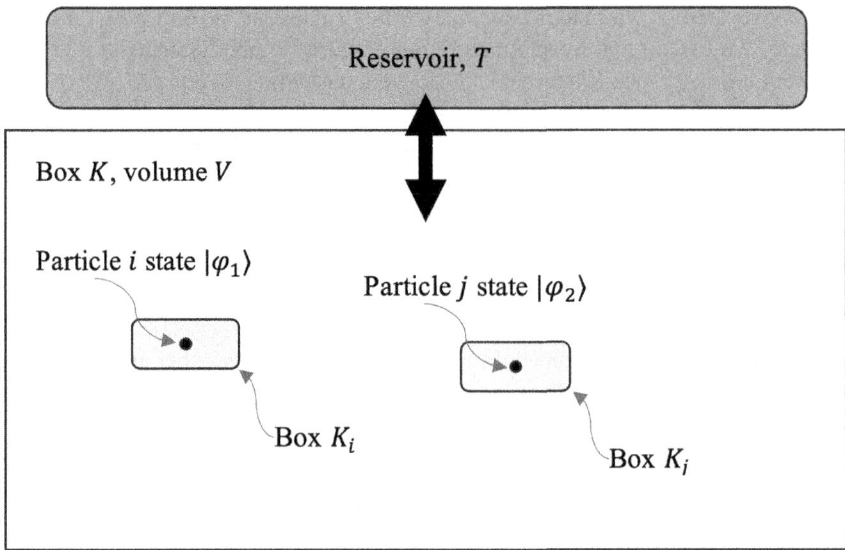

FIGURE 12.1 The starting point for the von Neumann thought experiment. Here the sizes of the individual boxes, K_n, are greatly exaggerated for clarity. There would also be a very large number, $\mathcal{N}$, of these boxes.

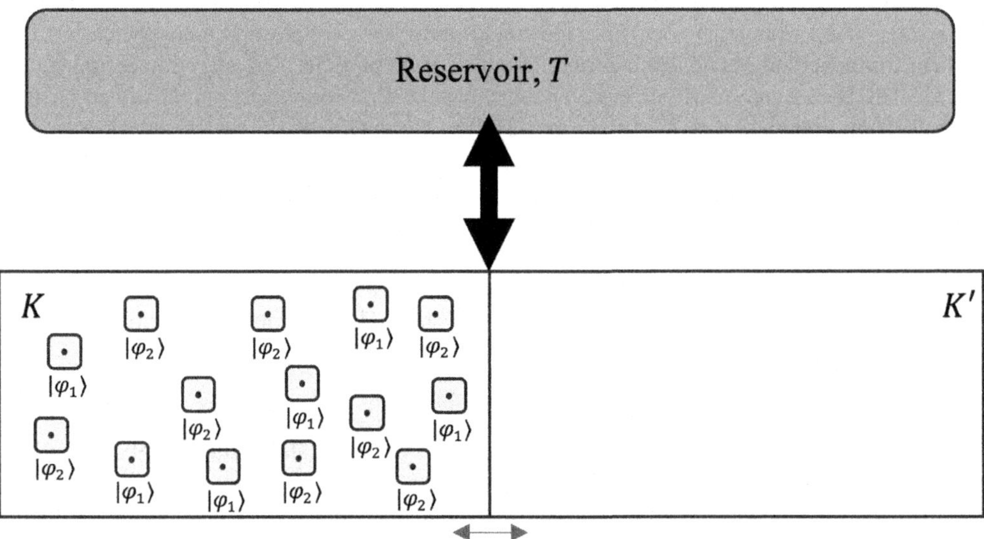

FIGURE 12.2 In Step 1, a second box of equal volume is connected to the first box. The partition between the two boxes can slide back and forth as illustrated.

$$\mathbb{D}_M = p_1|\varphi_1\rangle\langle\varphi_1| + p_2|\varphi_2\rangle\langle\varphi_2| \tag{12.6}$$

The remaining steps are designed to take this mixed ensemble on a reversible tour.

Step 1: Another box, K', of the same dimensions, and hence volume, is connected to K and to the thermal reservoir. This box is initially empty (Figure 12.2). The partition between the two boxes is moveable but completely impenetrable

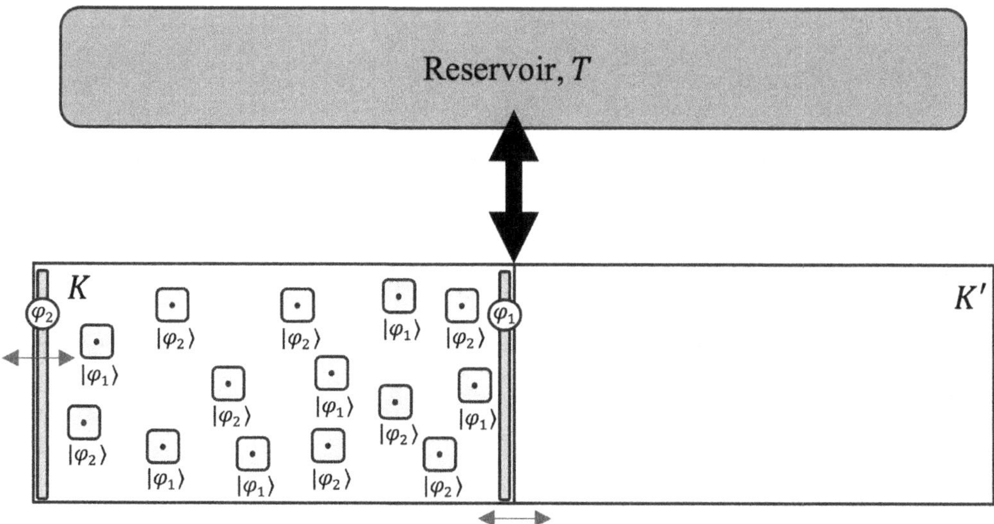

FIGURE 12.3 In this diagram, a fixed semi-permeable membrane has been added between K and K'. This membrane allows states $\varphi_1\rangle$ to pass while reflecting $\varphi_2\rangle$. A moveable semi-permeable membrane has also been placed at the far left-hand side of K. This one allows $\varphi_2\rangle$ to pass while reflecting $\varphi_1\rangle$.

Step 2: The moveable partition between the boxes is now supplemented by a fixed semi-permeable membrane of special design, which is placed on the K side of the partition. If one of the molecule boxes hits this membrane, a mechanism opens the box[9] and measures O for the particle within.[10] If the state is $|\varphi_1\rangle$, a slot opens in the membrane, allowing the (now closed) molecule box to pass through. If the state is $|\varphi_2\rangle$, no slot opens, and the molecule box bounces off without losing any energy. In his book, von Neumann goes to great lengths to prove that such a design is possible, and that its action on the "molecules" does not result in any work being done or energy transfer from the reservoir. He also shows that the membrane is consistent with the Second Law, provided the states are orthogonal, $\langle\varphi_1\,\varphi_2\rangle = 0$

In essence, the membrane acts as a semi-permeable barrier, allowing molecule boxes $|\varphi_1\rangle$ through, while reflecting $|\varphi_2\rangle$.

At the same time, another semi-permeable membrane is placed at the far left-hand side of the box K. This one is set to allow $|\varphi_2\rangle$ to pass while reflecting $|\varphi_1\rangle$ and is capable of movement (Figure 12.3).

Step 3: We now slide the partition and the moveable semi-permeable membrane from left to right as in Figure 12.4, moving them at the same rate so that the volume between them remains constant at V. In this way, we ensure that there is no work done and no temperature change.

As the partition slides, it opens a previous inaccessible volume to the molecule boxes, which will attempt to diffuse across. However, as the $|\varphi_1\rangle$-allowing membrane has remained in place, only molecule boxes of that state will pass into this new volume. At the same time, to the left of box K, the $|\varphi_2\rangle$-transmitting membrane is sweeping across reflecting $|\varphi_1\rangle$ to the right but allowing $|\varphi_2\rangle$ through to the left. Eventually, the sliding partition will reach the right-hand side of K'. By this time, the $|\varphi_2\rangle$-transmitting membrane will have nestled alongside its $|\varphi_1\rangle$ colleague. We have now separated $|\varphi_1\rangle$ from $|\varphi_2\rangle$ (Figure 12.5), *in a reversible manner*.

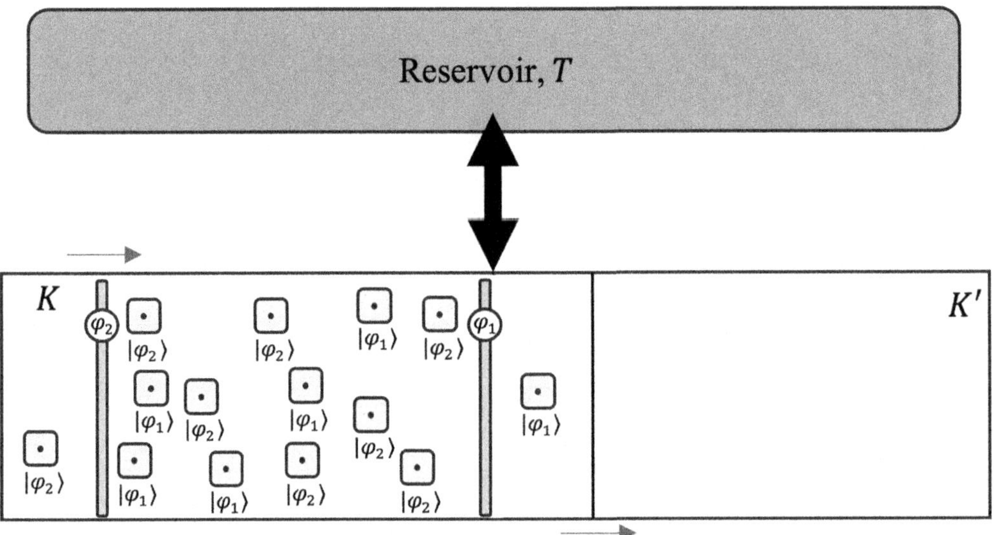

FIGURE 12.4 The partition between the boxes is now allowed to slide to the right. As this happens, the semi-permeable wall on the left tracks with it across K.

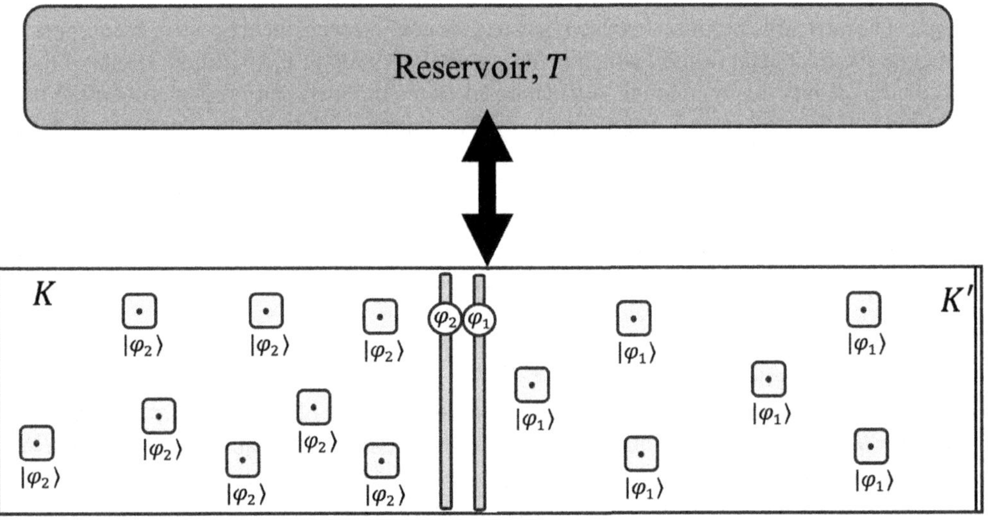

FIGURE 12.5 At the end of Step 3, we have segregated the two types of molecule box in a reversible manner.

Box K' now contains only molecule boxes of type $|\varphi_1\rangle$, while box K is full of $|\varphi_2\rangle$. However, K and K' do not contain equal numbers of molecule boxes, as we did not start with equal numbers in each state. So, the gases have different densities.

Step 4: Next, we compress the two large boxes so that the density in each is the same. As box K' contains $\mathcal{N}_1 = p_1\mathcal{N}$ molecule boxes, we need to reduce its volume to p_1V. At the same time, box K is squashed down to p_2V. Both compressions are carried out isothermally. Note that the total volume of K and K' is now $p_1V + p_2V = V$

The work done in the process is:

$$\mathcal{W} = \int_V^{p_1 V} \mathcal{P} dV + \int_V^{p_2 V} \mathcal{P} dV$$

$$= \int_V^{p_1 V} \frac{\mathcal{N}_1 k_B T}{V} dV + \int_V^{p_2 V} \frac{\mathcal{N}_2 k_B T}{V} dV$$

$$= k_B T \left(\mathcal{N}_1 \int_V^{p_1 V} \frac{dV}{V} + \mathcal{N}_2 \int_V^{p_2 V} \frac{dV}{V} \right)$$

$$= k_B T \left(\mathcal{N}_1 \ln\left(\frac{p_1 V}{V} \right) + k_B T \mathcal{N}_2 \ln\left(\frac{p_2 V}{V} \right) \right)$$

$$= k_B T \mathcal{N} \left(p_1 \ln(p_1) + p_2 \ln(p_2) \right) \tag{12.7}$$

As this takes place isothermally, the same amount of energy is transferred by heating the reservoir. {Note the use of the ideal gas equation here, which we have set up via the deployment of the molecule boxes trick.} Figure 12.6 shows the resulting situation.

Step 5: In this step, we open each box one by one and perform a unitary transformation on each particle that restores them to state $|\Psi\rangle$. Once again, von Neumann takes care to justify that this can be done reversibly. Indeed, we do not need to restore them to $|\Psi\rangle$, as long as they are all in the *same state* by the end of step 5

Step 6: All we need to do now is remove the semi-permeable membranes from between the boxes, and we will restore the initial conditions. Recall that the internal state has no thermodynamic relevance to the molecule boxes, so initial conditions are fully re-established as long as the internal states are all the same

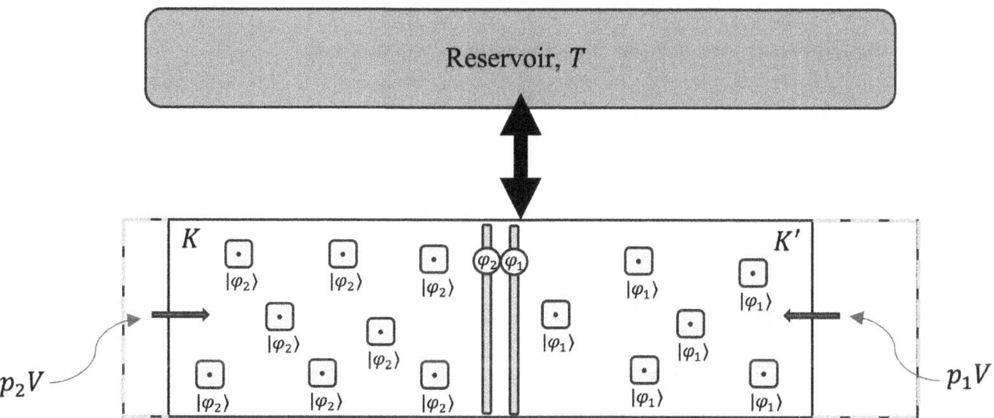

FIGURE 12.6 The two boxes are preferentially compressed so that the density is the same in each. The compression is carried out isothermally.

Having devised this series of steps, von Neumann then proceeded to analyse the entropy changes taking place, discovering a shortfall that could only be accounted for by introducing an entropy change specifically related to Process **1**.

12.3 VON NEUMANN ENTROPY

The first thing to say is that the overall classical entropy, $\mathcal{S}_C$, has not changed during our cycle.[11]

Each thermodynamic step was carried out in a reversible manner. Furthermore, the end result is an ensemble in exactly the same thermodynamic conditions as the start. As the phenomenological entropy is a function of state, $\Delta\mathcal{S}_C = 0$.

Looking at each of the steps in turn, the only point at which a change in entropy takes place is the isothermal compression, resulting in:

$$\Delta\mathcal{S}_C = k_B\mathcal{N}\left(p_1\ln\left(p_1\right)+p_2\ln\left(p_2\right)\right) \tag{12.8}$$

This leaves us with a puzzle. Overall, classical entropy does not change, but there is one process that brings about a decrease. Hence, there must be another, non-thermodynamic, step in here which compensates. As von Neumann has argued that all the Process **2** steps are reversible, and the various other manipulations with permeable membranes, etc., are also reversible, that only leaves the Process **1** taking place in step 0.

Now we hypothesize the *von Neumann entropy*,[12] $\mathcal{S}_{vN}$, categorized by:

$$\mathcal{S}_{vN} = -k_B\mathcal{N}\left(p_1\ln\left(p_1\right)+p_2\ln\left(p_2\right)\right) \tag{12.9}$$

in this case. This must be obtainable from the density matrix:

$$\mathbb{D}_M = p_1\left|\varphi_1\right\rangle\left\langle\varphi_1\right|+p_2\left|\varphi_2\right\rangle\left\langle\varphi_2\right| \tag{12.10}$$

so von Neumann proposed:

$$\mathcal{S}_{vN} = -k_B\mathrm{Tr}\left[D\times\ln\left(D\right)\right] \tag{12.11}$$

which, like the Gibbs entropy, morphs into an intensive version via:

$$\mathbb{S}_{vN} = \mathcal{N}\mathcal{S}_{vN} \tag{12.12}$$

Calculating this von Neumann entropy clearly involves computing the logarithm of a matrix. This is a mathematically dense and not wholly well-defined process, but fortunately it simplifies when the matrix is diagonal:

$$\ln\left(M\right)=\begin{pmatrix} \ln\left(M_{11}\right) & 0 & \cdots & 0 & 0 \\ 0 & \ln\left(M_{22}\right) & \cdots & 0 & 0 \\ 0 & 0 & \ln\left(M_{33}\right) & \cdots & 0 \\ \cdots & \cdots & \cdots & \cdots & 0 \\ 0 & 0 & 0 & 0 & \ln\left(M_{nn}\right) \end{pmatrix} \tag{12.13}$$

in other words, the log of a diagonal matrix is a matrix constructed from the logs of its diagonal elements. In this case, we have

$$\mathbb{D}_M = p_1|\varphi_1\rangle\langle\varphi_1| + p_2|\varphi_2\rangle\langle\varphi_2| = \begin{pmatrix} p_1 & 0 \\ 0 & p_2 \end{pmatrix} \tag{12.15}$$

so that:

$$\mathbb{S}_{vN} = -k_B\mathcal{N}\,\mathrm{Tr}\left[\begin{pmatrix} p_1 & 0 \\ 0 & p_2 \end{pmatrix} \times \begin{pmatrix} \ln(p_1) & 0 \\ 0 & \ln(p_2) \end{pmatrix}\right]$$

$$= -k_B\mathcal{N}\,\mathrm{Tr}\big(p_1\ln(p_1) + p_2\ln(p_2)\big) \tag{12.16}$$

According to von Neumann, if this entropy was introduced into the cycle with the Process **1** taking place in Step 0, it will compensate for the classical entropy change and explain why the total entropy change is zero.

Before we get into the debate that exists regarding the validity of von Neumann's argument, let's spend some time exploring $\mathcal{S}_{vN}$ and its various properties.

12.3.1 Unitary Transformations Acting on the Von Neumann Entropy

As we know, any density matrix can be diagonalized with a suitable unitary transformation, and as the trace is invariant under such transformations (**T6**), we are good to go. If we have a density matrix, D, expressed in its diagonal basis, the von Neumann entropy becomes:

$$\mathcal{S}_{vN} = -k_B\mathrm{Tr}\big[D \times \ln(D)\big] = -k_B\mathcal{N}\left\{\sum_K\sum_j D_{Kj}\ln\left(D_{jK}\right)\delta_{jK}\right\}$$

$$= -k_B\sum_K D_{KK}\ln\left(D_{KK}\right) \tag{12.17}$$

Of course, in this form, the elements are the probabilities that the various states contained within the density matrix will be manifest; hence, we get:

$$\mathcal{S}_{vN} = -k_B\sum_n p_n\ln\left(p_n\right) \tag{12.18}$$

which looks like something that we have seen before….

It is also worth considering the specific unitary transformation enacted by the time evolution operator, $\widehat{U}(\Delta t)$. The invariance of the trace under unitary transformations of a matrix is a very general property; it applies to time evolution as well. That being the case, *the von Neumann entropy does not change with time*. This exactly, and unsurprisingly parallels the issue with the Gibbs entropy that was fully discussed in Chapter 10. In von Neumann's words[13]:

> Although our entropy expression is, as we saw, completely analogous to classical entropy, it is still surprising that it is invariant under normal temporal evolution of the system (process 2), and increases only in consequence of measurements (process 1).
>
> Quote 12.2

This is a key issue that is explored more deeply in Section 12.4.3.

12.3.2 Purity of Mind, Purity of State

Strikingly, the von Neumann entropy is zero for a pure state. In its diagonal representation, a pure state density matrix will have only one non-zero element along the diagonal, the value of which is one. Every other element on the grid is zero. Generally, the log function gets upset if you try and feed it a zero; however, while we are taking the log of the zeroes along the diagonal, we are also multiplying by the density matrix ($D\times\ln(D)$) before we take the trace, so those same zeroes causing the log to misbehave will kill off the log in the product. That just leaves the single diagonal element with value 1 and $\ln(1)=0$. As the value of the trace is invariant under a unitary transformation, it does not matter which representation the density matrix is in; we can always say:

$$\mathcal{S}_{vN}=-k_B\mathrm{Tr}\left[D_{\mathrm{Pure}}\times\ln\left(D_{\mathrm{Pure}}\right)\right]=0 \tag{12.19}$$

This should be compared to our comment from Section 11.4.4, where we noted that a pure state has $\mathrm{Tr}\left[D^2\right]=1$. Both are measures of the degree of purity of the state.

Once again, it is worth remembering that superpositions also count as pure states. The only criterion is the extent of our knowledge about the state. If the state of the system is known with certainty, all is pure. Otherwise, we are dealing with a mixed state.

12.3.3 Von Neumann Entropy of the Mixed State

The only occasions when the von Neumann entropy is non-zero come about with a mixed state. In Section 11.5.2, we discussed various scenarios that generate mixed states, which we will return to shortly. For the moment, we are going to ponder the upper bound on the von Neumann entropy.

As always, we are considering an ensemble of systems with each representative sitting in a state randomly allocated from the collection $\{|\psi_n\rangle\}$ with probability $\{p_n\}$. This time, we also note that there are $\mathbb{N}$ states in the collection, $\{|\psi_n\rangle\}\ 1\le n\le\mathbb{N}$.

We can maximize the von Neumann entropy with our familiar technique, using a Lagrange multiplier, A, to incorporate a constraint on the total probability, $\Sigma\, p_n=1$ or equally $\Sigma\,\Delta p_n=0$.

As:

$$\mathcal{S}_{vN}=-k_B\sum_n p_n\ln\left(p_n\right) \tag{12.20}$$

the expression to maximize is:

$$d\mathcal{S}_{vN}=-k_B\mathcal{N}\sum_{n=1}^{\mathbb{N}}\left(\mathrm{d}p_n\ln\left(p_n\right)+\frac{p_n}{p_n}dp_n\right)+A\sum_{n=1}^{\mathbb{N}}dp_n$$

$$=-k_B\mathcal{N}\sum_{n=1}^{\mathbb{N}}\mathrm{d}p_n\left(\ln\left(p_n\right)+1-\frac{A}{k_B\mathcal{N}}\right) \tag{12.21}$$

As before, we can set the term within the brackets equal to zero:

$$\ln\left(p_n\right)+1-\frac{A}{k_B\mathcal{N}}=0 \tag{12.22}$$

Yielding:

$$p_n = e^{A/k_B \mathcal{N} - 1} = p \tag{12.23}$$

Evidently, the von Neumann entropy hits a maximum when the probability for each state is the same. This is a *maximally mixed state*. Furthermore, as each state has the same probability:

$$\sum_{n=1}^{\mathbb{N}} p_n = \sum_{n=1}^{\mathbb{N}} p = \mathbb{N}p = 1 \qquad \Rightarrow \qquad p = 1/\mathbb{N} \tag{12.24}$$

it is *determined by the number of states available*. Looping back to the actual value of the entropy, we get:

$$\mathcal{S}_{vN} = -k_B \sum_{n=1}^{\mathbb{N}} p\ln(p) = -k_B \mathbb{N} p\ln(p) = k_B \ln(\mathbb{N}) \tag{12.25}$$

The entropy of a maximally mixed state is the log of the "dimensionality" of the "space" of states.

12.3.4 Thermal States

An important mixed state is the ensemble in thermal equilibrium from Section 11.6. The appropriate density matrix is (11.153):

$$D_T = \frac{\sum_j e^{-\widehat{H}/k_B T} |j\rangle\langle j|}{\mathrm{Tr}\left[\sum_j e^{-\widehat{H}/k_B T} |j\rangle\langle j|\right]} = \frac{1}{Z}\sum_j e^{-\widehat{H}/k_B T} |j\rangle\langle j|$$

If we have access to the energy eigenstates, $|E_j\rangle$, this becomes:

$$D_T = \frac{1}{Z}\sum_j e^{-\widehat{H}/k_B T} \left|E_j\right\rangle\left\langle E_j\right| = -\frac{1}{Z}\sum_j e^{-E_j/k_B T} \left|E_j\right\rangle\left\langle E_j\right| \tag{12.28}$$

which we can plug into the von Neumann entropy (using **T1**):

$$\mathcal{S}_{vN} = -k_B \mathrm{Tr}\left[D_T \times \ln(D_T)\right]$$

$$\mathcal{S}_{vN} = -k_B \mathrm{Tr}\left[\left(-\frac{1}{Z}\sum_j e^{-E_j/k_B T}\left|E_j\right\rangle\left\langle E_j\right|\right) \times \left(\sum_m \left(\frac{E_m}{k_B T}\right)\left|E_m\right\rangle\left\langle E_m\right|\right)\right] + k_B \ln(Z)$$

$$= k_B \sum_j \frac{1}{Z}\left(\frac{E_j}{k_B T} e^{-E_j/k_B T}\right) + k_B \ln(Z)$$

$$= \frac{1}{T}\langle E\rangle + k_B \ln(Z) \tag{12.29}$$

In classical thermodynamics, $F = -k_B T\ln(Z)$, (5.48) so we can re-arrange this expression to get:

$$F = \langle E\rangle - T\mathcal{S}_{vN} \tag{12.30}$$

which is clearly an analogue of:

$$F = U - TS \tag{12.31}$$

in the classical case.

12.3.5 Product States

From our work in Section 11.5.3, we know that the density matrix of a combined state can be factored, if the states themselves can be factored. This happens when we have two systems which are uncorrelated, i.e., the behaviour of one has no impact on the other. In which case (**T7**):

$$D_{12} = D_1 \otimes D_2 \qquad \mathrm{Tr}\left[D_{12}\right] = \mathrm{Tr}\left[D_1 \otimes D_2\right] = \mathrm{Tr}\left[D_1\right]\mathrm{Tr}\left[D_2\right] \tag{12.32}$$

However, this is not enough to help us calculate the entropy; for that, we need to deal with the logs as well. The log of a tensor product breaks down as this:

$$\ln\left(D_1 \otimes D_2\right) = \ln\left(D_1\right)\otimes I_2 + I_1 \otimes \ln\left(D_2\right) \tag{12.33}$$

with I_2 being the unit matrix applied to system 2 and I_1 the unit matrix acting on system 1. For the von Neumann entropy, we now have the following:

$$\mathcal{S}_{vN}\left(D_{12}\right) = -k_B \mathrm{Tr}\left[D_{12} \times \ln\left(D_{12}\right)\right]$$

$$= -k_B \mathrm{Tr}\left[D_{12} \times \left(\ln\left(D_1\right)\otimes I_2 + I_1 \otimes \ln\left(D_2\right)\right)\right]$$

$$= -k_B \mathrm{Tr}\left[\left(D_1 \otimes D_2\right)\times\left(\ln\left(D_1\right)\otimes I_2 + I_1 \otimes \ln\left(D_2\right)\right)\right]$$

$$= -k_B \mathrm{Tr}\left[\left(D_1 \times \ln\left(D_1\right)\right)\otimes D_2 + D_1 \otimes \left(D_2 \times \ln\left(D_2\right)\right)\right]$$

$$= -k_B \left\{\mathrm{Tr}\left[\left(D_1 \times \ln\left(D_1\right)\right)\otimes D_2\right] + \mathrm{Tr}\left[D_1 \otimes \left(D_2 \times \ln\left(D_2\right)\right)\right]\right\} \qquad \text{(T2)}$$

$$= -k_B \left\{\mathrm{Tr}\left[D_1 \times \ln\left(D_1\right)\right]\mathrm{Tr}\left[D_2\right] + \mathrm{Tr}\left[D_1\right]\mathrm{Tr}\left[\left(D_2 \times \ln\left(D_2\right)\right)\right]\right\} \qquad \text{(T7)}$$

$$= -k_B \mathrm{Tr}\left[D_1 \times \ln\left(D_1\right)\right] - k_B \mathrm{Tr}\left[D_2 \times \ln\left(D_2\right)\right] \qquad \text{(D3)}$$

$$= \mathcal{S}_{vN}\left(D_1\right) + \mathcal{S}_{vN}\left(D_2\right) \tag{12.34}$$

demonstrating that the von Neumann entropy is *additive*: for two uncorrelated systems, the total entropy is the sum of their separate entropies. This is clearly a desirable attribute to gel with the physics of entropy generally.

12.3.6 Entangled State

Of course, two (or more) systems can be entangled, in which case their combined state cannot be factored and so neither can the density matrix. We can explore what happens in these situations if we first stop to define the *relative entropy* of two density matrices, D_1, D_2:

$$S_{vN}\left(D_1 \,||\, D_2\right) = k_B \mathrm{Tr}\left[D_1 \times \ln\left(D_1\right) - D_1 \times \ln\left(D_2\right)\right] \tag{12.35}$$

Using this measure, $S_{vN}\left(D_1 \,||\, D_2\right) \geq 0$, the equality being when $D_1 = D_2$.

Now, we can compare the density matrix for our entanglement of two systems, with the situation as if they were uncorrelated, i.e., $S_{vN}\left(D_{12} \,||\, D_1 \otimes D_2\right)$. As the state cannot be factored, we define D_1, D_2 as the reduced density matrices produced by tracing over the other system, respectively:

$$D_1 = \mathrm{Tr}_1\left[D_{12}\right] \qquad D_2 = \mathrm{Tr}_1\left[D_{12}\right] \tag{12.36}$$

The relative entropy definition gets us as far as:

$$\begin{aligned} S_{vN}\left(D_{12} \,||\, D_1 \otimes D_2\right) &= k_B \mathrm{Tr}\left[D_{12} \times \ln\left(D_{12}\right) - D_{12} \times \ln\left(D_1 \otimes D_2\right)\right] \\ &= k_B \mathrm{Tr}\left[D_{12} \times \ln\left(D_{12}\right)\right] - k_B \mathrm{Tr}\left[D_{12} \times \ln\left(D_1 \otimes D_2\right)\right] \\ &= -S_{vN}\left(D_{12}\right) - k_B \mathrm{Tr}\left[D_{12} \times \ln\left(D_1 \otimes D_2\right)\right] \end{aligned} \tag{12.37}$$

We can tackle the trace by looking back to (12.33):

$$\ln\left(D_1 \otimes D_2\right) = \ln\left(D_1\right) \otimes I_2 + I_1 \otimes \ln\left(D_2\right)$$

applying that to the problem:

$$\begin{aligned} k_B \mathrm{Tr}\left[D_{12} \times \ln\left(D_1 \otimes D_2\right)\right] &= k_B \mathrm{Tr}\left[D_{12} \times \left(\ln\left(D_1\right) \otimes I_2 + I_1 \otimes \ln\left(D_2\right)\right)\right] \\ &= k_B \mathrm{Tr}\left[D_{12} \times \ln\left(D_1\right) \otimes I_2\right] + k_B \mathrm{Tr}\left[D_{12} \times I_1 \otimes \ln\left(D_2\right)\right] \\ &= k_B \mathrm{Tr}\left[D_{12} \times \ln\left(D_1\right)\right] \mathrm{Tr}\left[I_2\right] + k_B \mathrm{Tr}\left[D_{12} \times \ln\left(D_2\right)\right] \mathrm{Tr}\left[I_1\right] \quad \text{(T7)} \\ &= k_B \mathrm{Tr}\left[D_{12} \times \ln\left(D_1\right)\right] + k_B \mathrm{Tr}\left[D_{12} \times \ln\left(D_2\right)\right] \end{aligned} \tag{12.38}$$

Re-inserting this back into our relative entropy brings us to:

$$S_{vN}\left(D_{12} \,||\, D_1 \otimes D_2\right) = -S_{vN}\left(D_{12}\right) - k_B \mathrm{Tr}\left[D_{12} \times \ln\left(D_1\right)\right] - k_B \mathrm{Tr}\left[D_{12} \times \ln\left(D_2\right)\right] \quad (12.39)$$

This is the moment to deploy our partial trace, which will only act on D_{12}. Recalling **PT1**:

$$\mathrm{Tr}\left[D_{12}\right] = \mathrm{Tr}_1\left[\mathrm{Tr}_2\left[D_{12}\right]\right] = \mathrm{Tr}_2\left[\mathrm{Tr}_1\left[D_{12}\right]\right]$$

we proceed to pick partial traces over system 1 and system 2 separately, depending on which system's density matrix is lurking inside the log:

$$S_{vN}\left(D_{12} \,||\, D_1 \otimes D_2\right) = S_{vN}\left(D_{12}\right) - k_B \mathrm{Tr}_1\left[D_1 \times \ln\left(D_1\right)\right] - k_B \mathrm{Tr}_2\left[D_2 \times \ln\left(D_2\right)\right]$$

$$= -S_{vN}\left(D_{12}\right) + S_{vN}\left(D_1\right) + S_{vN}\left(D_2\right) \quad (12.40)$$

As $S_{vN}\left(D_{12} \,||\, D_1 \otimes D_2\right) \geq 0$, we obtain:

$$S_{vN}\left(D_{12}\right) \leq S_{vN}\left(D_1\right) + S_{vN}\left(D_2\right) \quad (12.41)$$

When two systems are correlated (entangled) the entropy of their combined state is less than the sum of their separate entropies.

There is a clear message here. By taking the partial trace, we lose information, and *the entropy is larger as a result.*

12.3.7 Mutual Information

Thinking back to the relative entropy that we defined in equation (12.35):

$$S_{vN}\left(D_1 \,||\, D_2\right) = k_B \mathrm{Tr}\left[D_1 \times \ln\left(D_1\right) - D_1 \times \ln\left(D_2\right)\right]$$

which we then applied to our composite system, resulting in Equation (12.40) (reordered here for convenience):

$$S_{vN}\left(D_{12} \,||\, D_1 \otimes D_2\right) = S_{vN}\left(D_1\right) + S_{vN}\left(D_2\right) - S_{vN}\left(D_{12}\right)$$

we can see that in the special case of two uncorrelated, factorable, systems, $D_{12} = D_1 \otimes D_2$:

$$S_{vN}\left(D_1 \otimes D_2 \,||\, D_1 \otimes D_2\right) = 0 \quad (12.42)$$

Also, we noted that $S_{vN}\left(D_1 \,||\, D_2\right) \geq 0$, so we have a measure that is always positive and zero when the two systems are uncorrelated. This is also known as the *mutual information*:

$$I\left(D_{12}\right)=S_{vN}\left(D_{12}\,||\,D_1\otimes D_2\right)=S_{vN}\left(D_1\right)+S_{vN}\left(D_2\right)-S_{vN}\left(D_{12}\right)\geq 0 \tag{12.43}$$

which is *the amount of information present in* D_{12}, *which is not separately present in* D_1 and D_2. Bluntly, a correlated (entangled) pair of systems contains more information than the two systems did separately while uncorrelated. This is the reverse of the situation with the entropy. The von Neumann entropy for the entangled state is less than the sum of the separate entropies (12.41).

An interesting case comes about when the entangled systems are in a pure state. As we know, the entropy of a pure state is zero, hence $S_{vN}(D_{12})=0$. This makes the mutual information:

$$I_{\text{Pure}}\left(D_{12}\right)=S_{vN}\left(D_1\right)+S_{vN}\left(D_2\right) \tag{12.44}$$

Now, we turn over a card that we placed on the table during the previous chapter (Section 11.5.5). When the entangled systems are in a pure state, the eigenvalues for the reduced trace density matrices are the same. It follows that *their von Neumann entropies are equal*: $S_{vN}(D_1)=S_{vN}(D_2)$ and so:

$$I_{\text{Pure}}\left(D_{12}\right)=2S_{vN}\left(D_1\right)=2S_{vN}\left(D_2\right) \tag{12.45}$$

The total information in the entanglement is twice that held in each of the separate sub-systems.

It is important to note that the mutual information does not distinguish between quantum and classical contributions to the correlations between the systems. Indeed, this is an ongoing aspect of research in the field. We have already seen in Section 11.5.2 that mixed states generally contain classical probabilities when they arise due to "lossy" system preparation, but in the case of entanglement, what must be treated as a classical probability by one experimenter is a result of quantum probabilities from the perspective of another.

To develop this further, let's consider two separate cases.

Mixed states:

For simplicity, take an ensemble split between systems being in states $|0\rangle$ or $|1\rangle$. Each state is equally likely, but as we have no further information, we have a (maximally) mixed state scenario. The density matrix is:

$$D_M=\frac{1}{2}\begin{pmatrix}1 & 0\\ 0 & 0\end{pmatrix}+\frac{1}{2}\begin{pmatrix}0 & 0\\ 0 & 1\end{pmatrix}=\frac{1}{2}\begin{pmatrix}1 & 0\\ 0 & 1\end{pmatrix} \tag{12.46}$$

and the von Neumann entropy follows:

$$S_{vN}=-k_B\text{Tr}\left[D\times\ln(D)\right]=k_B\text{Tr}\left[\begin{pmatrix}1/2 & 0\\ 0 & 1/2\end{pmatrix}\times\begin{pmatrix}\ln(2) & 0\\ 0 & \ln(2)\end{pmatrix}\right]$$

$$=k_B\text{Tr}\left[\begin{pmatrix}1/2\ln(2) & 0\\ 0 & 1/2\ln(2)\end{pmatrix}\right]=k_B\ln(2) \tag{12.47}$$

Pure states:

In contrast, now take a pure state ensemble consisting of entangled pairs of systems in the state:

$$|\Psi\rangle = \frac{1}{\sqrt{2}}\{|0\rangle|1\rangle + |1\rangle|0\rangle\} \tag{12.48}$$

Writing:

$$|0\rangle|1\rangle = \begin{pmatrix} 0 \\ 1 \\ 0 \\ 0 \end{pmatrix} \qquad |1\rangle|0\rangle = \begin{pmatrix} 0 \\ 0 \\ 1 \\ 0 \end{pmatrix} \qquad |\Psi\rangle = \frac{1}{\sqrt{2}}\begin{pmatrix} 0 \\ 1 \\ 1 \\ 0 \end{pmatrix} \tag{12.49}$$

gives the density matrix:

$$D_P = \frac{1}{2}\begin{pmatrix} 0 & 1 & 1 & 0 \end{pmatrix}\begin{pmatrix} 0 \\ 1 \\ 1 \\ 0 \end{pmatrix} = \frac{1}{2}\begin{pmatrix} 0 & 0 & 0 & 0 \\ 0 & 1 & 1 & 0 \\ 0 & 1 & 1 & 0 \\ 0 & 0 & 0 & 0 \end{pmatrix} \tag{12.50}$$

Unfortunately, we can't proceed to calculate $\mathcal{S}_{vN}$ as this matrix has off-diagonal terms, and so the log is not well defined. A few moments with one of the many online matrix calculation tools available finds the unitary transformation that we need to diagonalize this matrix:

$$\mathcal{U} = \begin{pmatrix} 0 & 1 & 0 & 0 \\ 1 & 0 & -1 & 0 \\ 1 & 0 & 1 & 0 \\ 0 & 0 & 0 & 1 \end{pmatrix} \tag{12.51}$$

giving:

$$D_P' = \begin{pmatrix} 1 & 0 & 0 & 0 \\ 0 & 0 & 0 & 0 \\ 0 & 0 & 0 & 0 \\ 0 & 0 & 0 & 0 \end{pmatrix} \tag{12.52}$$

and hence:

$$\mathcal{S}_{vN} = -k_B \mathrm{Tr}\left[D_P' \times \ln\left(D_P'\right)\right] = -k_B \mathrm{Tr}\left[\begin{pmatrix} 1 & 0 & 0 & 0 \\ 0 & 0 & 0 & 0 \\ 0 & 0 & 0 & 0 \\ 0 & 0 & 0 & 0 \end{pmatrix} \times \begin{pmatrix} \ln(1) & 0 & 0 & 0 \\ 0 & \ln(0) & 0 & 0 \\ 0 & 0 & \ln(0) & 0 \\ 0 & 0 & 0 & \ln(0) \end{pmatrix}\right] = 0 \tag{12.53}$$

the painful logs of zero being killed off by the zeros on the diagonal of the first matrix in the product. The result is as we should expect from a pure state.

There is another, more long-winded, way to obtain the same density matrix:

$$D_P = |\Psi\rangle\langle\Psi| = \frac{1}{2}\{|0\rangle|1\rangle + |1\rangle|0\rangle\}\{\langle 0|\langle 1| + \langle 1|\langle 0|\}$$

$$= \frac{1}{2}\{(|0\rangle\langle 0|)(|1\rangle\langle 1|) + (|0\rangle\langle 1|)(|1\rangle\langle 0|) + (|1\rangle\langle 0|)(|0\rangle\langle 1|) + (|1\rangle\langle 1|)(|0\rangle\langle 0|)\}$$

$$= \frac{1}{2}\left\{\begin{pmatrix}1&0\\0&0\end{pmatrix}\otimes\begin{pmatrix}0&0\\0&1\end{pmatrix}+\begin{pmatrix}0&1\\0&0\end{pmatrix}\otimes\begin{pmatrix}0&0\\1&0\end{pmatrix}+\begin{pmatrix}0&0\\1&0\end{pmatrix}\otimes\begin{pmatrix}0&1\\0&0\end{pmatrix}+\begin{pmatrix}0&0\\0&1\end{pmatrix}\otimes\begin{pmatrix}1&0\\0&0\end{pmatrix}\right\}$$

$$= \frac{1}{2}\left\{\begin{pmatrix}1\begin{pmatrix}0&0\\0&1\end{pmatrix} & 0\begin{pmatrix}0&0\\0&1\end{pmatrix}\\ 0\begin{pmatrix}0&0\\0&1\end{pmatrix} & 0\begin{pmatrix}0&0\\0&1\end{pmatrix}\end{pmatrix}+\begin{pmatrix}0\begin{pmatrix}0&0\\1&0\end{pmatrix} & 1\begin{pmatrix}0&0\\1&0\end{pmatrix}\\ 0\begin{pmatrix}0&0\\1&0\end{pmatrix} & 0\begin{pmatrix}0&0\\1&0\end{pmatrix}\end{pmatrix}+\begin{pmatrix}0\begin{pmatrix}0&1\\0&0\end{pmatrix} & 0\begin{pmatrix}0&1\\0&0\end{pmatrix}\\ 1\begin{pmatrix}0&1\\0&0\end{pmatrix} & 0\begin{pmatrix}0&1\\0&0\end{pmatrix}\end{pmatrix}\right.$$

$$\left.+\begin{pmatrix}0\begin{pmatrix}1&0\\0&0\end{pmatrix} & 0\begin{pmatrix}1&0\\0&0\end{pmatrix}\\ 0\begin{pmatrix}1&0\\0&0\end{pmatrix} & 1\begin{pmatrix}1&0\\0&0\end{pmatrix}\end{pmatrix}\right\}$$

$$= \frac{1}{2}\left\{\begin{pmatrix}0&0&0&0\\0&1&0&0\\0&0&0&0\\0&0&0&0\end{pmatrix}+\begin{pmatrix}0&0&0&0\\0&0&1&0\\0&0&0&0\\0&0&0&0\end{pmatrix}+\begin{pmatrix}0&0&0&0\\0&0&0&0\\0&1&0&0\\0&0&0&0\end{pmatrix}+\begin{pmatrix}0&0&0&0\\0&0&0&0\\0&0&1&0\\0&0&0&0\end{pmatrix}\right\}$$

$$= \frac{1}{2}\begin{pmatrix}0&0&0&0\\0&1&1&0\\0&1&1&0\\0&0&0&0\end{pmatrix} \tag{12.54}$$

The advantage of this approach is that we can immediately see how to produce a reduced density matrix for one sub-system by tracing over the other. For example:

$$\begin{aligned}D_{P1} &= \mathrm{Tr}_2\left[D_P\right]\\ &= \mathrm{Tr}_2\left[\frac{1}{2}\left\{\begin{pmatrix}1&0\\0&0\end{pmatrix}\otimes\begin{pmatrix}0&0\\0&1\end{pmatrix}+\begin{pmatrix}0&1\\0&0\end{pmatrix}\otimes\begin{pmatrix}0&0\\1&0\end{pmatrix}+\begin{pmatrix}0&0\\1&0\end{pmatrix}\otimes\begin{pmatrix}0&1\\0&0\end{pmatrix}+\begin{pmatrix}0&0\\0&1\end{pmatrix}\otimes\begin{pmatrix}1&0\\0&0\end{pmatrix}\right\}\right]\end{aligned}$$

$$= \frac{1}{2}\left\{\begin{pmatrix}1&0\\0&0\end{pmatrix}\otimes 1I_2+\begin{pmatrix}0&1\\0&0\end{pmatrix}\otimes 0I_2+\begin{pmatrix}0&0\\1&0\end{pmatrix}\otimes 0I_2+\begin{pmatrix}0&0\\0&1\end{pmatrix}\otimes 1I_2\right\} = \frac{1}{2}\begin{pmatrix}1&0\\0&1\end{pmatrix} \tag{12.55}$$

Equally:

$$D_{P2} = \mathrm{Tr}_1\left[D_P\right]$$

$$= \mathrm{Tr}_2\left[\frac{1}{2}\left\{\begin{pmatrix}1&0\\0&0\end{pmatrix}\otimes\begin{pmatrix}0&0\\0&1\end{pmatrix}+\begin{pmatrix}0&1\\0&0\end{pmatrix}\otimes\begin{pmatrix}0&0\\1&0\end{pmatrix}+\begin{pmatrix}0&0\\1&0\end{pmatrix}\otimes\begin{pmatrix}0&1\\0&0\end{pmatrix}+\begin{pmatrix}0&0\\0&1\end{pmatrix}\otimes\begin{pmatrix}1&0\\0&0\end{pmatrix}\right\}\right]$$

$$= \frac{1}{2}\left\{1I_1\otimes\begin{pmatrix}0&0\\0&1\end{pmatrix}+0I_1\otimes\begin{pmatrix}0&0\\1&0\end{pmatrix}+0I_1\otimes\begin{pmatrix}0&1\\0&0\end{pmatrix}+1I_1\otimes\begin{pmatrix}1&0\\0&0\end{pmatrix}\right\}=\frac{1}{2}\begin{pmatrix}1&0\\0&1\end{pmatrix} \tag{12.56}$$

Once you are used to working with partial traces and matrices, there is a way of doing this "by eye" which is shown in Appendix PA1.7.

{Note in passing: both sub-systems are now in maximally mixed states (Section 12.3.3) as the probability of each state is the same and equal to 1/number of available states.}

The von Neumann entropy for either sub-state is also $k_B \ln(2)$ *as we have ended up with the same density matrix as in the mixed case*. It is very significant that tracing over one of the systems in a pure state produces a mixed state for the other. Recall what was said in Section 11.5.4: the process of generating a reduced density matrix for system 1 by forgetting about system 2 has lost us information about system 1.

However, for the mixed state, the entropy arose due to the classical probabilities needed to define the mixed ensemble. In the entangled case, there are classical probabilities from the perspective of someone experimenting on system 1 without having been granted information about experiments on system 2. We have obtained the reduced density matrix by wiping information to do with the system 2, but this information and probability is quantum from the perspective of that experimenter.

If we were just presented with the density matrix, *there would be no way to distinguish its two possible origins*. We can never know if the state is mixed because of uncertainty about the exact pure state or because the system is entangled with another system that lies outside of our control.

12.3.8 So, What is Probability Anyway?

Probability is one of those vexing topics that seems obvious intuitively but in practice is a lot harder to pin down. Somewhat, in fact, like entropy. It is possible that this in itself is telling us something.

Our first call on probability is to act as a tool to employ when we do not have the ability to calculate precisely. Perhaps there are too many molecules to keep track of (e.g., in a gas), or perhaps we cannot measure physical variables with sufficient precision to predict properly (e.g., with tossing a coin), or perhaps we do not have enough information (e.g., with a mixed state ensemble). In each case, information is available in principle, but in practice is inaccessible. A sufficient understanding of the mass distribution across a coin along with precise parameters to do with forces, air pressure, and gravitation would enable an exact prediction saying if a tossed coin would land heads or tails. Instead, we could toss a coin 100 times (preferably many more) and record 57 heads and 43 tails. From this data, we estimate a set of probabilities: $p_H = 0.57, p_T = 0.43$, but it is unclear what this *means*. At one level, we have simply summarized data that we already have. On the other hand, we hope to use this data to predict the future. One aspect of this probability is determined by the physical structure of the coin. It has two sides, after all. If we were to toss a six-sided die instead, that would affect the probability of an individual side coming up. In some sense, the probability is objective, but can we say the same for estimating the odds in a horse race, for example?

In quantum theory, we experience a new aspect of probability. The amplitudes/wave functions that we use to describe a state are in some ill-understood sense determinants of the probability of a result in a measurement. Here is genuine randomness, not covering for our lack of information. This is information *that does not exist.* Quantum measurement results cannot be predicted in principle as *there is no physics that exactly determines each outcome.* A superposition contains probability amplitudes of this form, and it is very striking that such a state, like any pure state, has zero von Neumann entropy. We only get an entropic value of this type if we are ignorant of the exact details of the state.

12.4 IS THE VON NEUMANN ENTROPY PHYSICAL?

Since the publication of von Neumann's book, especially since the 1960–1970s, there has been a burgeoning science of *quantum information theory*. This has been even further acceleration with the discovery of an algorithm which would run on a *quantum computer* and factorize large prime numbers with speed. As this is the key to decrypting encoded messages, this has generated funding and interest in the construction of quantum computers. The tools of quantum information theory will be key in the development and coding of such machines.

This research has also stimulated further work on the foundations of quantum theory, giving a different perspective on traditional issues (such as the measurement problem and entanglement) as well as enhanced experimental techniques allowing quantum systems to be isolated from their environment for longer and more effectively. The von Neumann entropy plays a central role in this work, but it is not the only measure of entropy used. Within the theory it is well established and has its place, but in a sense, that would still be the case whether it can be linked to the phenomenological (classical) entropy or not. That is still a question that is in dispute. It is now time to take up that debate, albeit in a limited manner.

12.4.1 Some Issues

The material in this section is drawn from several papers on the subject. Of these, we agree with the argument presented by Erin Sheridan[14] in that much of the critique misunderstands the role of the quantum gas thought experiment.

The validity of von Neumann's argument (Section 12.2) has been questioned based on the presumption that it is an attempt to justify a correspondence with the phenomenological thermodynamic entropy—what we have been calling $\mathcal{S}_T$. The use of the term "classical entropy" in von Neumann's text is a contributing factor to this confusion. However, given the reference to Gibb's ensemble in Quote 12.1 and the use of the phrase "completely analogous" in Quote 12.2, it seems more likely that von Neumann's aim here was the Gibbs entropy, not $\mathcal{S}_T$. Indeed, if the quantum gas cycle was von Neumann's argument for $\mathcal{S}_{vN} \Leftrightarrow \mathcal{S}_T$, it is puzzling that he went to such further trouble as we will see in Section 12.4.2. Also, when he gets to the conclusion of this further discussion, he uses a different phrase: "phenomenological thermodynamics of the real world" (Quote 12.4) indicating that he sees a difference between "classical entropy" ($\mathcal{S}_G$) and "phenomenological entropy" ($\mathcal{S}_T$). In justifying his need for further development, another striking phrase is used (Quote 12.3): "[we need to] investigate the precise analog of *classical macroscopic entropy*" [our emphasis], again suggesting that he is now after something more than he achieved with the quantum gas.

Other critical authors focus in on step three, where the expansion of the gas suggests an entropy change. This might be compensated by a change in the entropy of mixing, but as the distinction between the two sets of "molecules" is down to the quantum state within the molecule boxes, which by design is not thermodynamically "active", some feel that this is not entropically relevant. If we had two sets of molecules that were different colours[15] but otherwise identical, we might not expect any entropy change by mixing them. This is an interesting issue which, to a degree, begs the question.

We should give some weight to this argument but judge that this is not catastrophic if we view the von Neumann's discussion as a heuristic development of the expression $\mathcal{S}_{vN} = -k_B \mathrm{Tr}[D \times \ln(D)]$, which then goes on to prove its worth via its insight and application.

Whatever the status of von Neumann's 'box' argument, he certainly did not rest at its conclusion. For him the job was far from done. The rest of his development is interesting and illuminating, ultimately leading to an adapted form of entropy relevant to macroscopic observers.

12.4.2 Time for Change…

Von Neumann's motivation for developing $\mathcal{S}_{vN}$ was to explore the irreversibility of process **1**. His cyclic gas argument builds a case for $\mathcal{S}_{vN}$ being "analogous" to "classical entropy" (which we believe means $\mathcal{S}_G$) but there is one striking puzzle[16]:

> Although our entropy expression is, as we saw, completely analogous to classical entropy, it is still surprising that it is invariant under normal temporal evolution of the system (process **2**), and increases only in consequence of measurements (process **1**). In classical theory—where measurements in general played no role- its increase, as a rule, resulted from the ordinary mechanical evolution of the system. It is incumbent upon us to clear up this apparently paradoxical situation.

The "ordinary mechanical evolution" referenced in this quote is von Neumann's way of describing the time evolution of a system under the strictures provided by the laws of classical (Newtonian) mechanics. He is disturbed by the contrast between quantum and classical cases. The unitary time evolution of a quantum system is the completely deterministic quantum equivalent of the "mechanical evolution" in the Newtonian case. But, as has already been established, $\widehat{U}(\Delta t)$ leaves the density matrix, and hence $\mathcal{S}_{vN}$, unchanged. As the classical equivalent leads to entropy increase and the Second Law, where lies the root of the difference between the two situations?

To explore this puzzle, von Neumann works up a second thought experiment.

12.4.3 The Single Particle Gas

In this situation,[17] a lone particle is placed in a box which has a sliding partition across its centre. Initially, the partition is unable to move as it is wedged on both sides. In addition, the box is in thermal contact with a reservoir.

The experiment proceeds in one of two different ways.

- **Operation A** If the initial location of the particle is *completely unknown*, i.e., we don't even know which side of the box it is lurking, then the partition is simply removed. The particle is then free to diffuse where it will. The volume of the "gas" doubles; hence, the thermodynamic entropy increase is

$$\Delta\mathcal{S}_T = k_B \ln\left(\frac{V}{V/2}\right) = k_B \ln(2).$$

However, no work is done, and no energy is conducted from the reservoir. Hence, there is no corresponding entropy decrease in the reservoir and the universe's entropy increases. The process is irreversible. At the end of the experiment, we still do not know where the particle is, aside from within the box.

The key point here is that the simple time evolution (mechanical) of the system has brought about an entropy change.

Now, we consider the second way to conduct operations.

- **Operation B** If the location of the particle is *roughly known*, to the extent of being aware of which side of the partition it lies, then the wedge is removed on the *opposite side*. The particle's impacts on the partition cause it to slide towards the opposite wall of the box. Eventually, the partition meets the wall, and the volume of the gas has doubled. Calculable work has been done on the partition, $k_B T\ln(2)$, with a corresponding entropy increase of $k_B\ln(2)$. However, this time there has been a compensating decrease in the reservoir's entropy, due to the conduction of energy into the "gas". Overall, we have $\Delta S_T = 0$ making the process reversible. At the conclusion of this experiment, *we no longer have <u>any</u> awareness of the particle's location within the box.*

As we can see, at the end of both experiments, we have systems in the same situation, but one has brought about a thermodynamic entropy change while the other has not. Regarding operation B, von Neumann declares[18]:

> ... at the end of the process the molecule is again in volume V, but we no longer know whether it is left or right of the middle and there has been a compensating entropy decrease of $k_B\ln(2)$ (in the reservoir). That is, we have exchanged our knowledge for the entropy decrease $k_B\ln(2)$.

Now, von Neumann develops his point. If we know more about the particle, not just simply its crude initial location, either side of the partition, then we can calculate more effectively. Being privy to the initial momentum and a more precise initial location allows us to predict which side of the box the particle will be found in the future. At any time, we could carry out operation B. Hence, the entropy does not change[19]:

> if we knew all the properties (position and momentum) of the molecule before the diffusion process was initiated we could calculate at each subsequent moment whether it is on left or right, and entropy would not have changed. If, however, the only information at our disposal were the macroscopic information that the molecule was initially in the right (or left) half of the enclosure, then entropy would increase upon diffusion.

In conclusion, *simple time evolution does not bring about an entropy change in the classical case either*. The situation is congruent with quantum time evolution. Entropy changes result from incomplete information[20]:

> For a classical observer, who knows all coordinates and momenta, the entropy is therefore constant, and is in fact 0, since the Boltzmann "thermodynamic probability" is 1 – just as in our theory for [quantum] states $D = |\varphi\rangle\langle\varphi|$, since these again correspond to the highest possible state of knowledge of the observer, relative to the system.

[our addition, notation updated to our use. The "thermodynamic probability" is defined earlier in the text; it is the statistical weight in our terminology.]

In these comments, von Neumann establishes his *bona fides* as a "subjectivist" when it comes to classical entropy.

Regarding the Second Law, von Neumann is clear[21]:

> The time-variations of entropy are based then on the fact that the observer does not know everything – that he cannot find out (measure) everything that is measurable in principle. His senses allow him to perceive only the so-called macroscopic quantities.

This ushers in a new phase of the conversation. The central question now becomes "how are the microscopic observables of a quantum system connected with macroscopic quantities that are available to our inspection?" In von Neumann's words[22]:

> ... this clarification of the apparent contradiction mentioned at the outset [quantum vs classical time evolution and entropy] imposes upon us an obligation to investigate the precise analog of classical macroscopic entropy for quantum mechanical ensembles; i.e., the entropy as seen by an observer who cannot measure all quantities, but only a few special quantities, namely, the macroscopic ones.
>
> Quote 12.3 [our addition]

12.4.4 Simultaneous Position and Momentum Measurements

The aim is to find a way of dealing with the entropy of a system given an observer who can only measure macroscopic quantities, and even then, with limited accuracy. In an earlier chapter, von Neumann demonstrated that all limited-accuracy measurements of quantities such as position or momentum, can be replaced with absolutely accurate measurements of a different quantity which are functions of the originals $x \to f(x)$. These new quantities yield the classical macroscopic entropy, $\mathcal{S}_T(f)$.

As an example, von Neumann constructs an experiment which allows the conjugate microscopic quantum observables position, x, and momentum, p, of a particle to be measured via macroscopic observables. This is illustrated in Figure 12.7. There are two aspects to the experiment:

Position measurement: To achieve this, a photon is directed through a slit system (to collimate its path) towards a target particle. The photon is scattered from the particle towards a photographic plate via a further set of collimating slits. The photon produces a macroscopically detectable spot on the plate, the location of which (X') compared to where it should have ended up if had not been scattered by the particle, gives information about the particle's initial location.[23] The particle recoils as a result.

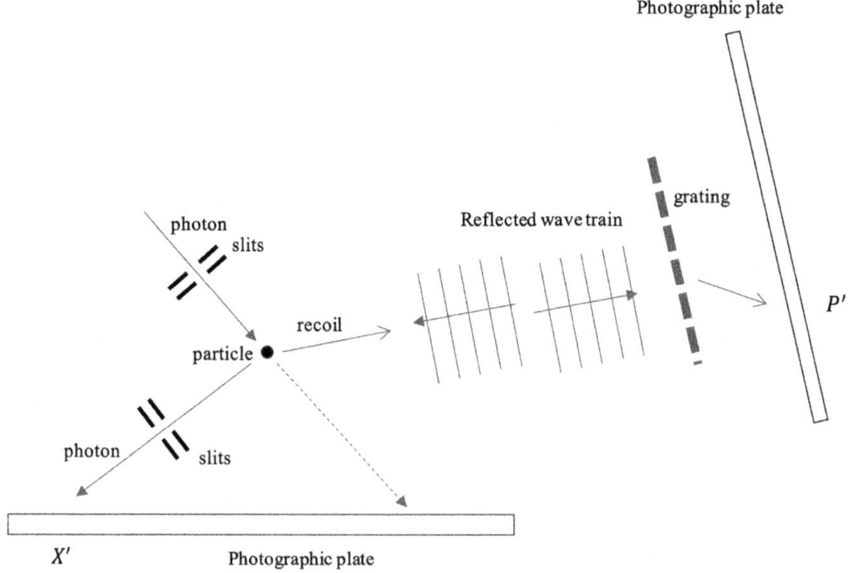

FIGURE 12.7 An experiment to measure the microscopic position and momentum of a particle.

Momentum measurement: A wave train is directed in the path of the recoiling particle. Some of the waves are reflected and Doppler shifted due to the velocity of the particle. These waves are diffracted by a grating and then impact on a photographic plate. The location of the second spot, P', tells us about the angle through which the grating deflects the wave and hence the shifted wavelength. In turn, this yields the velocity/momentum of the particle

While $\{x,p\}$ are not simultaneously measurable, being conjugate and so subject to the uncertainty principle $\Delta x\Delta p \sim h$, $\{X',P'\}$ *can* both be measured at the same time. At the classical level, $\{X',P'\}$ can be measured to arbitrarily high precision, but this does not yield uncertainty violating measures of $\{x,p\}$. The precision to which we can determine x is limited by the wavelength of the scattering photon, which must be less than the width of the slits, or diffraction will occur. Equally, the *coherence length* of the wave train that is Doppler shifted turns out to be the limiting factor to the momentum determination.[24] Together, these factors ensure that uncertainty is not violated.

12.4.5 Projectors

Von Neumann uses the schema for simultaneous position and momentum measurements at the macroscopic level as an exemplar for general macroscopic measurements. He proceeds in the following manner.

Along with the operators $\{\hat{x},\hat{p}\}$ at the quantum level, we can define $\{\hat{X}',\hat{P}'\}$ as operators for the macroscopic quantities. Let's make the eigenvalues of $\hat{P}$ the collection $\{\lambda_n\}$ and the eigenstates $\{|\phi_n\rangle\}$. It's important to remember that these are eigenstates of *macroscopic measurements* that we make. In his book, von Neumann goes into detail in establishing the intuitively appealing idea that all macroscopic measurements can take place at the same time. In quantum terms, *none of them are conjugate*, hence *all the operators commute* allowing $\{|\phi_n\rangle\}$ to be the collection of eigenstates for *all* macroscopic measurements.[25]

> It is reasonable to attribute to this result a general significance, and to view it as disclosing a characteristic of the macroscopic method of observation. According to this, the macroscopic procedure consists of replacing all possible operators A,B,C,..., which as a rule do not commute with each other, by other operators A′,B′,C′,... (of which these are functions to within a certain approximation) which do commute with each other.

The next step is to re-work the notion of measurement in line with common practice these days in foundational quantum theory. The idea of measuring the observable P' can be recast as a question: "does $P' = \lambda_N$?" selecting a specific value to be the subject of the question. In turn, λ_N corresponds to one (or more, assume one for the moment….) eigenstate $|\phi_N\rangle$, making the question "is the system in state $|\phi_N\rangle$?". This question can be constructed in terms of an operator, $\hat{Q}_N$. Back in Section 11.4.5, we discussed the idea that the density matrix was a "selection observable", which is exactly what we need now. Consider a system that might be in one of the states $\{|\phi_n\rangle\}$. We select one and construct the question "is the system in state $\{|\phi_N\rangle\}$?" via the operator:

$$\hat{Q}_N = |\phi_N\rangle\langle\phi_N| \tag{12.57}$$

There are now two options if we act on a state with this operator:

$$\hat{Q}_N|\phi_M\rangle = |\phi_N\rangle\langle\phi_N|\phi_M\rangle = 0 = 0|\phi_M\rangle \tag{12.58}$$

or:

$$\hat{Q}_N|\phi_N\rangle = |\phi_N\rangle\langle\phi_N|\phi_N\rangle = |\phi_N\rangle = 1|\phi_N|\rangle \tag{12.59}$$

making the eigenvalues of this observable $q_N = \{0,1\}$ or {no, yes}. An operator that looks for a specific state, or collection of states, is known as a *projection operator*, as they "project out" or "select" their states from the full collection.

Thinking back to our discussion of compound operators in Section 11.1.5, any function of an eigenvalue, $f(o_i)$ can be obtained from an eigenstate by applying the same function of the appropriate operator, $f(\widehat{O})$. This allows us to uncover an important property of our projection operators. If the eigenvalues are to be $q = \{0,1\}$, then the function $f(q) = q^2 - q$ must vanish in all instances. Applying the same idea to the operator function:

$$f\left(\widehat{Q}\right) = \widehat{Q}^2 - \widehat{Q} = 0 \qquad \widehat{Q}^2 = \widehat{Q} \tag{12.60}$$

shows us that a projection operator must be idempotent (as is the case with pure state density matrices).

In his detailed argument, von Neumann develops this approach further by showing how to combine primitive questions into compound ones. Ultimately, we end up with a collection $\{\widehat{Q}_N\}$ such that[26]:

> The $\{\widehat{Q}_N\}$ correspond to all macroscopically answerable questions … i.e., to all discriminations between alternatives in the system under investigation that can be carried out macroscopically. They are all commutative. We can conclude … that $1-\widehat{Q}_N$ belongs along with $\widehat{Q}_N$ to the set of all projectors associated with macroscopically answerable questions (propositions), … It is reasonable to assume that every system **S** admits of only a finite number of such questions; i.e., only a finite number of such operators: $\{\widehat{Q}_N\}$.

[notation changed to coincide with that used in this book]

Von Neumann makes another crucial point regarding these projectors, almost as a casual aside[27]:

> It should be observed, additionally, that the $\{\widehat{Q}_N\}$. —which are the elementary building blocks of the macroscopic description of the world—correspond in a certain sense to the cell division of phase space in classical theory.

[notation changed to coincide with that used in this book]

In other words, transforming the theoretical measurements of quantum observables (some of which do not commute), into macroscopic measurements of observables at that scale, all of which do commute, is tantamount to a coarse-graining mechanism. Ultimately, this is why von Neumann is able to produce an entropy that can be said to correlate with the classical thermodynamic or phenomenological entropy.

Meanwhile, back to our projection operators.

We should expect some "degeneracy" amongst the collection of eigenstates, i.e., more than one state $|\phi_N\rangle$ with the same eigenvalue as our original measurement, λ_N. Otherwise, the results of our macroscopic measurements would *completely and unambiguously determine the microscopic state*, which we know is very much not the case.[28] For a question posed as "is $P' = \lambda_N$?", we move into "is the system in a state from the collection $\{|\phi_N\rangle\}$ all of which have $P' = \lambda_N$?" to a projector $\widehat{Q}_N$ which will have more than one eigenstate with $q_N = 1$.

To cope with this, we introduce s_n, following von Neumann's notation, to indicate the dimensionality of the "space" of eigenstates with the eigenvalue $q_N = 1$. More prosaically, s_N counts the number of eigenstates projected out of the collection by $\widehat{Q}_N$, which makes:

$$\mathrm{Tr}\left[\widehat{Q}_n\right] = s_n \tag{12.61}$$

Any situation where $s_N = 1$ (unlikely) would correspond to a single eigenstate, $|\phi_N\rangle$, and the projection operator would then be the density matrix for that state, $\hat{Q}_N = |\phi_N\rangle\langle\phi_N|$. In other situations:

$$\hat{Q}_n = \sum_{\text{projected states}}^{\#=s_n} |\phi_n\rangle\langle\phi_n| \tag{12.62}$$

In general, $s_N > 1$, and most probably $s_N \gg 1$.

12.4.6 Macroscopic Entropy

Having spent some time setting up the macroscopic measurement scenario, von Neumann then poses a crucial question: if a state change takes place at the quantum level, what is the largest compensatory entropy change that can be determined at the macroscopic level, given the array of possible measurements? In his words[29]:

> Now...what entropy does the mixture U have for an observer whose indecomposable projections are $[\hat{Q}_1, \hat{Q}_2, \ldots]$? Or, more precisely: How much entropy can such an observer maximally obtain by transforming U into V—i.e., what entropy increase (or decrease) can he (under suitable conditions and the most favorable circumstances) produce in external objects as compensation for the transition U → V?

[notation changed to coincide with that used in this book]

The first move in answering this question is to propose an ensemble of the following form (with the nature of the $\{x_n\}$ to be established in the argument that follows):

$$\mathbb{D} = \sum_n x_n \hat{Q}_n \tag{12.63}$$

To guarantee that this is a validly normalized ensemble, we must ensure that $\mathrm{Tr}[\mathbb{D}] = 1$. Assuming all the s_n to be finite, $\mathbb{D}$ must contain s_1 occurrences of the eigenvalue x_1, s_2 occurrences of the eigenvalue x_2, etc. Its trace is then:

$$\mathrm{Tr}[\mathbb{D}] = \Sigma\, s_n x_n = 1 \tag{12.64}$$

Also, $\mathrm{Tr}[\mathbb{D}\times\ln(\mathbb{D})]$ must be built from s_1 values $x_1\ln(x_1)$, s_2 values $x_2\ln(x_2)$, and so on. This makes the von Neumann entropy of the ensemble:

$$\mathcal{S}_{vN} = k_B \mathrm{Tr}[\mathbb{D}\times\ln(\mathbb{D})] = k_B\, \Sigma\, s_n x_n \ln(x_n) \tag{12.65}$$

To put a further constraint on the x_n, consider the product $\hat{Q}_M \times \mathbb{D}$:

$$\hat{Q}_M \times \mathbb{D} = \sum_n x_n \hat{Q}_M \times \hat{Q}_n = x_M \hat{Q}_M \tag{12.66}$$

using the properties of the projectors.

Following on:

$$\text{Tr}\left[\hat{Q}_M \times \mathbb{D}\right] = x_M \text{Tr}\left[\hat{Q}_M\right] = x_M s_M \tag{12.67}$$

Re-arranging and using (12.61):

$$x_M = \frac{\text{Tr}\left[\hat{Q}_M \times \mathbb{D}\right]}{s_M} = \frac{\text{Tr}\left[\hat{Q}_M \times \mathbb{D}\right]}{\text{Tr}\left[\hat{Q}_M\right]} \tag{12.68}$$

Taking this value back to the entropy produces:

$$\mathcal{S}_{vN} = k_B \sum_n s_n x_n \ln\left(x_n\right) = k_B \sum_n \text{Tr}\left[\hat{Q}_n \times \mathbb{D}\right] \ln\left(\frac{\text{Tr}\left[\hat{Q}_n \times \mathbb{D}\right]}{\text{Tr}\left[\hat{Q}_n\right]}\right) \tag{12.69}$$

which is the result we require. Von Neumann then proposes that the macroscopicly determined entropy for *any* ensemble D is:

$$\mathbb{S}_{vN} = k_B \sum_n \text{Tr}\left[\hat{Q}_n \times D\right] \ln\left(\frac{\text{Tr}\left[\hat{Q}_n \times D\right]}{s_n}\right) = k_B \sum_n \text{Tr}\left[\hat{Q}_n \times D\right] \ln\left(\frac{\text{Tr}\left[\hat{Q}_n \times D\right]}{\text{Tr}\left[\hat{Q}_n\right]}\right) \tag{12.70}$$

This is the entropy that he believes is the quantum equivalent of the thermodynamic entropy.

In justification for this assertion, he starts by pointing out how:

$$k_B \sum_n \text{Tr}\left[\hat{Q}_n \times D\right] \ln\left(\frac{\text{Tr}\left[\hat{Q}_n \times D\right]}{\text{Tr}\left[\hat{Q}_n\right]}\right) \geq \text{Tr}\left[D \times \ln(D)\right] \tag{12.71}$$

the equality being the case $D = \mathbb{D} = \sum_n x_n \hat{Q}_n$. The argument to prove this inequality does take some extended manipulation, so we ask that this be taken on trust and refer the reader to von Neumann's book. In the same section, he is also able to show that $\mathbb{S}_{vN} > 0$, unlike $-k_B \text{Tr}\left[D \times \ln(D)\right]$, which we know is zero for a pure state.

Better than both these encouraging properties, *this entropy changes with time*. The unitary time evolution operator, $\hat{U}(\Delta t) = e^{-i\hat{H}\Delta t/\hbar}$, acts on the density matrix, transforming $\text{Tr}\left[\hat{Q}_n \times D\right]$ into:

$$\text{Tr}\left[\hat{U} \times D \times \hat{U}^{\dagger} \times \hat{Q}_n\right]$$

If the $\widehat{Q}_n$ commute with $\widehat{H}$, the Hamiltonian operator,[30] this can be rearranged:

$$\mathrm{Tr}\left[\widehat{U}\times D\times\widehat{U}^{\dagger}\times\widehat{Q}_n\right]=\mathrm{Tr}\left[D\times\widehat{U}^{\dagger}\times\widehat{Q}_n\times\widehat{U}\right] \qquad \textbf{T5}$$

$$=\mathrm{Tr}\left[D\times\widehat{U}^{\dagger}\times\widehat{U}\times\widehat{Q}_n\right] \qquad \text{commutation}$$

$$=\mathrm{Tr}\left[D\times\widehat{Q}_n\right]$$

So, the issue hinges on the various $\widehat{Q}_n$ commuting with $\widehat{H}$. According to von Neumann, this is almost self-evidently not the case[31]:

> But all $\widehat{Q}_n$—i.e., all macroscopically observable quantities—are in no way all commutative with $\widehat{H}$. Indeed, many such quantities—for example, the center of gravity of a gas in diffusion—change appreciably with t; i.e., $\mathrm{Tr}[D\times\widehat{Q}_n]$ is not constant. Since all macroscopic quantities do commute, $\widehat{H}$ is never a macroscopic quantity; i.e., the energy cannot be measured macroscopically with complete precision. This is plausible without additional comment.

[notation changed to coincide with that used in this book]

Of course, changing with time does not necessarily mean that $\mathbb{S}_{vN}$ *increases*, as we would require. Interestingly, von Neumann does not attempt to prove this in his book. His comment runs[32]:

> Since the macroscopic entropy [$\mathbb{S}_{\mathrm{vN}}$] is always time variable, the next question to be answered is this: Does it behave like the phenomenological thermodynamics of the real world; i.e., does it predominantly increase? This question is answered affirmatively in classical mechanical theory by the so-called Boltzmann H-theorem. In that, however, certain statistical assumptions—namely, the so-called "disorder assumptions"—must be made. In quantum mechanics it was possible for the author to prove the corresponding theorem without such assumptions.[33] Since the detailed discussion of this subject, as well of the ergodic theorem closely connected with it (see the reference cited in [footnote 30], where this theorem is also proved) would go beyond the scope of this volume, we cannot report on these investigations. The reader who is interested in this problem can refer to the treatments in the references.
>
> Quote 12.4 [our addition]

The key point here is that *this* is the form of entropy that von Neumann believes corresponds to the thermodynamic entropy. Again, we see that entropy comes about due to some loss or obscuring of information, in this case via the coarse graining inherent in the macroscopic measurements.

12.5 CONCLUSIONS

We can think of the whole universe as being described by a single wavefunction evolving in time. It's a pure state with entropy zero. Everything is entangled with everything else. As it becomes practically impossible to exploit that entanglement … we coarse grain. Coarse graining in quantum mechanics means tracing over unmeasurable components. This increases the entropy and moreover turns a pure state into a mixed state. In this way, classical probabilities emerge from a completely deterministic quantum system.

> In summary, von Neumann entropy lets us understand both the information loss by measurement and by losing entanglement.
>
> Matthew D. Schwartz[34]

The sifting quantum boxes argument reads as a heuristic justification for von Neumann's "entropy" expression, $\mathcal{S}_{vN}$. As it stands, $\mathcal{S}_{vN}$ is useful and informative within the context of quantum information theory. It also helped von Neumann explore the difference between his process **1** and **2** evolutions of quantum states. It is possible that the term "entropy" is unhelpful applied to $\mathcal{S}_{vN}$. Mind you, *entanglement entropy* is even worse… Perhaps if we thought of $\mathcal{S}_{vN}$ as an *information measure* leading to *an entropy at the macroscopic level*, things would be clearer.

His second thought experiment throws light on how the von Neumann "entropy" remains frozen in time, even though the equivalent classical time evolution does increase entropy. He argues that the classical increase in entropy comes about due to our lack of information having burgeoning consequences.

From our perspective, developing a conceptual understanding of entropy, it is significant that $\mathcal{S}_{vN} = 0$ for a pure state. Even though pure states do not have the level of information about a system that the classical state would have, there is no entropy to mine. The von Neumann entropy only becomes non-zero when information *about the state*, not *within the state*, is absent. Our lack of knowledge is the significant factor. As a subjectivist, von Neumann is aware of this. He also points out that while we can *calculate* using quantum observables, we never directly *measure* them. Measurement is restricted to macroscopic quantities, which somehow need to be mapped to the quantum level observables. Effectively this is done via a form of coarse graining, leading to an entropy formulation which behaves in a much more amenable manner. Arguably, von Neumann does not suggest that $\mathcal{S}_{vN}$ maps into $\mathcal{S}_T$ in the thermodynamic limit, it is the macroscopic $\mathbb{S}_{vN}$ that he is lobbying for.

Suppose that the most optimistic hopes of the quantum cosmologists are correct, and that the business of the entire universe is governed by one uber-wave function. This would be the ultimate entanglement. It would also be a pure state. Presumably, the time evolution of this wave function would be complicated, but unitary. Consequently, the Gibbs/von Neumann entropy would be constant and further, given the pure state, zero. Practically, this wave function would not be of much use. To conduct experiments on (much!) smaller sub-systems within the entanglement, we would have to trace over vast amounts of information that we could not know or measure anyway. This is the quantum version of course graining. It also turns what was a pure state into a mixed state, introducing entropy.

And then there is the measurement problem. What could conceivably bring about state collapse on a wave function of universal scale? How does a web of possibilities turn into an instance of actuality? A universal quantum gravity wave function would describe a superposition of different space–time geometries (see Chapter 15). Presumably, some collapse leads the observed version, but the mechanism is far from clear. Perhaps some modification to the Schrödinger equation is needed with retains a random element, but accounts for state collapse, given a suitable trigger. In which case, unitary time evolution might well be an approximation which holds true up to the moments of state collapse. Penrose has suggested that the superposition of space–time geometries can be maintained until the difference in their gravitational energies passes a threshold,[35] at which time "collapse" into one of the alternatives takes place. In such circumstances, von Neumann's process **1** gains a physical mechanism, which might shed further light on entropy. If something like Penrose's suggestion turns out to be correct, the random element will remain, and entropy will arise as a result.

The most popular solution to the measurement problem among quantum cosmologist is the Many Worlds Interpretation, first suggested by Hugh Everett III.[36] Everett believed that the measurement problem was down to our not taking the quantum predictions seriously enough. He suggested that unitary time evolution was the only progression in play. All branches of a superposition continued into the future and state collapse never happens. Our observation of a single measurement result comes about as we are confined to a single branch of the superposition which spans parallel evolving worlds. In this grand conception, every measurement triggers the evolution of another parallel universe or world. Initially, the two worlds exist "within touching distance", but as a measured system gets increasingly entangled with its environment so the worlds diverge to the point where they can no longer coherently interact. Hence, we only observe one branch outcome. There are consequences for our perceptions of continuity of consciousness, but these fascinating ideas are not germane to us here. More relevant are the implications for probability. If each branch of the superposition exists, albeit separated from us, what are we to make of the Born rule? Were we to take "God's eye view" across the branching worlds, all possibilities would be present, as per an ensemble, but without the relative frequency of outcome that we bake into our collection of systems. In simple terms, if we approach a measurement with the quantum prediction of a 1:3 chance of a certain result, what can we take that to mean if all results come about at the same time? Various defences of probability have been made by Many Worlds theorists, notably David Deutch[37] who constructs an argument from decision theory. Whatever the merits of these particular thoughts, the discussion of Section 12.3.8 is re-enforced. We can't claim to have a full understanding of entropy without some view of the nature of probability.

Arguably the measurement problem is a key unresolved issue in science, a fact that is often brushed under the carpet, at least while pragmatists "shut up and calculate". It is entirely possible that we lack fundamental concepts that will point us in the right direction. In fact, the existence of a debate over the interpretation of quantum theory may, in itself, indicate that the last conceptual words have not yet come to light. In which case, it would be premature to close any consideration of entropy and state collapse (process **1**).

NOTES

1 von Neumann, J. *Mathematical Foundations of Quantum Mechanics* (R. T. Beyer, Trans.): Princeton University Press, 2018, p. 234.
2 Lev Davidovich Landau 1908–1968, Nobel Prize in Physics 1962, awarded for the mathematical theory of superfluidity. See also: Landau, L. (1927). "Das Daempfungsproblem in der Wellenmechanik". *Zeitschrift für Physik.* **45** (5–6): 430–464
3 John von Neumann, 1903–1957, Hungarian-American mathematician, physicist, computer scientist, engineer, and polymath.
4 That is, the manner in which coherent superpositions collapse in defiance of ordinary unitary time evolution.
5 *Mathematische Grundlagen der Quantenmechanik*, Springer 1932.
6 The generalization shown here from the definition in Section 10.2.1 is due to us now having to consider matrices as well, hence $*$ is replaced by $\dagger$.
7 As per 1, page 233.
8 As per 1, page 235.
9 Or an especially trained Demon, rented from Chapter 14.
10 Recall that after Step 0, the particles within the molecule boxes are all in eigenstates of O. Hence this measurement will not disturb the state of the particle within.
11 The reader should note the implicit caginess of this statement. Any judgement of von Neumann's argument hinges on being satisfied that he has successfully argued the equivalence of his form of entropy, to be revealed shortly, and the thermodynamic or phenomenological entropy. Much has been made of this debate in the physics-philosophy of science press. In his paper Erin Sheridan has, in my {our} view successfully argued that the gas cyclic experiment equates the von Neumann entropy with the Gibbs entropy and that further points need to be made to join up with the thermodynamic entropy, which are also in von Neumann's book. As this discussion will be opened up later in the chapter, we are using 'classical entropy' at this stage to cover all bases.
12 This is also known as the entanglement entropy, but more correctly that is a specific example of the von Neumann entropy applied to entangled states.
13 von Neumann, J. *Mathematical Foundations of Quantum Mechanics* (R. T. Beyer, Trans.): Princeton University Press, 2018, p. 259.
14 Sheridan "A Man Misunderstood: Von Neumann did not claim that his entropy corresponds to the phenomenological thermodynamic entropy" https://doi.org/10.48550/arXiv.2007.06673, last accessed February 2024.

15 Were such a thing possible.
16 As per 12.
17 The version presented here is a simplified version of the one in von Neumann's text. There are obvious parallels with Joule expansion as discussed in Chapter 5.
18 As per 12.
19 As per 12.
20 As per 12.
21 As per 12.
22 As per 1, page 260.
23 By selecting positions for the two sets of slits, we are effectively asking the question "did the particle start somewhere near where the original path and the final path of the photon intersect". However, from a collection of such slits, we can effectively perform a co-ordinate measurement.
24 The frequency of a wave train is only precisely defined if the wave train is infinite. A wave train of finite length can only be constructed by from a summation of waves of differing frequency. This is the Fourier theorem.
25 As per 1, page 262.
26 As per 1, page 264.
27 As per 1, page 265.
28 Von Neumann argues this somewhat more rigorously in the following way (As per 1, p265): "*If all the* $s_n = 1$ ... *then (each)...* $\widehat{Q}_n = \phi_n\rangle\langle\phi_n$, *and because* $\widehat{Q}_1 + \widehat{Q}_2 + \ldots = 1$ *the* $\{\phi_n\rangle\}$ *would form a complete orthonormal set. This would mean that macroscopic measurements would themselves make possible a complete determination of the state of the observed system. Since this is ordinarily not the case, we have in general* $s_n > 1$, *and in fact* $s_n \gg 1$". [updated with our terminology]
29 As per 24.
30 As the exponential of the matrix is defined in terms of the power series $e^{-i\widehat{H}\Delta t/\hbar} = 1 - \frac{i\widehat{H}\Delta t}{\hbar} + \frac{1}{2!}\left(\frac{i\widehat{H}\Delta t}{\hbar}\right)^2 + \ldots$, commuting with $\widehat{H}$ is sufficient to guarantee commuting with $\widehat{U}$.
31 As per 1, footnote 204.
32 As per 1, page 269.
33 Here von Neumann references the paper: von Neumann "Proof of the ergodic theorem and the H-theorem in the new mechanics", *Z. Phys.* 57, 30–70, 1929.
34 From https://scholar.harvard.edu/files/schwartz/files/6-entropy_0.pdf, last accessed March 2024.
35 Most likely something called the *Planck mass*. It is possible to combine the fundamental constants of the universe to give a 'natural' mass scale, known as the Planck mass: $M_P = \sqrt{\hbar c/G} \sim 2.2 \times 10^{-8}\,\text{kg}$.
36 H Everett, Relative State Formulation of Quantum Mechanics, *Reviews of Modern Physics*, 1957, 29: 454–462.
37 D. Deutsch, Quantum Theory of Probability and Decisions, *Proc. R. Soc.*, London, February 1999

Part III

13 Life

Living organisms obtain energy from fuel in the form of food, yet are fundamentally different from heat engines. Chemical reactions are described, first qualitatively in terms of molecular rearrangement and then quantitatively from a thermodynamic perspective through a derivation of the Law of Mass Action. All chemical reactions, including those important in a biological context, proceed in a direction which reduces free energy. After briefly reviewing one important example, muscle contraction, we argue that the conversion of high-free-energy food to low-free-energy waste inevitably increases the entropy of the environment, both at the level of a single organism level and also for entire ecosystems.

13.1 CHEMISTRY

So far in the book, we have, where possible, linked our development and discussion of entropy to human experience—of all branches of physics (with the possible exception of mechanics), thermodynamics most lends itself to this viewpoint. The most intimate human experience of all is that of being alive, and by now, we are ready to apply what we've learned to living organisms. We know that engines—entities that do useful work—require fuel for their operation to supply the energy needed. For living things, the fuel is called food. What if we treat the human body as an engine? As we've already seen in Chapter 2, an important way of describing what we eat uses a unit of energy, the large Calorie, equivalent to 4184J. Particularly energy-rich foodstuffs are carbohydrates such as sugar. Our *metabolism*, the process which governs energy transfer within our bodies, combines a molecule of sugar, such as glucose, with six molecules of oxygen from the air we breathe; the resulting chemical reaction produces six molecules of carbon dioxide and six of water (eventually released back to the atmosphere when we exhale) and, crucially, releases energy. For every mole of glucose consumed this way, 2.8 MJ of energy are released[1]—it seems little surprise we get such a buzz from sweet treats!

A moment's thought, however, should convince us that things must be more complicated. Typically, we measure the energy released by combining glucose with oxygen by burning it in a calorimeter; the heat released can be deduced from the resulting temperature rise in the apparatus, very much in the same spirit as Joule's groundbreaking experiments with paddle wheels discussed in Chapter 2. However, identifying *metabolism* with *combustion* and the body with a heat engine is far too naive. A heat engine performs work by operating between two reservoirs at different temperatures — the great triumph of classical thermodynamics, reviewed in Chapter 4, was to relate the engine's efficiency to the ratio of those temperatures. One thing we all know about the human body is that it operates at a constant temperature of roughly 37°C; indeed, departures from this norm are interpreted as a sign that something is not right. There are no reservoirs within

DOI: 10.1201/9781003121053-17

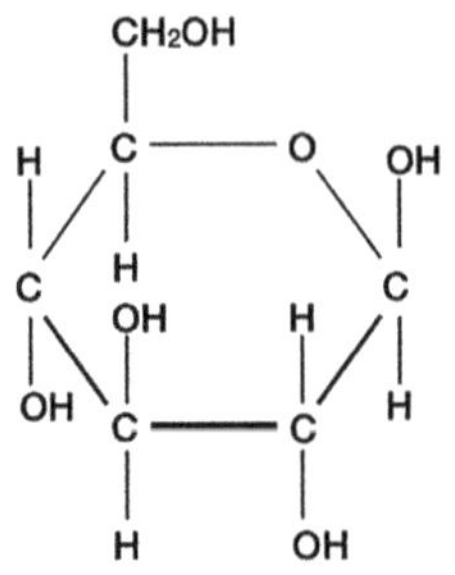

FIGURE 13.1 Diagram showing the relative arrangement of atoms in a glucose molecule, with chemical bonds represented by straight lines. The bond lengths and angles are not to scale.

the body where heat produced during, say, the compression stroke of a piston can be dumped. Another important observation is that there are no significant pressure differences within the fluids of the body, so that the identification of useful work with the area of a closed cycle in the *p, V* diagram cannot be made. This chapter will present a (necessarily) simplified overview of how thermodynamics applies to the processes within *living* systems—we will find, again, that entropy plays a key role in the discussion.

Let's begin by re-expressing the reaction between glucose and oxygen in the manner used by chemists:

$$C_6H_{12}O_6 + 6O_2 \rightarrow 6CO_2 + 6H_2O. \tag{13.1}$$

The letters C, H, and O stand for carbon, hydrogen, and oxygen atoms. They are grouped in combinations representing molecules—so, for instance, a glucose molecule consists of 6 carbon, 12 hydrogen, and 6 oxygen atoms. This isn't the most general or detailed way of representing glucose since it communicates nothing of the arrangement of the atoms within the molecule (see Figure 13.1), but it will serve for now. The molecular formulae for oxygen, carbon dioxide, and water already appeared in Table 3.1 of Chapter 3. The arrow suggests that the reaction has a direction, in other words that glucose and oxygen combine to form carbon dioxide and water. If you check carefully, you'll see that on each side of the equation totals of 6 C atoms, 12 H atoms, and 18 O atoms are represented. In other words, in the course of the chemical reaction, no atoms are created or destroyed, but instead are merely rearranged into different molecular groups.[2] In contrast, the number of molecules involved is *not* conserved: there are 7 *reagent* molecules on the left hand and 12 *product* molecules on the right.

Molecules are stable configurations of atoms bound together by attractive forces resulting from the intermingling and subsequent rearrangement of negatively charged electrons orbiting positively charged atomic nuclei. A faithful and quantitative account of this process requires quantum mechanics. We frequently visualize these interactions by a line joining the interacting atoms representing a *chemical bond*, e.g., in the schematic depiction of a glucose molecule shown in Figure 13.1.

Each bond contains a pair of electrons occupying a lower energy quantum state than those found in isolated atoms, so that energy is needed to break the bond between the atoms. When a chemical reaction forming a different set of molecules takes place, the participating electrons are redistributed into new bonds with different energies—so chemical reactions such as (13.1) are accompanied by a net energy change, which can be positive or negative. This (extremely) broad-brush picture accounts for the 2.8MJ release.

13.2 MASS ACTION

From a thermodynamic perspective, the key feature of chemical reactions, namely that molecules are assemblies of integer numbers of atoms and that during the course of a reaction, numbers of molecular species may change, but the number of atoms remains fixed, can be incorporated into our machinery as follows. Let's consider an isolated system contained in a fixed volume V, with entropy

$$S = S\left(U, V, N_1, \ldots, N_s\right). \tag{13.2}$$

Here N_i, with $i = 1,\ldots,s$ are the number of molecules A_i, where each i represents a distinct molecular species. Since the system is isolated with a fixed volume U and V are constant, but the N_i may vary as a consequence of chemical reactions, changes in entropy satisfy

$$dS = \sum_i dN_i \left.\frac{\partial S}{\partial N_i}\right|_{U,V,N_j,j\neq i}. \tag{13.3}$$

In equilibrium, entropy is maximal, so $dS = 0$. The term multiplying each factor of dN_i in the sum is related to the *chemical potential* μ_i of the ith molecular component:

$$\mu_i = -T\frac{\partial S}{\partial N_i}. \tag{13.4}$$

Chemical potential was first encountered in Section 8.3.3. Like temperature, it is a *thermodynamic control parameter*: its relation to the number of molecules present is analogous to that between temperature and the molecules' energy. From a theoretical perspective, it plays a central role in the description of systems where the number of particles present may vary, for instance, as a consequence of a chemical reaction. It's a less familiar concept than temperature merely because our kitchens and workshops, while amply equipped with thermometers as described in the opening chapter, don't contain many chemical potentiometers.

Now, let's write a general chemical reaction among our s molecules in a less familiar way:

$$\nu_1 A_1 + \nu_2 A_2 + \cdots + \nu_s A_s \rightarrow 0. \tag{13.5}$$

Because molecules are composed of whole numbers of atoms, and because atoms are neither created nor destroyed in chemical reactions, the numbers ν_i must be integers. However, because all the terms are now written on the same side of the reaction, the ν_i for the reagent molecules must be taken as negative; e.g. for the reaction (13.1): $\nu_{\text{glucose}} = -1$; $\nu_{O_2} = -6$; $\nu_{CO_2} = \nu_{H_2O} = 6$. Every time a reaction occurs, the molecular species populations N_i (the total numbers in the bucket) change as follows: $dN_{\text{glucose}} = -1$; $dN_{O_2} = -6$; $dN_{CO_2} = dN_{H_2O} = 6$, i.e. $dN_i = \nu_i$. Assuming a constant reaction temperature, therefore, Equations (13.3,4) can be recast in equilibrium as

$$\sum_i \nu_i \mu_i = 0. \tag{13.6}$$

This equation holds for any chemical reaction. To go further, we need more insight into chemical potential μ_i. Let's recall definitions (5.36, 5.37), viz. $T = (\partial S/\partial U\,|_V)^{-1}$, $p = T\partial S/\partial V\,|_U$ from Chapter 5:

$$S = S(U,V,N_i) \Rightarrow dS = \frac{\partial S}{\partial U}dU + \frac{\partial S}{\partial V}dV + \sum_i \frac{\partial S}{\partial N_i}dN_i \tag{13.7}$$

i.e.

$$dS = \frac{dU}{T} + \frac{pdV}{T} - \frac{1}{T}\sum_i \mu_i dN_i. \tag{13.8}$$

We deduce a modified version of equation (4.17) (previously derived as (9.44)):

$$dU = TdS - pdV + \sum_i \mu_i dN_i. \tag{13.9}$$

In this form μ_i can be interpreted as the increase in energy of a system when a single molecule of type A_i is added. Now recall that in systems at constant temperature and volume (such as the human body), it is more convenient to reexpress thermodynamic relations in terms of the free energy $F = U - TS$, $dF = dU - TdS - SdT$, so that

$$dF = -SdT - pdV + \sum_i \mu_i dN_i = \sum_i \mu_i dN_i, \tag{13.10}$$

where the second equality results from $dT = dV = 0$. The equilibrium condition $dF = 0$, which follows from (13.6) corresponds to a *minimum* of the free energy. Chemical potential is related to free energy by

$$\mu_i = \frac{\partial F}{\partial N_i}. \tag{13.11}$$

What we gain from these gymnastics is the ability to deploy the power of statistical mechanics set out in Chapter 5 once we recall the alternative definition $F = -k_B T \ln Z$, with Z the partition function. For now, assume the collection of molecules can be treated using the statistical description developed for an ideal gas:

$$F = F(T,V,N_1,\ldots,N_s) = \sum_i F_i(T,V,N_i) \tag{13.12}$$

with F_i the free energy of the A_i molecular species. F can be split up like this into the sum of individual components whenever the interactions between the components, especially when well-separated, can be neglected. Hence:

$$F_i(T,V,N_i) = -k_B T \ln Z_i(T,V,N_i); \qquad Z_i(T,V,N_i) = \frac{1}{N_i!}[Z_i(T,V,1)]^{N_i} \tag{13.13}$$

Again, the second equality in (13.13) results from the ideal gas assumption of vanishingly small interactions between different molecules and dilute occupation of quantum states, this time of the same species, and the factor $N_i!$ in the denominator asserts the indistinguishability of states in which two identical particles are exchanged, as discussed in Physics Appendix A1. Using Stirling's approximation for $\ln N!$ we arrive at

$$F_i(T,V,N_i) = -k_B T N_i \left[\ln Z_i(T,V,1) - \ln N_i + 1\right] \tag{13.14}$$

Hence,

$$\mu_i = \left.\frac{\partial F}{\partial N_i}\right|_{T,V,N_j,j\neq i} = -k_B T\left[\ln Z_i(T,V,1) - \ln N_i\right]. \tag{13.15}$$

Now substitute this result into the equilibrium condition (13.6):

$$-k_B T \sum_i \nu_i \left[\ln Z_i(T,V,1) - \ln N_i\right] = 0;$$

$$\text{i.e.,} \sum_i \nu_i \ln N_i = \sum_i \nu_i \ln Z_i(T,V,1), \tag{13.16}$$

then exponentiate both sides to derive (using the product symbol defined in (MA.21))

$$\prod_i N_i^{\nu_i} = \prod_i \left[Z_i(T,V,1)\right]^{\nu_i} \equiv K(T,V), \tag{13.17}$$

where we have defined a dimensionless *equilibrium constant* $K(T,V)$.[3] Each chemical reaction has its own equilibrium constant.

Equation (13.17) is sometimes known as the *Law of Mass Action*, and it enables us to predict the relative equilibrium concentrations of the different reagents and how this varies as external conditions are changed. First, we need to rewrite our prototype reaction (13.1) in a more nuanced way:

$$C_6H_{12}O_6 + 6O_2 \rightleftharpoons 6CO_2 + 6H_2O. \tag{13.18}$$

The arrows pointing both to right and to left signify that the reaction can actually proceed in either direction; equilibrium is attained only once the leftward rate is equal to the rightward—under such conditions, the reagent concentrations will remain unaltered over time. The Law of Mass Action (13.17) permits these concentrations to be predicted from first principles, for instance, using the ideal gas expression for the partition function of a single molecule from Equation (7.47):

$$Z_i(T,V,1) = V\left(\frac{2\pi m_i k_B T}{h^2}\right)^{\frac{3}{2}} Z_i^{int}. \tag{13.19}$$

In this expression, m_i is the mass of molecule A_i and Z_i^{int} is the internal partition function of the molecule, which implicitly contains all the information about chemical bond energies referred to earlier and is also sensitive to the molecule's rotations and vibrations, which need to be taken into account for all save monatomic species. In principle, this information can be calculated using quantum mechanics or deduced from experimental data from the molecule's absorption of infrared and microwave radiation of particular wavelengths.

To illustrate the law in action, consider an idealized reaction $AB \rightleftharpoons A + B$ so that $\nu_{AB} = -1$, $\nu_A = \nu_B = 1$. We'll assume, for simplicity: that A, B are sufficiently simple molecules that they have no internal degrees of freedom so that $Z_A^{int} = Z_B^{int} = 1$; that the chemical bond connecting A to B in the AB molecule requires an energy ε_{AB} to break[4]; and that higher rotational and vibrational states of

AB require such large energies to excite that their population is negligible at temperatures of interest. We infer $Z_{AB}^{int} = e^{\varepsilon_{AB}/k_BT}$. Note that all species are allotted a common reference energy so that the energy of the AB molecule in comparison to a widely separated A, B pair is *negative*, i.e. $-\varepsilon_{AB}$. The Law of Mass Action (13.17) together with (13.19) then yields

$$K(T,V) = \frac{N_A N_B}{N_{AB}} = V\left(\frac{2\pi k_B T}{h^2}\right)^{\frac{3}{2}}\left(\frac{m_A m_B}{m_{AB}}\right)^{\frac{3}{2}} e^{-\frac{\varepsilon_{AB}}{k_BT}}. \tag{13.20}$$

Suppose a fraction x of the AB molecules originally present dissociate to form equal numbers of A and B. The concentration of a reagent R is written $[R]$ and quoted in units of moles per litre.[5] If we denote by $[AB]_0$ the concentration of AB in the limit $x \to 0$, then the equilibrium constant

$$K = \frac{[A][B]}{[AB]} = \frac{(x[AB]_0)^2}{(1-x)[AB]_0}. \tag{13.21}$$

Figure 13.2 uses relations (13.20,21) to plot x as a function of both T and $[AB]_0$. As might be anticipated, at low temperatures (to the right of the plot as drawn) x is practically zero, but rises monotonically with T (i.e. to the left) as the thermal energy available to break the AB bond increases, reaching unity (i.e. complete dissociation) at temperatures of $O(10^3 K)$ for all but the most concentrated systems. However, the upward curvature of lines of constant T show x also has an interesting dependence on $[AB]_0$; it is possible to reduce the dissociated fraction x merely by adding more AB molecules to the system, or alternatively to increase x by removing them (for instance via another chemical reaction).

In general, such predictive power is hard to achieve unless the reaction involves gases comprising simple molecules. However, the Law of Mass Action (13.17) is more widely applicable. It can also be deduced from a completely different standpoint, namely reaction kinetics. Consider a reaction of the form $pA + qB \rightleftharpoons rC + sD$. We've already seen that the equilibrium state is invariant in time and yet not static: in equilibrium, the reaction is proceeding at the same rate *in either direction*. Let's specify $p = 1, q = 2$, and consider the reaction proceeding from left to right. Since a single A molecule is involved, whatever the detailed mechanism, the reaction rate r_{LR} must surely be proportional

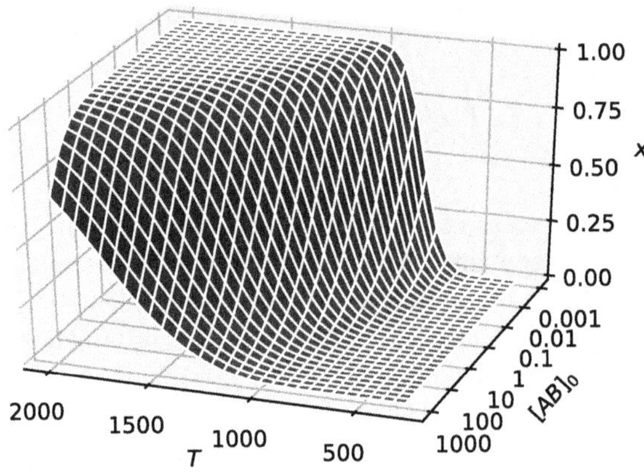

FIGURE 13.2 Plot of the fraction x of dissociated AB molecules as a function of T and $[AB]_0$.

Note: The plot assumes $m_A = m_B = \frac{1}{2}m_{AB} = 100\ u$, and bond energy $\varepsilon_{AB} = 1.5\ eV$.

to the concentration of A, i.e. $r_{LR} \propto N_A \propto [A]$. The reaction also requires 2 B molecules; in this case the reaction rate must depend on the *square* of B's concentration,[6] i.e. $r_{LR} \propto [B]^2$. Hence

$$r_{LR} = c[A][B]^2, \tag{13.22}$$

where c is a constant. However, we can also apply similar reasoning to deduce

$$r_{RL} = c'[C]^r[S]^s, \tag{13.23}$$

where now by keeping the species counts r, s unspecified we are being as general as possible. In equilibrium $r_{LR} = r_{RL}$, a condition known as *Detailed Balance*.[7] We deduce

$$\frac{[C]^r[D]^s}{[A]^p[B]^q} = \frac{c'}{c} = K, \tag{13.24}$$

a particular instance of (13.17).

The next big step is to apply the same formalism to situations out of equilibrium. Away from equilibrium using (13.14) a change in free energy following from changing molecule numbers $\Delta N_i = \nu_i \Delta R$, with ΔR the number of discrete chemical reactions accompanying the change is written (we have assumed the thermodynamic limit so $\ln N_i - 1 \approx \ln N_i$):

$$\Delta F(N_i) = -k_B T \sum_i \left[\ln Z_i(1) - \ln N_i\right] \Delta N_i = -k_B T\, \Delta R \sum_i \nu_i \left[\ln Z_i - \ln N_i\right]. \tag{13.25}$$

We've suppressed the functional dependence of all quantities on T, V which, as ever, are assumed constant. For just a single reaction $\Delta R = 1$ and we deduce the free energy change

$$\Delta F_{LR} = -k_B T \sum_i \nu_i \left[\ln Z_i - \ln N_i\right], \tag{13.26}$$

i.e.,

$$\frac{\Delta F_{LR}}{k_B T} = \sum_i \left[\ln N_i^{\nu_i} - \ln Z_i^{\nu_i}\right] \Rightarrow e^{\frac{\Delta F_{LR}}{k_B T}} = \prod_i \frac{N_i^{\nu_i}}{Z_i^{\nu_i}} = \frac{\widetilde{K}}{K} \tag{13.27}$$

where K is the equilibrium constant defined in (13.17), and $\widetilde{K} = \prod_i N_i^{\nu_i}$ specifies the molecular concentrations actually present.

Let's spell out in detail what happens next. First, if $\widetilde{K} = K$ then the system is already in equilibrium, and no further change happens. In this case Equation (13.27) tells us $\Delta F_{LR} = 0$, which we knew already from (13.10). Next, if the free energy change $\Delta F_{LR} < 0$ then $\widetilde{K} < K$, in other words there is a deficit of product over reagent molecules compared to equilibrium.[8] The system will attempt to correct this, approaching equilibrium by chemical reactions going from left to right, in other words, the change in free energy $\Delta F = +\Delta F_{LR} < 0$. By contrast if $\Delta F_{LR} > 0$ then $\widetilde{K} > K$ corresponding to an excess of product over reagent molecules. The system now approaches equilibrium by chemical reactions going from right to left, in other words, the change in free energy $\Delta F = -\Delta F_{LR} < 0$.

In all cases, therefore, the chemical reactions which take place result in a *reduction* of free energy, working to approach the equilibrium state where F is a minimum.

To sum up, chemical thermodynamics dictates that reactions always occur in a direction which reduces free energy, whose value depends on conditions such as temperature and reagent concentration. Free energy and its transfer define the appropriate thermodynamic language for chemistry in the same way that energy transfer is the language appropriate for mechanics and heat transfer for engines.[9] It can be conceptualized as a form of energy capable of doing work at constant temperature, in so-called *chemical engines.* A biological cell in a living organism epitomises such an engine.[10]

13.3 CHANNELLING FREE ENERGY IN BIOLOGICAL SYSTEMS

Let's flesh this out by returning to our original example. In the metabolism of a real organism, the reaction (13.1) in which glucose fuel is consumed to yield water and carbon dioxide products actually proceeds over many many diverse and separate stages, so that the free energy it makes available is directed towards the task in hand. At each stage reagents with relatively high free energy are transformed into products with lower free energy. There is a long list of tasks needing free energy input within a living organism, which includes: the synthesis of important biomolecules such as proteins, lipids, polysaccharides, and nucleic acids such as DNA from simpler precursor molecules such as amino acids and sugars; the transport of substances across membranes from regions of low concentration to places where they are required but where the concentration is already high[11]; maintenance of constant body temperature in varying external environments; and mechanical work, such as that performed by muscles. In Chapter 14, we will see that the accurate information transfer needed during the biosynthesis of DNA and proteins, the key step in cell replication, also requires a supply of free energy. Clearly, some agency is needed to direct and transport the free energy to where it's needed. In most living cells this role is performed by a molecule called *adenosine triphosphate* (ATP), whose structure is visualized in Figure 13.3.[12]

ATP is a large and relatively complex molecule—at the left-hand side of Figure 13.3 you can see three phosphate groups, labelled α, β, and γ, each consisting of an atom of phosphorus bonded to four oxygens, some of which carry an excess of electrons and are hence negatively charged. The "terminal" phosphate group γ is particularly susceptible to being removed in a reaction known as *hydrolysis*:

$$ATP^{4-} + H_2O \rightarrow ADP^{3-} + HPO_4^{2-} + H^+ \qquad (13.28)$$

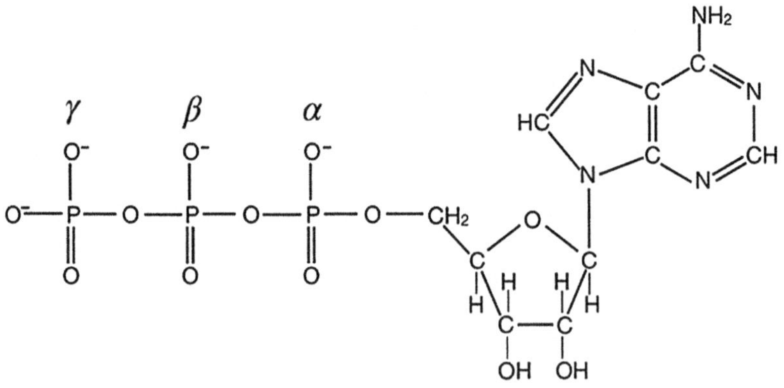

FIGURE 13.3 Molecular structure of ATP. The symbol P stands for atoms of phosphorus.

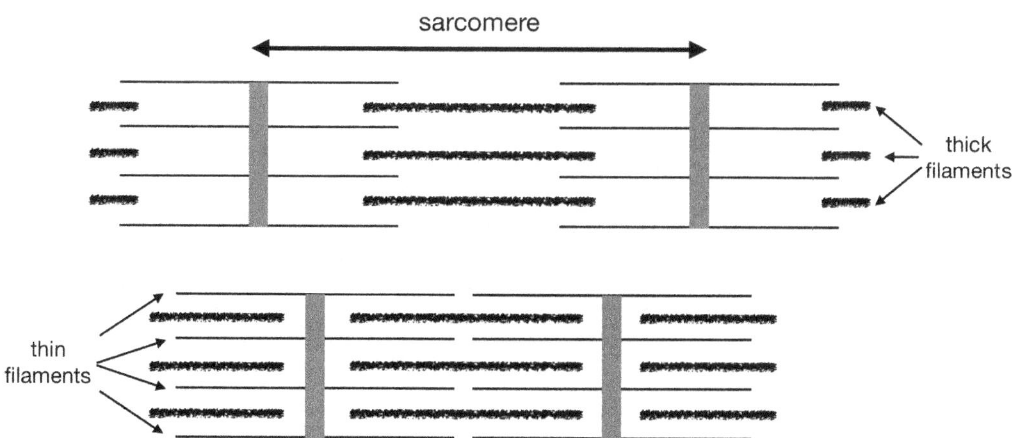

FIGURE 13.4 Schematic depiction of muscle contraction.

Here, ADP is adenosine diphosphate, basically, the same molecule as depicted in Figure 13.3 but with just the α,β phosphate groups left. Whenever this happens free energy ΔF is released. As we have seen, the precise amount depends on temperature and reagent concentration; under typical conditions[13] found in a cell $\Delta F = -51.9 \text{ kJ mol}^{-1} = -0.538 \text{ eV}$. Actually this is something of an intermediate value; there are some super high-energy phosphate compounds produced when glucose is first broken down, which can pass on their free energy in the form of a phosphate group by combining with ADP to form ATP, and some lower free-energy "phosphate acceptor" products of biological importance formed when ATP hydrolyses. ATP lies roughly in the middle of the free energy spectrum, in the thick of the action.[14]

The bioenergetics of living cells is a complex subject to which we can't hope to do justice here. However, it is helpful to look at one aspect in a little more detail precisely because it mirrors the discussion of heat engines in earlier chapters. Organisms do work when muscle fibres contract under tension, which causes the point of application of a force to move. In vertebrates, muscles are composed of two distinct kinds of filament: one set, which is thicker, is mainly composed of rod-like molecules called *myosins*, and then there are thinner filaments made from twisted strands of *fibrous actin.* The two sets of strands are arranged in parallel in repeating units, about 2.5 μm long, called *sarcomeres*, only partially overlapping when the muscle is in a relaxed state. As shown in Figure 13.4, when the muscle contracts the thin filaments are pulled into the spaces between the thick ones, and the fibre shortens in length.

As shown diagrammatically in Figure 13.5, each myosin has a large molecular group acting as a "head" at one end. Once the myosin molecules in the thick filaments hydrolyse ATP, the head assumes a "cocked" position, and is then able to bind weakly to the adjacent actin (see (d)—(e)—(a) (b) in Figure 13.5). At this stage, both ADP and phosphate (P_i) groups remain bound to the myosin. Next, the P_i detaches, initiating a reconfiguration of the myosin molecule so that the head springs back, setting in motion a power stroke (c) generating a force of approximately 10^{-12} N; in so doing it pulls the actin fibre approximately 10^{-8} m to the left, shortening the sarcomere and causing the muscle to contract. At the end of the power stroke the ADP group is released, but the myosin remains tightly bound to the actin until the next ATP binds, at which point the myosin detaches and completes the cycle by returning us to (d). In this way the free energy released during hydrolysis is used to perform mechanical work. The process is controlled by the concentration of calcium Ca^{2+} ions present in the muscle, which bind to a protein called *troponin* attached every so often to the actin filaments, unblocking further binding sites and thereby strengthening the myosin–actin binding just prior to the power stroke, which in turn accelerates the subsequent release of P_i and then ADP. When muscle

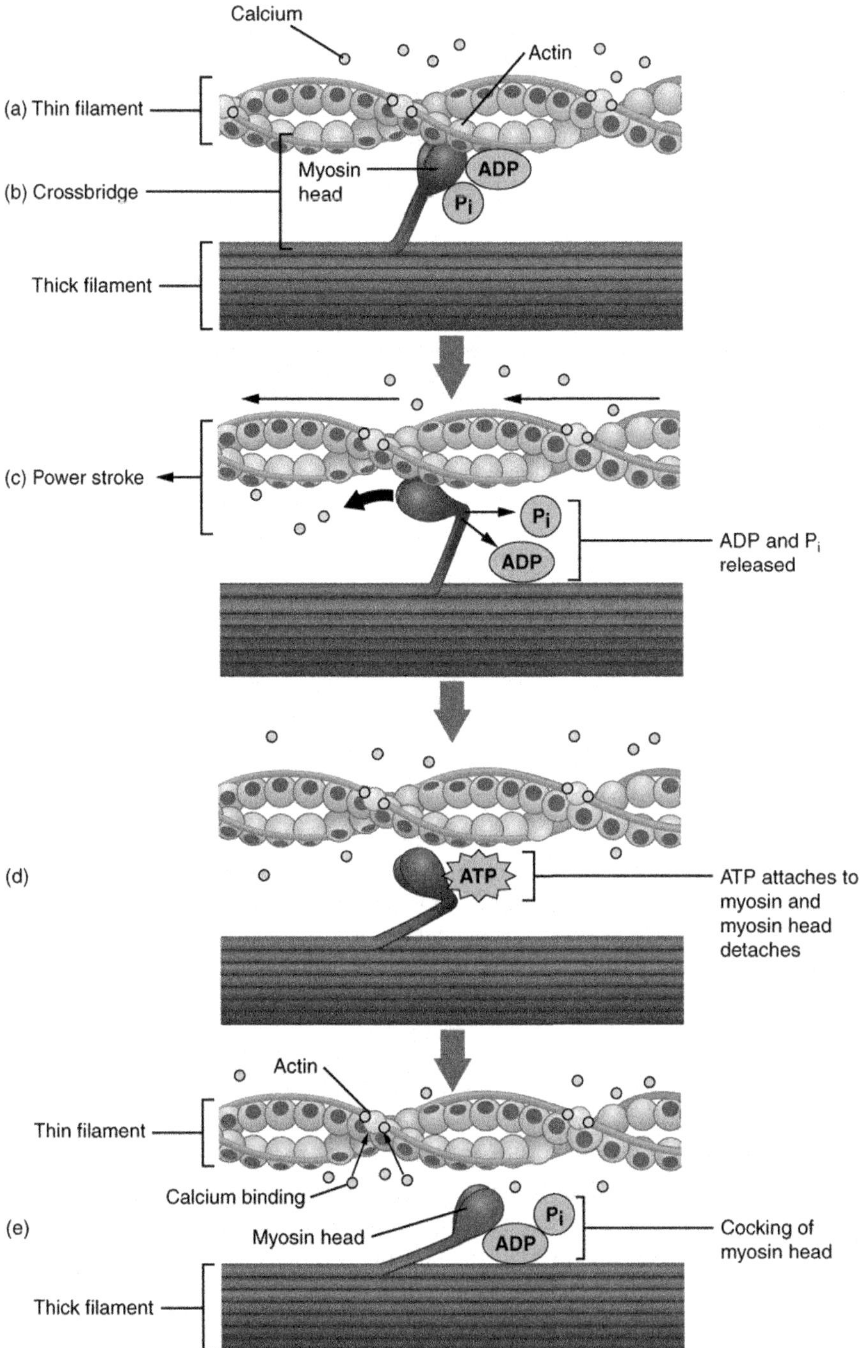

FIGURE 13.5 Chemomechanical cycle responsible for muscle contraction.[a]

a The figure is taken from Version 8.25 from the textbook *Anatomy and Physiology* (OpenStax, published 18 May 2016, https://openstax.org/details/books/anatomy-and-physiology-2e). A helpful animated visualization can be found at https://valelab4.ucsf.edu/external/moviepages/moviesMolecMotors.html

contraction is no longer needed, it is allowed to relax by transport of the excess Ca^{2+} away from the muscle fibre under the action of a different enzyme, again drawing free energy from the local supply of ATP. While no heat flow takes place, this *chemomechanical* process shares a key feature with the heat engines of Part 1; it is a cycle.

13.4 STAYING ALIVE

After this briefest of glances under the hood, let's return to our basic reaction (13.1) for glucose combining with atmospheric oxygen to form CO_2 and water. We now know the correct metric for the useful work that can be done as a result of taking on this fuel is free energy rather than the internal energy quoted earlier as $2.801\ \text{MJ mol}^{-1}$. Under standard conditions of 25°C and 1 atm pressure, the free energy change for (13.1) is $2.855\ \text{MJ mol}^{-1}$. Given the thermodynamic relation $\Delta F = \Delta U - T\Delta S$, we deduce an entropy change[15]

$$\Delta S = \frac{\Delta U - \Delta F}{T} = +183\text{JK}^{-1}\text{mol}^{-1}. \qquad (13.29)$$

Reassuringly, the sign of ΔS is consistent with the Clausius inequality $\Delta S \geq 0$ expressed in Equation (4.24). Our metabolism processes the high-free-energy foodstuff glucose into low-free-energy high-entropy products, releasing free energy used to get on with the important business of staying alive. The reaction products in (13.1) are high in entropy because 6 extra moles of gas are formed: following the Sackur–Tetrode relation (7.51) this gives an entropy increase of $6R\ln\left[CVT^{\frac{3}{2}}\right]$, where $C = \left(e^{\frac{5}{2}}/6N_A\right) \times (2\pi\bar{m}k_B/h^2)^{\frac{3}{2}}$, with $\bar{m} = m_{H_2O}m_{CO_2}/m_{O_2}$.[16]

It's not hard to guess where this entropy ends up—eventually it is returned to the environment as we exhale, or excrete, the reaction products CO_2 and H_2O.[17] If entropy, which remember is a function of state, were to remain contained within the body then over time the organism (viewed as a thermodynamic system) would necessarily change as a consequence, countering our natural sensation that from day to day, by-and-large our bodies remain in the same condition over reasonably long periods, at least certainly over the timescales in which biochemical reactions take place. The technical term for this staying the same is *homeostasis.* Of course, this cannot be the whole story—over the course of a lifetime organisms grow, mature, reproduce, lose or gain weight, become ill, age, and ultimately die. None of these essential processes, lying outside the scope of this book, are captured within the simple picture being developed. Our point is to stress that the process of living *generates* entropy, as schematically depicted in Figure 13.6, starting with the breakdown of high free-energy low-entropy

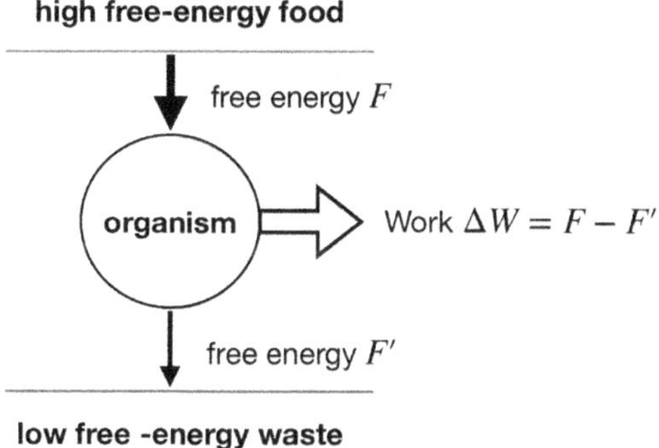

FIGURE 13.6 Representation of free energy flow through a living organism.

fuel—we have discussed foodstuffs in terms of large complex organic molecules such as sugars, fats and protein, appropriate for *heterotrophic* cells, but there is also low-entropy fuel available in the form of sunlight, exploited by *autotrophic* cells to synthesise the organic biomolecules needed for their living using atmospheric carbon dioxide.[18] The entropy released by processing the fuel is released back to the environment, with the outcome that the state of the organism is stable over time, while the entropy of the universe increases. This is the riposte to simple-minded arguments that complicated structures such as living organisms, and complex processes such as evolution, are fundamentally incompatible with the Second Law. From a thermodynamic perspective, we are forced to conclude, echoing John Donne, *No man is an island, entire of itself.*

We can also view the biosphere as a whole in thermodynamic terms. Broadly, autotrophic cells use sunlight-powered photosynthesis of CO_2 and H_2O sourced from the environment to make a living, while heterotrophs use the products of autotrophs plus atmospheric O_2 for their metabolism, releasing CO_2 and H_2O back into the environment as waste products. Since atoms are neither created nor destroyed, it's possible to analyse the whole process in cyclic terms—for instance, the "fast" carbon cycle just described is estimated to circulate some 200Gtonnes of carbon between the atmosphere, the biosphere, and the oceans each year. There is another "slow" carbon cycle of roughly 10^{-2} to 10^{-3} the volume, which is worked through over longer periods and is now jealously monitored for signs of secular behaviour, such as the weathering of carbon-containing rocks and the steady increase in atmospheric CO_2 associated with man-made emissions. Other cycles important for life are associated with nitrogen and water.

It's tempting to associate such cycles with the thermodynamic cycles studied extensively in Chapter 4, where the performance of a heat engine was understood in terms of the flow of heat through the system from a hot reservoir to a cold one, periodically restoring the engine to its original state while generating an inevitable increase in the entropy of the environment. Is there an analogue either for a single organism or the biosphere as a whole, in which the living process is understood as the flow of indivisible atoms through an open yet stable system, converting low-entropy fuel into high-entropy waste? Could we plot such a cycle using, say, μ and N as coordinates rather than p,V or T,S ? There are a couple of important distinctions worth noting.

We have seen in this chapter that the many many tasks required to keep life going are accomplished through highly controlled free energy transfers with very specific pathways, each governed by very specific enzymes. It is by no means clear how to capture such complexity with a simple metric such as the area of a closed path in (μ_i, N_i)-space—for one thing the cycle must take place in a multi-dimensional space generated by the thousands of different molecular species involved. Perhaps the closest we can come to encapsulating the metabolic cycle is an adaptation of Figure 4.4, which highlights the similarities and distinctions between the operation of a heat engine and a heat pump. Figure 13.6 shows the free energy F entering an organism as food and F' leaving as waste. Since $F = U - TS$ we can write

$$\Delta F = F' - F = \Delta U - T\Delta S - S\Delta T + \sum_i \mu_i \Delta N_i \approx \Delta U - T\Delta S \tag{13.30}$$

where the last approximate equality arises because the process is isothermal and the molecular composition of the organism is constant over time. It is also apparent that $-\Delta F$ equates to the useful work ΔW that results. Now, for the organism U is a function of state which we expect changes relatively little as a consequence of homeostasis, i.e., $\Delta U \approx 0$. We conclude[19]

$$-\Delta F = \Delta W \approx T\Delta S, \tag{13.31}$$

which links work done to entropy production. For an adult human, the free energy cost of homeostasis is about 6.3 MJ per day.

By now, we've already drifted towards thinking in talking about "low-entropy" food and "high-entropy" waste. It highlights a view of life is as a constant struggle against a return to equilibrium. In the mid-1960s James Lovelock[20] was working as a consultant engineer–scientist working for NASA on the *Viking* programme of space probes destined to explore the Martian surface. Along with many other scientists on the project he was concerned with the question of how to identify signs of life on Mars. His proposed solution was to look for signs of "entropy reduction", in other words for bulk signs of a system maintained far from thermodynamic and chemical equilibrium by the presence of life processes. As an example he pointed to the Earth's own atmospheric composition, in principle accessible to a suitably equipped planetary scientist based on Mars. Taking into account incoming solar radiation and the presence of both oceans and land masses, he argued the simultaneous presence of both oxygen and methane (CH_4), which one might expect normal atmospheric chemistry to rapidly convert to CO_2 and H_2O evidenced a disequilibrium state which could only arise via the presence of some living process dumping 500Mtonnes of methane into the atmosphere annually, coupled with a complementary process replacing the oxygen consumed when methane is oxidized. While Lovelock's original thinking focussed on the Solar System, the atmospheric composition of extrasolar planets can now be probed each time an exoplanet's disk passes in front of its star.[21]

The subsequent work of Lovelock and his collaborators has helped to change the way we view our world and our interaction with it—it's also entered the culture: in Cixin Liu's masterful *Three Body Problem* sci-fi trilogy,[22] an observer from a remote, advanced and indifferent civilization identifies the presence of "low-entropy entities", dubbed "Star-Pluckers", based on analysis of gravitational wave signals received from the direction of the Solar System—and then initiates a process for their elimination.

Our brief analysis of life in thermodynamic terms has by no means answered the basic question "What is Life?"[23] but has managed to highlight fresh ways to understand entropy and interpret its production. As we'll see in Chapter 15, the Second Law is often viewed in a rather negative light, as a harbinger of increased disorder and decay over time, leading ultimately and inevitably to things shutting down and turning off as universal entropy increases inexorably. Here, by contrast, we've seen the production of entropy by living organisms is inextricably linked with the creation and maintenance of order and complexity, both at the level of individual organisms and perhaps also on a planetary scale. Entropy is what we make. At body temperature 37°C, each of us is producing just over 20000 JK^{-1} on a daily basis.

NOTES

1 For those paying careful attention to what they eat a less alarming way of quantifying this is 3.72 Cal/g. A 1kg bag of sugar contains about $5\frac{1}{2}$ moles of glucose.

2 A chemist would regard this statement as pedestrian in the extreme, but we are particle physicists …

3 At first sight it seems strange to refer to something expressed as a function of T and V as "constant", but remember we are working under conditions in which these quantities don't vary.

4 Typical bond energies between carbon atoms in an organic molecule range from 3 to 6 eV.

5 In chemistry this is known as the *molar concentration.* To convert to SI units use $[AB] = N_{AB}/1000N_AV$.

6 Think about waiting for a bus, given the probability p that one will arrive in the next minute. Assuming bus arrivals are completely uncoordinated (not the bus company's ideal, but you know what rush hour traffic is like…), the probability of two buses arriving in the next minute is p^2.

7 This is very reminiscent of discussion of the Fermi Master equation in Section 10.3.1.

8 Strictly it's moot which we regard as reagents and which as products, so we'll stick with the convention that products live on the right-hand side of the reaction, have $\nu_i > 0$, and hence appear in the numerator of K.

9 At this point, we need to come clean: professional chemists and biochemists actually work with the Gibbs free energy $G = F + pV$, the appropriate thermodynamic potential for systems in which it is T and p held constant rather than T and V. The difference matters when some of the reagents are gases. We've chosen to ignore this important distinction in the name of simplicity, but if you want to do accurate work you should toe the line…

10 This point of view is presented in *Principles of Biochemistry*, A.L. Lehninger (Worth, 1982).

11 The similarity of this transport phenomenon to heat flowing from a cool place to a hot place is more than coincidental; the Second Law dictates that (free) energy, quantified through the Law of Mass Action, is needed to make it happen.

12 The key role played by ATP in free energy transfer throughout living organisms was first identified by biochemist Fritz Lipmann (1899–1986).

13 Biochemists consult a table of ΔF° values for reactions under standard conditions: unit molar concentrations of all reagents, temperature 298 K, pH 7.0, pressure 1 atm. You can then use the Law of Mass Action to calculate ΔF for the prevailing conditions.

14 In all cases reactions involving ATP hydrolysis are catalysed by *kinase* enzymes highly specific to the particular reaction needed at that time and location.

15 As alluded earlier, a professional would instead specify $T\Delta S = \Delta H - \Delta G$, where the *enthalpy* $H = U + pV$. In this reaction since 6 moles of excess gas are produced, the difference $\Delta H - \Delta U = \Delta G - \Delta F = 6RT = 14.9$ kJ mol^{-1}; the same result for ΔS is obtained in either case.

16 Once again we have ignored complications associated with internal molecular excitations such as rotation.

17 If the organism is warm-blooded, then further entropy transfer can also take place through heat flow.

18 As you might guess, autotrophs are relatively self-sufficient while heterotrophs have to subsist on the products of other cells.

19 You can also think of ΔF as the free energy change of the organism's environment.

20 James Lovelock (1919–2022) had a remarkable career spanning wide-ranging scientific and technical interests, including work for British Intelligence and developing an early form of microwave oven for reviving cryogenically frozen lab hamsters! He is best known as the originator of the *Gaia* hypothesis which views Earth as a self-regulating living system.

21 See for instance the ARIEL space mission: https://arielmission.space/

22 *Death's End*, Cixin Liu, translated by Ken Liu (Head of Zeus, 2016).

23 *What is Life?* (Cambridge University Press, 1944) is the title of a short but influential book written in 1944 by the theoretical physicist Erwin Schrödinger (1887–1961), one of the founders of quantum mechanics. The book sets out for the first time many of the ideas reviewed in this chapter, but unfortunately uses the confusing term *negative entropy* in place of free energy.

14 Information

Many aspects of contemporary life revolve around data, yet there is an important distinction between data held in various media and transmitted through various channels and the resulting information communicated. It is possible to relate information to the average surprise we feel encountering the next symbol, expressed quantitatively via the Shannon entropy measured in bits per symbol. The rate and accuracy of information transmission are maximized if all transmitted symbols have an equal probability of occurring, corresponding to a low entropy coding. The formal similarity between the Shannon and Gibbs expressions for entropy suggest a deeper connection, explored in a series of thought experiments featuring a sentient and intelligent entity called Maxwell's Demon. Although the Demon presents an apparent challenge to the Second Law, it can shown to be satisfied either because of physically irreversible processes needed by the Demon for measurement or logically irreversible processes needed for memory erasure. A biological example of reversible computation highlights the importance of entropy generation for accurate gene transcription.

14.1 DATA VERSUS INFORMATION

We live in a data-driven age. If the 19th and 20th centuries could be characterized by growing public awareness of the importance of energy, in terms of both its supply and its applications, as set out in Chapter 2, then surely data is the dominant theme of the 21st century. Prior to the advent of personal computing and communication devices, probably few of us regularly used the term. Now data is a central, even commodified aspect of our lives—we carefully shop around for the best value when acquiring a mobile phone. Broadband access is now considered a household necessity on a similar footing to water and power, and for many of us, it's a stressful problem when the WiFi's down. Constant improvements in the way data are acquired, processed and stored have revolutionized practice in the natural, medical, and social sciences[1]; "data scientist" is now a respectable profession, equipped to tackle questions beyond the capacity of previous generations either to ask or to answer[2]; the associated techniques and analyses are rapidly being assimilated into the education of younger researchers. On the flip side, we're also much more conscious of the swathes of personal data concerning each of us—our health, purchasing preferences, political and social outlook, hobbies—held by both public agencies and private companies and have a growing awareness of our rights concerning data privacy. We are constantly affronted by personalized advertising targeted at us through our tablets and internet browsers—except for when it's promoting the very thing we were looking for....[3]

DOI: 10.1201/9781003121053-18

In this chapter, we will take an apparent excursion from the book's main theme to review some basic issues about data, including how it's quantified and communicated, but we will also try to draw a useful distinction between *data* and *information*. On first reading the two terms are synonymous, but actually that can't be the case. If data encapsulates a statement or fact about the world, then are all facts equally as valuable, or are some inherently more interesting? There are well-worn examples to suggest not: "Man bites dog" is a more arresting headline than "Dog bites man", and anyone who's ever completed a jigsaw puzzle will confirm that the hardest part to complete involves the pieces of the sky, which all look the same. If one is fortunate enough to be in a close personal relationship with another, and that person has fallen into the annoying habit of completing our sentences for us, then perhaps it's time to question the information content of one's utterances and think harder before speaking! These nostrums communicate the obvious truth that we are more struck by data that surprise us or provide some contrast with what's gone before.

What has all this to do with entropy? Remarkably, there is a branch of science called *Information Theory* in which apparently meaning-laden nuances such as we've listed are addressed and even quantified. Developed in the 1940s, information theory now underpins all modern techniques for fast and accurate communication of data. More remarkable still, the formulae at the theory's heart bear a striking resemblance to those we have already seen in thermodynamics describing entropy. We will need to understand whether this is coincidental or whether there is some deep-lying link between thermodynamics and information. The journey will prove to have some unexpected, even diabolical, turns, but in order to get started, let's consider a message of particular importance to this book, namely the Clausius statement of the Second Law of Thermodynamics:

It is impossible to devise an engine which, working in a cycle, shall produce no effect other than the transfer of heat from a colder to a hotter body.

If we were to read this sentence out loud, then the information inherent in it would be communicated through the complex modulation of pressure variations causing the recipient's eardrums to vibrate, which arrive either via coherent travelling oscillations of an elastic medium called air, which we refer to as *sound*, or perhaps as voltage modulations in an electrical circuit comprising a phone line. The resultant waveform could in principle be recorded and plotted to form a visual representation of the message, and we might guess it would be incredibly complicated, occupying several pages. Fortunately, there is another medium known as the Roman alphabet which can effectively communicate virtually all the same information[4] using just 52 letter symbols[5] plus some punctuation marks. In order to simplify our presentation, we're going to use just upper-case letters and ignore all punctuation so that the Clausius statement becomes

IT IS IMPOSSIBLE TO DEVISE AN ENGINE WHICH WORKING IN A CYCLE SHALL PRODUCE NO EFFECT OTHER THAN THE TRANSFER OF HEAT FROM A COLDER TO A HOTTER BODY

Despite the insistence of generations of over-zealous English teachers, this stripped-down alphabet of 27 symbols still communicates very effectively![6] The Second Law can thereby be encapsulated using 148 symbols, occupying just a little over two lines of text.

14.2 CODED MESSAGES

The Roman alphabet is an effective means of communication if you write on a blackboard, send a letter through the post, or perhaps consult a text in a bookshop or library, but what if the message is more urgent? Information is sent much more rapidly through transmission lines or processed in digital electronic circuits, by encoding everything using just two symbols, 0 and 1, which could correspond, say, to two clearly distinct voltages transmitted through a conducting wire. The resulting *binary code* was introduced in Chapter 5. Each symbol is known as a binary digit or *bit*, and the

TABLE 14.1
Naive Encoding of Roman Alphabet Using 5 Bits per Symbol

A	00000	J	01001	S	10010
B	00001	K	01010	T	10011
C	00010	L	01011	U	10100
D	00011	M	01100	V	10101
E	00100	N	01101	W	10110
F	00101	O	01110	X	10111
G	00110	P	01111	Y	11000
H	00111	Q	10000	Z	11001
I	01000	R	10001	space	11010

amount of information it carries can be thought of as sufficient to answer a binary question, i.e. one requiring a yes-or-no answer.[7] A string of p bits can therefore supply 2^p distinct answers, or in other words communicate 2^p distinct messages. The most straightforward way to encode the alphabet in binary starts from the observation that 5 bits is enough to encode each of the 27 alphabet symbols we need since $2^5 = 32 > 27$. A possible encoding is shown in Table 14.1.

With this encoding the Clausius statement takes the following form:

```
01000100111101001000100101101001000011000111101110100101001001000000
01010110010011010100110111011010000110010010101010001001000100110100
00000110111010001000110100110010000110100100110101011000111010000001
00011111010101100111010001010100100001101001101101001000011011101000
00011010000101100000010010110010011010100100011100000010110101111010
01111100010111000011101000001000100110100110101110110100010000101001
01001000001010011110100111010011001110010010001110101001100111000000
11011101010011001110010011010100111000100000011011001000101001001000
11101001110001011101000111001000000010011110100010110001011100110011
01000000110100001001110010110001100100100011101010011011101101000000
110100011101110100111001100100100011101000001011100001111000    (14.1)
```

740 bits are required to convey the same message as 148 Roman letters—exactly what we'd expect from coding using 5 bits per symbol.

Is this the most efficient way of encoding the alphabet? The fact that 5 of the 32 available 5 bit codes remain unused is hardly a problem—if we wished these could be used for the most essential punctuation symbols. However, a moment's thought should convince you it's sub-optimal because the same number of bits is used for each letter, regardless of whether the letter appears frequently (such as **E**, **T**, or **A**) or only relatively rarely (such as **J**, **X**, or **Z**). Taking the letter's frequency into account can result in improvements, as appreciated as early as the 1830s in the famous Morse Code[8] shown in Table 14.2, which encodes letters in terms of short "dots" (.) and long "dashes" (-), each transmitted manually using an electrical contact controlled with a mechanical key by a skilled operator.

As shown in Table 14.2, **E**, **T**, and **A** are respectively coded by the short sequences ".", "-",and ".-", while the rarer **J**, **X**, and **Z** have longer codes ".- – -", "-..-", and "- -..". In practice, the operator must also include uniform gaps between successive letters (and still longer gaps between words) to resolve the ambiguity between, say, **TTE** and **G**. We have to represent this gap by a third symbol / to distinguish -/-/./ from --./. In information theory terms, Morse is thus a three-symbol code. With this important stipulation, the Second Law in Morse Code reads

TABLE 14.2
Morse Code Using Dots, Dashes, and a / Denoting Gaps Between Letters

A	.-/	J	.---/	S	.../
B	-.../	K	-.-/	T	-/
C	-.-./	L	.-../	U	..-/
D	-../	M	--/	V	...-/
E	./	N	-./	W	.--/
F	..-./	O	---/	X	-..-/
G	--./	P	.--./	Y	-.--/
H	..../	Q	--.-/	Z	--../
I	../	R	.-./	space	/

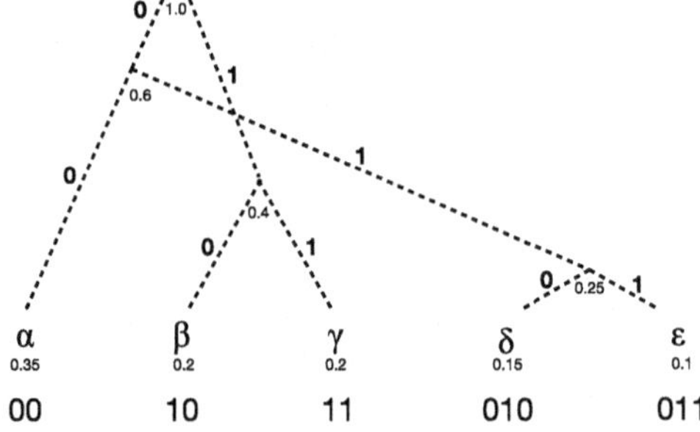

FIGURE 14.1 Huffmann coding of a five letter alphabet (α, β, γ, δ, ε). The decimal numbers are the relative frequencies of the letters, and the resulting binary code is shown beneath.

*../-//../...//../--/.--./---/.../.../../-.../.-.././/-/---//-../../...-/../..././/.-/-.//./-./--./../-././/.--/..../../-.-./....//.--/---/.-./-
.-/../-./--.//../-.//.-//-.-./-.---.-./.-.././/.../..../.-/.-../.-..//.--./.-./---/-../..-/-.-././/-./---//./..-./..-/./-.-./-//---/-/..../
./.-.//-/..../.-/-.//-/...././/-/.-./.-/-./.../..-././.-.//---/..-.//..../././.-/-//..-./.-./---/--//.-//-.-./---/.-../-.././.-.//-/---//.-//
..../---/-/-/./.-.//-.../---/-../-.--* (14.2)

At 456 symbols, the resulting compression is impressive, but this has come at the cost of introducing a third symbol, so Morse cannot be classified as a binary encoding. Physical limitations to the accurate transmission of signal pulses through a long wire disfavour three-symbol over binary code in modern applications,[9] though Morse's solution was remarkably efficient for the time and remained in use throughout much of the 20th century.[10]

An efficient binary encoding which both resolves the ambiguity of when one letter ends and the succeeding one starts and also takes into account the relative frequency of letters' appearance, is provided by *Huffmann Coding*.[11] As a simple example, we outline how to Huffmann code a simple alphabet containing five letters α, β, γ, δ, and ε, given the information that their relative frequencies of appearance in meaningful text are in the ratio 0.35:0.2:0.2:0.15:0.1. The rule, illustrated in Figure 14.1, is to write out the letters in a line, identify the two letters with the lowest frequencies (which may be equal), and connect them with a "wishbone". The wishbone is now treated as a compound letter and assigned a frequency which is the sum of its components. This process of identifying and then combining the two lowest frequency letters is repeated as many times as necessary until a single overarching wishbone is reached. The left-hand arm of each wishbone in the resulting *binary decision tree* is then assigned a 0 symbol and its right-hand counterpart a 1. Starting from the

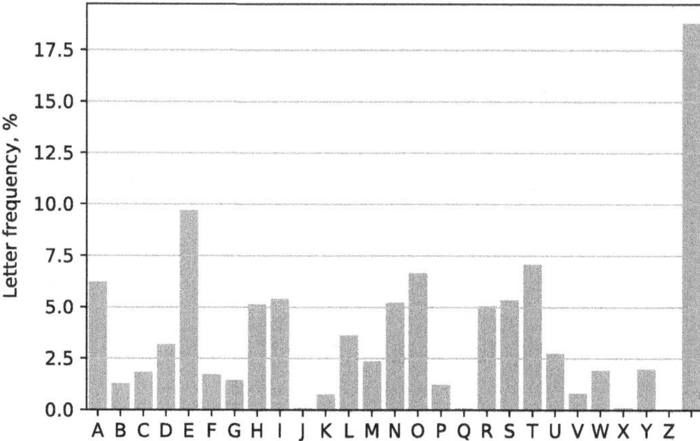

FIGURE 14.2 Relative letter frequencies in Shakespeare.

TABLE 14.3
Huffmann Coding of the Roman Alphabet

A	1001	J	0000111101	S	1010
B	101111	K	0101110	T	0010
C	10110	L	00111	U	01010
D	01111	M	000010	V	00001110
E	0001	N	1000	W	0000110
F	001101	O	0100	X	0101111
G	101110	P	001100	Y	010110
H	01110	Q	000011111	Z	0000111100
I	0110	R	00000	space	11

apex, there is a unique downward path to each letter in the alphabet, and traversing it generates the bit sequence corresponding to its Huffmann encoding.

The resulting binary code has two important properties: firstly, because fewer binary decisions are needed to arrive at the more frequently-used letters α, β, and γ, these are coded using fewer bits than the less common δ and ε; secondly, no letter's code forms the first part of a different. letter's code, so that successive letters can be concatenated into a compact form ignoring spaces without introducing ambiguity in any subsequent decoding. This *prefix property* is a direct consequence of the binary decision tree underpinning the code.

In order to Huffmann code the whole Roman alphabet, we need relative letter frequencies pertinent to, say, the English language. This has to be obtained through empirical analysis of a large body of text.[12] Fortunately, nowadays, it is straightforward to obtain digital text files of many major works of literature,[13] so with a nod to C.P. Snow, we present in Figure 14.2 relative letter frequencies based on the complete works of William Shakespeare. The height of the bar above each letter represents that letter's frequency, and we quickly pick out our old friends **E**, **T**, and **A**. The tallest bar on the right corresponds to the space character, which at 18.8% implies Shakespeare's average word length is just over five letters. Using this data and the algorithm summarized in Figure 14.1 yields the Huffmann coding shown in Table 14.3. The desired variation in bits per letter depending on relative frequency is manifest: the space character requires just two bits, **E**, **T**, and **A** four, while **J** and **Z** now each need no fewer than 10 bits!

The Huffmann-coded Second Law now needs only 611 bits:

```
01100010110110101011011000001000110001001010101001101011110011100011
10010010011011110001000011100110101000011110011000110001100010111001
10100000011100001100111001101011001110110000110010000000010111001101
00010111011011010001110011110110010110101100011100011110100111010010
01110011111001100000000010001111010101011000011110000100110001001101
01101000110110001011010000100111000010000011001001110100110001100100
11100001110010000000100110001010001101000100000110100001101110111 0000
11001001011001101000000010000001011100111101100100001110111100010000
11001001001110011101110010000100010000100000011101111010001111010110
```

(14.3)

14.3 SHANNON ENTROPY

In 1948 Claude Shannon[14] laid the foundations of Information Theory by specifying how to quantify the information contained in a message. Any crossword enthusiast will confirm that the information communicated by a single letter is related to the inverse of the frequency with which it occurs in regular use—think how much more helpful it is to find a **J** or an **X** in the intersection between an across and a down clue, compared with the **A**s, **E**s, and **T**s we so often have to work with! Shannon defined the "surprise" factor of the ith letter in terms of its relative frequency p_i by[15]

$$h_i = \log_2\left(\frac{1}{p_i}\right). \tag{14.4}$$

Here $\log_2$ is the logarithm with respect to base 2, i.e. if $y = \log_2 x$ then $x = 2^y$.[16] We can interpret y as the number of bits needed to represent x in binary notation, so it is natural to measure h_i in *bits*. Equation (14.4) specifies that the smaller the p_i, the larger the resulting surprise h_i. Now when considering the whole alphabet of possibilities we can define the expected value of the surprise in terms of a probability-weighted sum over all i letters:

$$H = \sum_i p_i h_i = \sum_i p_i \log_2\left(\frac{1}{p_i}\right) = -\sum_i p_i \log_2 p_i. \tag{14.5}$$

Shannon termed H the *entropy* of the information conveyed by the alphabet, measured in bits/symbol, by mathematical analogy with the Gibbs expression $-k_B \sum_i p_i \ln p_i$ for thermodynamic entropy given in (5.46). While we have approached it by thinking about letters in an alphabet, the definition (14.5) holds for any quantity specified by a discrete probability distribution, and with care can be extended to also cover *analogue* (i.e. continuously varying) quantities. Shannon entropy is maximized by maximizing uncertainty, implying a uniform distribution in which all the p_i are equal so that $H = \log_2\left(\frac{1}{p_i}\right)$,[17] and is reduced below this by any degree of non-uniformity, eventually vanishing in the limit of complete certainty, i.e. that only one possibility is encountered.

Figure 14.3 shows the surprise of each letter in the alphabet based on the relative Shakespearean frequencies of Figure 14.2. The distribution of h_i among the letters is much more uniform than p_i; note that the obvious standouts **J**, **Q**, **X**, and **Z** were previously invisible in Figure 14.2. We have also shown the letter scores allocated in Scrabble® augmented by addition of three; the correlation

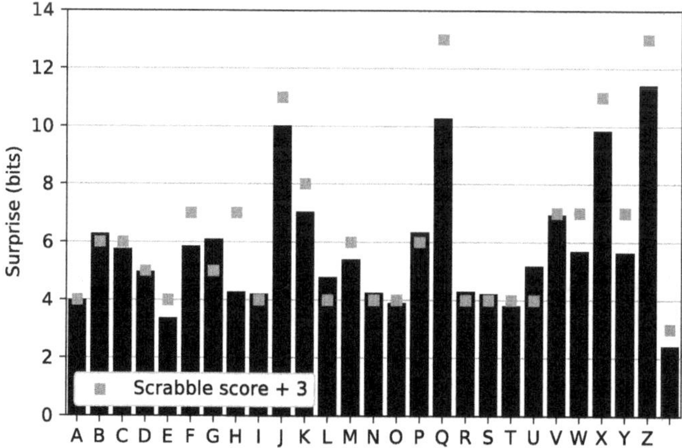

FIGURE 14.3 The Shakespearean surprise h_i..

TABLE 14.4
Entropies of Some Famous Works of Literature

Title	Letters	Entropy (bits/symbol)
Shakespeare's works	4977812	4.098
King James Bible	3972426	4.042
Ulysses	1424632	4.148
Moby Dick	1146092	4.124
Les Misérables (Eng)	3004898	4.104
Les Misérables (Fre)	2893118	3.993

is surprisingly good, although we feel that use of **H** is significantly over-rewarded![18] Using (14.5) the entropy of Shakespearean English is calculated to be about 4.1 bits/symbol, to be compared with the maximum entropy 4.755 bits/symbol for a 27 letter alphabet. It's amusing to repeat this exercise with other famous works of literature, as summarized in Table 14.4. There is remarkably little variation, although the difference between the original French and the English translation of *Les Misérables* is possibly significant.[19] Even analysis of a much shorter text sample, such as the Clausius statement of the Second Law, yields 4.013 bits/symbol.

14.4 SOURCE CODING THEOREMS

What raises this from an amusing parlour game to a powerful communication tool is Shannon's *Source Coding Theorem*. Suppose the source, i.e. the message we want to send, has entropy H bits per symbol, and the channel, i.e. the means of communication (be it a telegraph line, fibre optic cable, satellite relay, or whatever) has a *capacity* of C bits per second. It is then possible to encode the output of the source to achieve a maximum average communication rate of C/H symbols per second. For instance, the Huffmann coding (14.3) encoded the 148 symbol Clausius statement using 611 bits, i.e. using 4.13 bits per symbol, very close to our previous estimates of the source entropy and hence the theoretical maximum rate.[20] Compare this with the 5 bits per symbol needed by the naive coding (14.1), or the $\frac{456}{148} \times \log_2 3 = 4.88$ bits per symbol achieved with Morse code (14.2).

Efficient source encoding has become an essential, if invisible, aspect of our digital lives. A typical mobile phone camera produces digital colour images consisting of 2448×3264 pixels, each specified by 1 of 256 levels of brightness in each of red, blue, and green. A naive encoding of the image therefore requires 3 bytes per pixel (where 1 byte equals 8 bits) so the entire image apparently consumes almost 24 MB of hard drive. However, in practice it is much more efficient to store the differences in intensity values between adjacent pixels, which usually have a dynamic range significantly smaller than 256, and then use an encoding algorithm such as Huffmann coding. A typical photo might then fit into less than 2 MB of the phone's memory, and can be rapidly reconstructed whenever it is viewed on screen. In a related example, it is more efficient to code English text not simply one letter at a time, but as consecutive groups of two or more. This reflects the fact that letters in English text are *correlated* so that it is possible to predict the likelihood of the subsequent letter based on knowledge of the preceding one—e.g. if we see a **Q** then the following letter is almost certainly **U**. Analysing Shakespeare using *digram* frequency (i.e. letter pairs such as **TA** or **GH**) yields an estimate for $H \approx 3.8$ bits/symbol, implying that a Huffmann coding employing 27^2 different digram codes yields significantly faster communication according to the Source Coding Theorem. In fact for n-gram coding the potential gains keep rising until $n \approx 10$ by which time $H \approx 1.8$ bits/symbol. This implies that of the 27^{10} conceivable 10-grams only a tiny fraction $2^{1.8\times10} \approx 260000$ occur with any reasonable (i.e. non-vanishing) probability,[21] which makes n-gram coding a practicable proposition.[22] At some point the fact that meaning is conveyed by words and sentences rather than mere letters asserts itself. If we use $n = 13$, then there exists an encoding in which both **MAN BITES DOG** and **DOG BITES MAN** are represented as Huffmann bit strings—the former having fewer bits than the latter! The payoff is that as n rises the frequency analysis needed to estimate H accurately requires ever larger blocks of source text; indeed, Shannon's proof of the source coding theorem assumes that messages are very long.

The Source Coding Theorem applies to the ideal case of a *lossless* channel, in which every bit of the encoded message is transmitted without error. In reality, errors due to noisy interference, perhaps a simple misreading of a semaphore signal sent by a remote correspondent at dusk, or a Morse code error resulting from tired operator fingers, are a fact of life. In general, transmissions relying on electrical signals are susceptible to corruption from noise due to thermal or even (at sufficiently low temperatures) quantum effects. The thermal noise power in the circuit is[23]

$$\mathcal{N} = k_B T \mathcal{W}, \tag{14.6}$$

where $\mathcal{W}$ is the noise bandwidth[24] measured in Hz. Electrical signals must be transmitted with a power exceeding $\mathcal{N}$ in order to be reliably decoded. It's possible to mitigate noise by including redundancy in the transmitted message, i.e. extra bits which enable transmission errors to be detected and corrected. A simple approach is to divide the message into sequences of n^2 bits, and after each insert an extra $2n$ bits chosen to reflect the *parity*, i.e., treat the sequence as square array with n rows and n columns, and then assign a parity bit whose value depends on whether the sum of the n bits in each row/column is odd or even. With no transmission error, these extra bits are completely determined and therefore carry no information, but if there is a single bit error, then such an encoding will identify both the row and the column within the $n \times n$ square, and it can be readily corrected.[25] Information Theory handles this new complication by distinguishing the source message X from the output message Y, each with their own entropy $H(X), H(Y)$ defined using (14.5). Clearly Y should somehow depend on X, but in the presence of noise that relation is *stochastic*, meaning that it is described in general by a joint probability distribution $p(x_i, y_j)$ governing the likelihood of a transmitted symbol x_i being received as y_j. This permits a particular symbol x_i to be received as any of several y_j and *vice versa*, which describes the situation when errors are present. The joint distribution has an associated entropy

$$H(X,Y) = -\sum_{i,j} p(x_i, y_j) \log_2 p(x_i, y_j), \tag{14.7}$$

leading to the definition of an important quantity called the *mutual information*[26]:

$$I(X,Y) = H(X) + H(Y) - H(X,Y). \tag{14.8}$$

If source and output are completely independent so that $p(x_i, y_j) = p(x_i)p(y_j)$, then

$$\begin{aligned} H(X,Y) &= \sum_j p(y_j)\left[\sum_i p(x_i)\log_2(p(x_i))\right] + \sum_i p(x_i)\left[\sum_j p(y_j)\log_2(p(y_j))\right] \\ &= H(X) + H(Y), \end{aligned} \tag{14.9}$$

where in the second step we use $\Sigma_i\, p(x_i) = \Sigma_j\, p(y_j) \equiv 1$; therefore $I(X,Y) = 0$ and no information is transmitted. By contrast, if the channel is lossless, then $p(x_i, y_j) = p(x_i) = p(y_j)$ if and only if $i = j$, and is otherwise zero; now $I(X,Y) = H(X,Y) = H(X) = H(Y)$, the maximum information possible. Mutual information enables a more general and powerful definition of the channel capacity:

$$C = \max_{p(X)} I(X,Y), \tag{14.10}$$

i.e. the channel capacity is achieved by finding the source distribution $p(X)$ which maximises the mutual information between source and output. Shannon's *Noisy Channel Coding Theorem* then states that if the source entropy $H(X) \le C$, there exists a coding such that the output Y contains an arbitrarily small error frequency. If $H(X) \ge C$, then a coding can be found such that $H(X \mid Y) \le H(X) - C$, where the *equivocation* $H(X \mid Y)$ defined by

$$H(X \mid Y) = H(X) - I(X,Y) \tag{14.11}$$

quantifies the error rate. The proof of the Noisy Channel Coding Theorem is beyond the scope of our presentation, except to note that again it relies on the length of the transmitted messages being arbitrarily long. Its importance in delimiting the possibilities for error-free transmission of say, TV or wireless broadband signals should be obvious.

14.5 A DEMONIC CHALLENGE TO THE SECOND LAW

What's in a name? In other words, how seriously should we take the fact that the central concept of Information Theory shares the same name as that of thermodynamics, viz. *entropy*? Let's tackle this by first setting out the evidence in favour of a deeper link: the mathematical congruence between the definitions (5.45) and (14.5) for Gibbs and Shannon entropies, respectively, which holds up to a trivial rescaling resulting from converting between natural and base-two logarithms; the fact that each quantity, while at first sight having a rather abstract, elusive meaning, actually leads to considerable conceptual and technical simplifications in the treatment of real-world devices such as heat engines and transmission lines; and the importance of the large-n or *thermodynamic* limits as required ingredients of arguments leading to precise statements.[27] However, there is a sense in which the key ideas are aligned in the opposite direction. In statistical thermodynamics, our concern is to count indistinguishable microstates that all yield the same macrostate characterized by a few bulk variables, whereas the whole point of Information Theory is that the messages being counted are distinguishable because otherwise, they wouldn't tell us anything. Finally, what about the dimensionful factor of k_B in (5.45); in other words, how do we translate between Joules per Kelvin and bits per symbol?

The quest to forge a link was started in 1871 by James Clerk Maxwell in his *Theory of Heat*. Maxwell wished to argue that the universal entropy increase predicted by the Second Law was not dictated by fundamental dynamical principles, but rather resulted from a sequence of microscopic processes whose direction was statistically overwhelmingly likely to lead to entropy increase. This point of view continues to be held by most physicists and corresponds to the pragmatic point of view outlined in Chapter 5. Maxwell envisioned a microscopic being endowed with senses, intelligence, and preferred pronouns he/him/his, subsequently dubbed a *Demon* by his colleague Kelvin. The Demon controls a valve connecting two adjacent vessels containing gas. Initially, the gases are in equilibrium and have the same temperature. The molecules comprising the gas are moving with a range of speeds described by the Maxwell-Boltzmann distribution first seen in Figure 3.2. The Demon is sufficiently nimble and perceptive that on viewing a gas molecule to the left of the valve moving faster than average towards him, he is able to briefly open the valve with negligible energy expenditure to allow it to transfer to the right-hand vessel; similarly, on viewing a molecule to the right approach with a slower speed he can allow it to pass through to the left. Over time, the excess kinetic energy carried from left to right results in a redistribution of molecular velocities as a result of collisions. The molecules in the right vessel have a larger mean square speed than those to the left; in other words, the gas in the right vessel is now at a higher temperature, and that in the left at a lower temperature. Since this process can continue as long as the Demon chooses, heat flows from a cooler to a hotter body, in direct contravention of the Second Law.[28]

Maxwell's Demon has by now entered physics culture on a similar footing as Schrödinger's Cat, also posited as a means to highlight the apparent absurdity of some all-encompassing physical principle. If anything, he is a more compelling figure, active rather than passive and exuding a dark glamour; his actions can be perceived as either malevolent in undermining a cherished physical law or as benevolent in enabling a heat engine to operate between the left and right-hand reservoirs, yielding useful work for no apparent energy expenditure.[29] The Demon has not, to date, made the same impact in general culture.[30] We've learned much through conceptual attempts over the subsequent hundred years to confound the Demon and restore the primacy of the Second Law.

An important "physicist's objection" was raised by Brillouin[31] in 1951, who pointed out that in order for the Demon to act he must be able to see the molecules coming, which requires him to use a torch emitting light of frequency f exceeding that of the thermal black body background radiation (introduced in Section 9.4.2) in the vessel at temperature T_0, i.e.

$$hf \gg k_B T_0, \tag{14.12}$$

where Planck's constant h relates the light frequency to an energy scale. In order to achieve this the torch's filament has temperature $T_1 \gg T_0$. Suppose the total energy radiated during the demonic action is E, leading to a decrease of filament entropy $\Delta S_f = -E/T_1 \simeq -k_B$. The Demon detects a molecule by absorbing a single photon scattered into his eye, which increases his entropy by

$$\Delta S_d = \frac{hf}{T_0} = k_B b \Rightarrow b \gg 1, \tag{14.13}$$

where the second inequality follows from (14.12). Note also that $|\Delta S_d| \gg |\Delta S_f|$. Following the Demon's action directly influencing the microscopic arrangement of the enclosed gas, its entropy $S_g = k_B \ln\Omega$ is decreased because the number of indistinguishable microstates is reduced by an amount ω, presumed much smaller than Ω, i.e.

$$\Delta S_g = k_B \ln\left(\frac{\Omega - \omega}{\Omega}\right) \simeq -k_B \frac{\omega}{\Omega}, \tag{14.14}$$

where we have used the small-x approximation (MA.17c). We conclude that the entropy change of the entire system

$$\Delta S_d + \Delta S_g + \Delta S_f = k_B\left(b - 1 - \frac{\omega}{\Omega}\right) > 0; \tag{14.15}$$

the Second Law thus appears safe provided the Demon himself is subject to the laws of physics.

14.6 REVERSIBLE COMPUTING

Remarkably, further insight comes from a different direction, associated with a discipline now known as Computer Science. A key feature of Brillouin's argument is that the torch filament is not in thermal equilibrium with either the gas or the Demon himself; in other words, the measurement process underlying the Demon's actions is posited to be inherently irreversible. However, first Landauer (1961) and then Bennett (1982)[32] argued that this assumption is unwarranted, and in fact, the measurements needed by the Demon can be performed with arbitrarily little dissipation using theoretical model devices performing error-free computations. To get a flavour for what this means we need to introduce a slightly different demonic scenario. Maxwell's original is a "temperature demon", in which the system of Demon and left and right vessels is thermally isolated. We can also consider a "pressure demon" who works with a system in contact with a heat bath to convert energy transferred in the form of heat into mechanical work via a moveable piston. In 1929, using insight from Information Theory, Szilard[33] took this line of reasoning to its logical conclusion by considering a gas consisting of just a single molecule.

Figure 14.4, adapted from Bennett's 1982 paper,[34] shows both the operation of a Szilard engine and the corresponding states of the Demon's "mind" invoked in Bennett's analysis, which can be considered in one of three states according to the Demon's knowledge of the molecule's whereabouts: a standard reference state **S**, and states **L** or **R** depending on which side of a partition the molecule is observed or known by the Demon to lie. The blobs in the grids to the right of Figure 14.4 simultaneously represent both the allowed range of the molecule's horizontal coordinate and the state of the Demon's mind.

In the analysis of heat engines in Chapter 4, we stressed the importance of working in a cycle in which the engine starts and finishes in the same state—recall this stipulation is built into the Second Law. We now describe the Szilard cycle in detail. At stage (a), the molecule wanders freely through the vessel, and the Demon doesn't know where it is, so his mind is in the standard state **S**. In (b), the Demon inserts a partition halfway across the cylinder, which traps the molecule on one side or the other—in Figure 14.4, we have chosen this to be the left-hand side, but it's important to keep in mind that another equally likely sequence in which **L** and **R** are interchanged is possible. In (c), the Demon measures the molecule's position, and accordingly, his internal state changes from **S** to **L**. Using this information, in (d), the Demon replaces the partition with a moveable piston on the right-hand side of the cylinder, and uses it to extract isothermal work $-\Delta W$ from the molecule[35] by allowing it to repeatedly drive the piston to the right, with the molecule's kinetic energy replenished by heat ΔQ transferred from the heatbath. By (e) the piston is back at its original position, the volume accessible to the molecule has increased by a factor of 2, and we deduce from the considerations of Chapter 4 that the amount of work done is $k_B T\ln 2$. However, the Demon's internal state is still **L**. Hence, the cycle is not yet complete—this is only done by stage (f) once the state is restored to **S**. This final step is not so innocent as it seems—the memory erasure that occurs when going from (e) to (f) reduces the Demon's personal phase space from {**L**,**R**} to **S**, i.e. by a factor of 2.

If the Demon is a physical entity, this necessarily reduces his entropy by an amount $k_B\ln 2$ which can only be consistent with the Second Law if there is a corresponding entropy increase somewhere

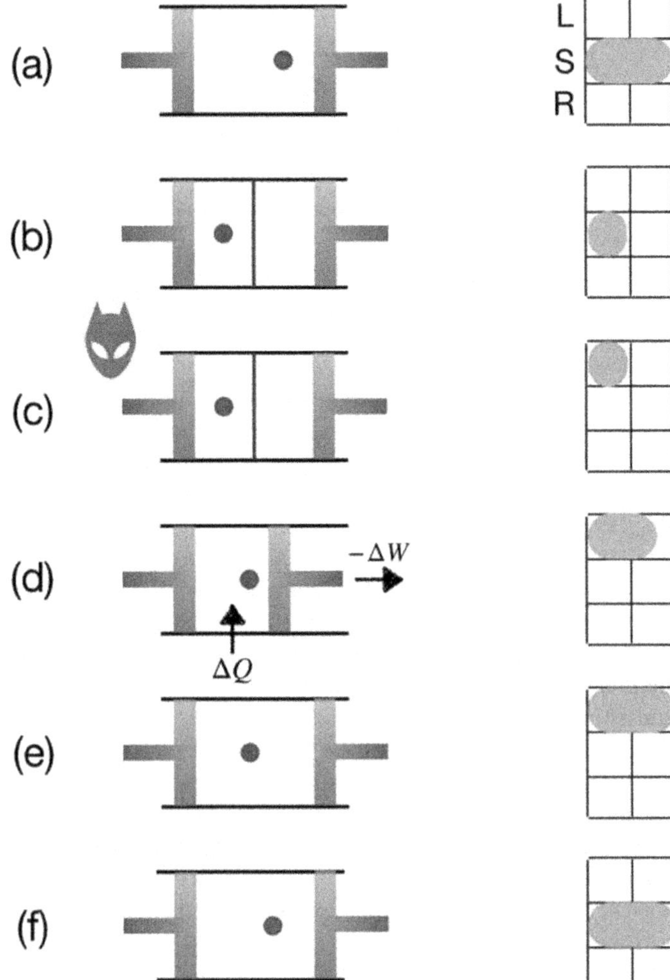

FIGURE 14.4 Six stages in the operation of a Szilard engine, following Bennett.

else. In Information Theory terms, erasing the memory reduces the uncertainty in the Demon's internal state by precisely 1 bit. It is very tempting to posit a relation between thermodynamic and information theory entropies:

$$S = \left(k_B \ln 2\right) H \simeq 0.693 k_B H. \tag{14.16}$$

What physical form could the Demon's memory device take? A straightforward way to store one bit mirrors the Szilard engine itself; simply store a molecule in a nearby constant temperature cylinder with a partition ensuring that it is either **R**, or **L**, which can be regarded as the **S** state. At the end of the cycle remove the partition, causing an irreversible Joule expansion with associated entropy increase $k_B \ln 2$ and then reset the memory by compressing the one-molecule gas into the left half of the box. It is important that the erasure procedure has to be the same regardless of whether the memory is in **L** or **R** state; the physically irreversible Joule expansion therefore implements the logically irreversible erasure. The isothermal compression requires work $k_B T \ln 2$, and results in an entropy $k_B \ln 2$ transferring from the gas + cylinder to the heatbath in the form of an equivalent amount of heat, as described in Chapter 4, leaving the memory in exactly the same reference state **S** as at the start. The Szilard cycle is thus *thermodynamically* reversible in the same sense as the Carnot cycle, implying that we can employ $\Delta Q = S\Delta T$, but the key step of memory erasure is *logically* irreversible. The heat $\Delta Q = k_B T\ln 2$/bit dissipated in the environment during memory erasure, anticipated in the conversion between S and H in (14.16), is known as the *Landauer limit*.

Let's take stock. For a reversible cycle as proposed by Bennett, each cycle of the Szilard engine delivers an amount of work $k_B T\ln 2$ as a consequence of the one-molecule gas absorbing heat from the reservoir. To close the cycle, exactly the same amount of work is needed to effect memory erasure, and the resulting entropy production is manifested as heat transfer back to another reservoir. For a reversible process the memory reservoir must share the same temperature as the engine reservoir, so the heat transferred again equals the work put in. This mirror arrangement makes it clear we are not getting something for nothing as a result of the Demon's action! It's a little reminiscent of the setup in Figure 4.3 in which two Carnot cycles working between two reservoirs at different temperatures in opposite senses were used to demonstrate the universal relation between heat transfer and absolute temperature. Note though that in a Szilard cycle everything takes place at a common temperature. Other erasure procedures can be formulated: Bennett's paper gives an example more closely resembling existing computer technology, in which bits are stored via the direction of magnetization of a small ferromagnetic sample, and measurement is a copying process employing a transverse magnetic field to induce a second bit to adopt the same orientation. Regardless of the detailed mechanism, no work is performed in a reversible cycle once memory erasure is taken into account. This differs essentially from Brillouin's analysis, which identified entropy increase with an irreversible measurement process. Another interesting solution in this category due to Pierce[36] uses a version of Szilard's engine in which the information about the molecule's location is transmitted via a radio signal. It turns out that due to thermal noise of the form (14.6) in the radio receiver, the minimum energy needed to reliably transmit one bit again coincides with the Landauer limit.

The reinterpretation of thermodynamic entropy in terms of information loss when going from a low-entropy state to a high-entropy state has proved to be very fruitful, and we'll see in subsequent chapters how it's spawned further ideas which remain at the forefront of modern theoretical physics. On the face of it, the fundamental Landauer identification of entropy production with memory erasure seems entirely reasonable—we hardly need reminding that deleting a file is an irreversible act (unless we've taken the trouble to make a backup), somehow increasing our ignorance of the world. In the limit that the Demon operates reversibly it also points to a fascinating intrusion of thermodynamics into the theory of computation. Remember, though, that it's only a thought experiment, whose interpretation ultimately depends on the extent to which we deem the Demon to be a physical entity subject to the same physical laws it purports to undermine. Perhaps an entity or device could be programmed to act as a Demon with arbitrarily little dissipation, but we have still to discuss the difficulties of performing quasistatic expansions and compressions of a one-molecule gas, where work is transmitted via sporadic, hard-to-predict collisions with the piston, making the problem of maintaining reversibility by carefully varying the load on the piston a formidable one.

While reversible computation and memory erasure close to the Landauer limit remain at the very limits of what can be achieved in experiments with artificial devices, there is a key biochemical process where the same concepts can be usefully applied, namely the process in a cell nucleus whereby an RNA strand containing the genetic information needed for the synthesis of proteins is copied from a DNA molecule, as described in Bennett's 1982 paper from which Figure 14.5 is adapted. The process is catalysed by an enzyme called RNA polymerase represented by rectangular box in Figure 14.5, a large molecule with sites in which a segment of the DNA molecule is held in proximity to the RNA strand as it is constructed link by link. At each stage of the build, the enzyme causes one of four nucleotide pyrophosphates (ATP, GTP, CTP, or UTP—we already met ATP in Chapter 13) from the surrounding solution to form a strong chemical bond with the previous

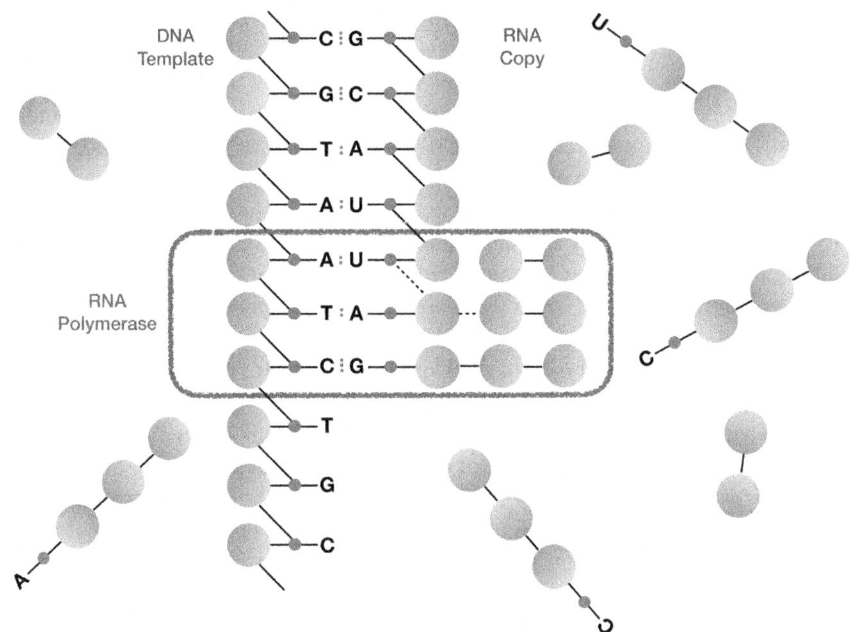

FIGURE 14.5 Cartoon of the synthesis of a complementary RNA strand from D via the action of RNA polymerase. As drawn the process is happening from the top downwards. Circles denote phosphate groups, solid lines strong (covalent) chemical bonds, dashed lines bonds in the process of forming or breaking, and the small groups of two/three vertical dots the highly specific hydrogen bonding between nucleotide base pairs. The nucleotides are represented by the bold letters **A**, **C**, **G**, **T**, and **U**.

nucleotide in the chain, at the same time jettisoning its pyrophosphate tail (i.e. the groups labelled γ, β in Figure 13.3). As a consequence of the many weaker bonds between various locations on the polymerase, DNA and RNA, the building process is highly specific: the **C** nucleotide (cytidine) forming part of the RNA *must* pair with **G** (guanosine) on the DNA; similarly **G** on RNA with **C** on DNA, **A** (adenosine) on RNA with **T** (thymidine) on DNA, and **U** (uridine) on RNA with **A** on DNA. This is how the coding sequence is copied from the DNA to the RNA.[37] Once the nucleotide is bound to form a link in the RNA the enzyme shifts forward one notch along the DNA in preparation for copying the next nucleotide.

As discussed in Chapter 13, the chemical reactions involved in RNA synthesis are reversible; their direction and rate are controlled by the local concentrations of both nucleotide (e.g. ATP) and free (PP) pyrophosphates through the Law of Mass Action. The reaction rate has been measured in live bacteria to run forward at roughly 30 nucleotides per second, dissipating $\Delta F \approx 20k_BT$of free energy per nucleotide. However, by varying the reagent concentrations the reaction can be slowed considerably, with the synthesis now taking almost as many steps backward as forward along the way, and a much reduced free energy cost per forward step as determined by Equation (13.27). Bennett refers to this limit as a *molecular Brownian computer*[38] executing what is in effect a one-dimensional random walk along the DNA strand. As the concentrations are tuned towards chemical equilibrium (i.e. $\Delta F = 0$), the reaction rate becomes very small, and the computation close to being reversible. However, the DNA to RNA transcription has a small intrinsic error rate η due to some RNA synthesis steps proceeding without catalysis. As the reversible limit is approached, using the approximation (MA.17b) we find the ratio $R = (1 - e^{-\Delta F/k_BT})^{-1} \simeq k_BT/\Delta F$ of total synthesis steps to net forward steps needed grows large, making the transcription prone to a high error rate $R\eta$. In order to keep errors infrequent to ensure accurate transcription, we have to work in a regime where

$\Delta F/k_BT$ is large. This underlines the assertion made in our discussion of metabolism in Chapter 13 that the accurate information transfer needed for vital processes in living organisms requires a supply of free energy. In order to reproduce we first need a good meal...

NOTES

1 Let's note immediately that in this context digitized images are also data.
2 An obvious example is the proliferation of statistical data accompanying professional sport, both in the training of athletes, analysis of their subsequent performance, and in the commentary through which most of us follow it. Sport is an intensely competitive business, with clearly identified winners and losers. If my team's data scientists are better than your team's, and we win, then go figure how much to pay them...
3 One of the authors is constantly amused by emails from a purchasing website recommending him his own books...
4 Of course, in vocal communication, there is enormous meaning communicated non-verbally via vocal emphasis and variation of tempo and pitch, but this should be less relevant for a scientific statement which has been crafted to be as unambiguous and universal as possible.
5 Upper and lower case letters count as distinct symbols.
6 Note that the space between words " " is also counted as a symbol; were it not, then there would be scope for confusion between, say, ANDREA DWORKIN GLOOM and AND READ WORKING LOOM, although an alert reader might be able to resolve this ambiguity based on context.
7 Some authors advocate highlighting the distinction between information and data by quantifying the former in *bits* and the latter in *binary digits*. We have some sympathy with this point of view but suspect the ship has already sailed...
8 Named after Samuel Morse (1791–1872), one of the inventors of single-wire telegraphy.
9 Encodings requiring several symbols are more susceptible to noise as transmission power is reduced.
10 One of the authors learned about Morse Code from his mother, who was a radio operator in the Women's Auxiliary Air Force, and the other from his father, who was a Morse operator in the run-up to the Normandy landings, both serving during the Second World War.
11 Developed by MIT information theory student David Huffmann in 1952.
12 Apparently, Morse addressed this issue not by analysis of text but by counting the contents of the different compartments of a printer's type box.
13 See for instance https://www.gutenberg.org/ebooks/
14 Claude Shannon (1916–2001) was a mathematician, electrical engineer, and computer scientist working at Bell Telephone Laboratories.
15 Writing the frequency using the symbol p_i emphasizes that it is the same thing as a probability obeying the two key conditions $p_i \geq 0$ and $\Sigma_i\, p_i = 1$.
16 To convert $\log_2$ to the natural logarithm ln, use the identity (MA.7), viz: $\log_a x \equiv \ln x/\ln a$.
17 This is equivalent to the discussion of the microcanonical ensemble in Chapter 7.
18 Scrabble was invented by Alfred Mosher Butts in 1938, who based the scores on a frequency analysis using sources such as the *New York Times*.
19 For reference we have treated e.g. é, è, and ê as the same letter.
20 Huffmann coding has the property that the average number of bits per symbol is never greater than $H+1$
21 For n-gram coding, the definition of the "entropy of English" is $H_n = -\frac{1}{n}\Sigma_i\, p_i \log_2 p_i$.
22 For instance, the trigram **TQX** and all subsequent n-grams containing it will never occur and so contribute zero to H: these can be eliminated from further consideration at an early stage. Even 96 of the 729 conceivable Shakespearean digrams never occur.
23 This result is due to Harry Nyquist (1889–1976), a physicist and electronic engineer working at Bell Telephone Laboratories.
24 The bandwidth quantifies the range of frequencies present: a signal of bandwidth $\mathcal{W}$ and duration $\mathcal{T}$ can be faithfully reconstructed from $2\mathcal{W}\mathcal{T}$ discrete samples.
25 The optimal choice of n is determined by the average transmission error rate.
26 The resemblance to the *mutual information* defined in Chapter 12.3.7 is more than coincidental; note however that Information Theory deals in classical probabilities, with no analogue of quantum entanglement.
27 Another technical requirement for the application of Information Theory is that message sources are *ergodic*; that is, symbol frequencies obtained by analysis over a large ensemble of sources are equal to those obtained from the time-averaged analysis of a single long message.
28 There are clear congruences with the thought experiment introduced by von Neumann, as discussed in Section 12.2.
29 Maxwell's Demon could power a *perpetual motion machine of the second kind*, i.e., one whose operation violates the Second Law. Guess which law is violated by perpetual motion machines of the first kind!
30 He does make a cameo appearance in Thomas Pynchon's novel *The Crying of Lot 49* (Jonathan Cape, 1967).

31 Theoretical physicist Léon Brillouin (1889–1969) made important contributions to quantum theory and condensed matter physics. His paper *Maxwell's Demon Cannot Operate: Information and Entropy I* J.App.Phys. **22** (1951) 334 clearly shows the influence of Shannon's ideas.

32 Physicists Rolf Landauer (1927–1999) and Charles Bennett (b. 1943) both worked at IBM Research. Bennett is also one of the pioneers of quantum cryptography.

33 During the Second World War, physicist Leo Szilard (1898–1964) was a key advocate for developing nuclear weapons, leading to the Manhattan Project. Famously (and chillingly), in London in 1933, he conceived of the nuclear chain reaction while waiting for the lights to change so he could cross the street.

34 C.H. Bennett, *The Thermodynamics of Computation — A Review,* Int.J.Theo.Phys. **21** (1982) 905.

35 For consistency with Chapter 2, we keep the convention that ΔW is the work done *on* the molecule.

36 J.R. Pierce, communications engineer and sci-fi author (1910–2002), in *An Introduction to Information Theory: Symbols, Signals and Noise* (Dover Books 2nd ed., 1980).

37 In turn, successive sequences of three nucleotides on the RNA strand code one of twenty amino acids, which are sequentially synthesised into proteins via the agency of a structure called a *ribosome*. The process employs the famous "Genetic Code", e.g. the triplet **GCU** codes for alanine.

38 Brownian motion is the random motion of optically-visible particles suspended in a medium, due to collisions with invisible molecules. It was discovered by botanist Robert Brown in 1827, and is one of the earliest pieces of empirical evidence supporting atomism.

15 Death

The Second Law expresses a tendency for all things to approach thermal equilibrium over time; we sense this when experiencing radiative cooling on a cold clear night. We trace back the origin of the various low-entropy fuels needed to make our lives viable, concluding that gravitational clustering of matter in the past to promote the onset of thermonuclear reactions is key. Our experienced "Arrow of Time" suggests the universe has evolved from an initial state with much lower entropy. We present a simplified cosmology consistent with a flat universe in which visible matter is composed of hydrogen atoms, while entropy is dominated by photons, outnumbering baryons by a factor of approximately a billion. This simple model describes the universe back to a few minutes after the Big Bang, but on its own, it cannot address fundamental questions such as why there is more matter than antimatter present today. Looking forward, the inexorable march towards universal equilibrium seems to predict a grim "Heat Death" for us all: we review one proposed strategy to keep going predicated on clocking down our processing speed and frequent hibernation.

15.1 EQUILIBRIUM WITH THE UNIVERSE

Can you remember the last time you stepped outside on a clear cloudless night to look up at the sky and enjoy the majestic spectacle on offer? Was it years ago, or maybe just last week? Perhaps you were lucky enough to be in the countryside, far from sources of artificial light pollution, and were able to discern many many stars, far more numerous and faint than those depicted in simple maps or planispheres; perhaps you even made out the luminous sheen of the Milky Way, our own galaxy in profile, straddling the sky from horizon to horizon. It's a breathtaking sight. Perhaps for a moment, you were able to override the illusion of the heavens being spread over the inner surface of a hemisphere like continents on a globe and felt truly immersed in the cosmos. Perhaps you also recall feeling rather chilly.

Figure 15.1 shows the black body spectrum for electromagnetic radiation in thermal equilibrium, characterized by temperature T. It was derived using the quantum statistical properties of bosons in Section 9.4.2 and plotted in Figure 9.1 as a function $u(f,T)$ of frequency f; here instead Figure 15.1 shows $u(\lambda,T)$ as a function of wavelength λ. The two plots convey the same information. The distribution contains radiation of arbitrarily long wavelength, but is cut off rather sharply as $\lambda \to 0$. The area under each curve $\int_0^\infty u(\lambda,T)\,d\lambda$ is the energy density of thermal radiation—clearly this quantity increases rapidly with temperature, indeed:

$$\int u(\lambda,T)\,d\lambda = \frac{\pi^2 k_B^4 T^4}{15\hbar^3 c^3} \propto T^4. \tag{15.1}$$

DOI: 10.1201/9781003121053-19

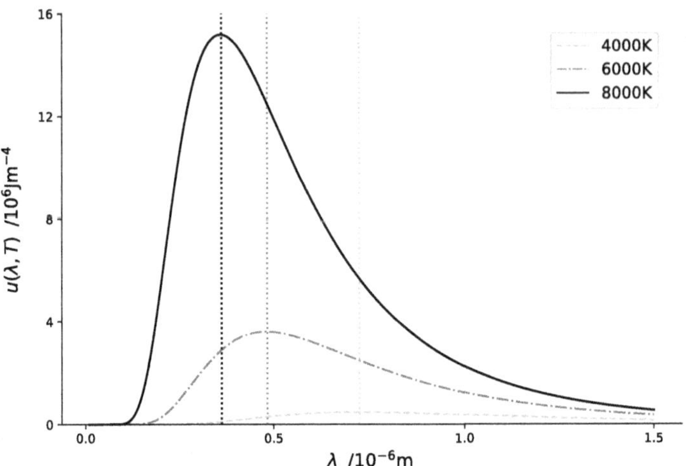

FIGURE 15.1 Representative black body spectra $u(\lambda,T)$ plotted as a function of wavelength.

The vertical dotted lines in Figure 15.1 show the wavelength for which the energy density is maximum, as described by Wien's Displacement Law[1]:

$$\lambda_{max} \simeq \frac{3}{T}\text{mm}. \tag{15.2}$$

The lines move to the left with increasing T; given that visible light falls in the range $0.4\ \mu\text{m} < \lambda_{vis} < 0.7\ \mu\text{m}$ (accommodated by the horizontal axis in Figure 15.1), we deduce that the characteristic colour of glowing objects evolves from red through orange to yellow then bluey-white as they get hotter.

It feels cold outside on a clear night because your skin radiates heat in the form of infrared radiation, with wavelength roughly 0.01 mm for skin temperature 300K using (15.2). When indoors we're not particularly conscious of this because we're bathed in radiation from surfaces such as walls or furniture at similar temperatures, so the net heat transfer is small. Outside, far from sheltering buildings or trees, your body is largely open to the sky, which radiates back at a much lower temperature; the 2.7K characteristic of the so-called *cosmic microwave background* (CMB). Hence your body loses heat through radiative cooling at a constant rate[2]; your feeling cold is a reflection of the struggle you're engaged in to achieve thermal equilibrium with the universe.

Really, this is just another instance of equilibration between bodies governed by the Laws of Thermodynamics we teased out in the opening few chapters, the extra gloss due to one of the bodies being the universe as a whole. This chapter aims to develop this universal perspective. Let's start by moving indoors, removing our scarves, and warming up after our cosmic excursion. It's warm inside because the house is heated, perhaps by a coal fire or by central heating powered by burning oil or gas. The energy released results from a chemical reaction between hydrocarbon molecules in the fuel and oxygen molecules in the air. In Chapter 13, we discussed chemical processes from a thermodynamic perspective and can use the language we developed to analyse burning. Free energy is invariably reduced when burning fossil fuels because the products are gases, in this case, carbon dioxide (CO_2) and water (H_2O). Gases are inherently high-entropy forms of matter because there are many many more ways to distribute small fast-moving molecules compared to the possible configurations of a single larger hydrocarbon molecule. The energy released from the rearrangement of the chemical bonds is transformed into molecular kinetic energy, resulting in the gases becoming hotter. Sometimes, to a good approximation, this process is quasistatic—an expanding gas could drive a piston so that a heat engine produces work as discussed at length in Chapters 2 and 4, or perhaps a mass of warm air rises and moves coherently to transfer heat throughout a larger space in a process known as *convection*. Eventually, however, the heat produced does the job we want, warming us up by flowing from a hotter place to a cooler place by conduction through our skin and internal organs—we'll breath in warmer air, and enjoy a cup of tea!—all in strict accordance with the Second Law. Because we've come in from the cold, our skin temperature is initially less than that in the room, so this process cannot be quasistatic, and as a consequence of our stargazing, the entropy of the universe increases.

15.2 LOW-ENTROPY FUEL SUPPLIERS

The First Law tells us that energy is neither created nor destroyed but can be interconverted between different forms, such as the examples discussed in Chapter 2. The fossil fuels used to heat our homes and power electricity generation were produced as a result of the decay of living plant material, exposed over many millions of years to heat and pressure within the Earth's crust and the anaerobic digestive action of microorganisms, to form low-entropy fuels such as coal, petroleum, and natural gas. The plants' organic material was originally rich in carbohydrates such as glucose (see Figure 13.1), which formed as a result of a chemical reaction called *photosynthesis*, which builds carbohydrate molecules from CO_2 and H_2O found in the atmosphere, producing oxygen O_2 as a by-product. Photosynthesis is powered by sunlight, which at a microscopic level consists of a stream of quanta called photons. The photons which reach us from the Sun are all travelling in the same direction, the ones active in photosynthesis having wavelengths in the range 400–700 nm,[3] comparable with the wavelength λ_{max} predicted by Wien's Law (15.1) if the surface temperature of the Sun is approximately 6000K.

As a consequence of photosynthesis the plant grows, reemitting infrared radiation into its environment appropriate to the ambient temperature, which bears away a much higher entropy than that which arrived with the incident sunlight for two reasons. Firstly, rather than forming a well-collimated sunbeam, the plant's thermal radiation is emitted in all directions; secondly, the amount of energy carried off by a photon is given by

$$E_{photon} = h\frac{c}{\lambda}, \tag{15.3}$$

where c is the speed of light and h the fundamental *Planck constant*.[4] Since infrared photons have a longer wavelength than those of sunlight, by a factor of roughly 20 according to equation (15.2), then for a given energy transfer consistent with homeostasis, there must be roughly 20 times more photons reradiated as absorbed. Both factors give the reemitted radiation a much higher entropy, for the same reason as the entropy production associated with exhaust gases in burning: there are simply many more ways to arrange the low-energy photons which travel independently in random directions. With so many more microstates available to explore, $S = k_B \ln\Omega$ must increase. We should therefore classify sunlight as another low-entropy source of energy, whose consumption increases the world's entropy in accordance with the Second Law. As an aside, the reemitted IR radiation is much more effectively absorbed by the Earth's atmosphere and for the most part does not succeed in escaping back to space. While we still feel a chill on clear nights, the main effect is to heat the air; the resultant *greenhouse effect* helps maintain a comfortable ambient temperature across much of the earth's surface.[5] Sunlight is also behind burgeoning power generation technologies such as the photovoltaic cells in solar panels, which generate a voltage whenever they absorb photons, and wind turbines — the world's weather is, after all, driven by solar radiation falling with different intensity on different parts of the globe in different seasons, producing local variations in pressure and moisture content.

For other sustainable energy sources entropy production arises through different means. Tidal power generation exploits bulk motions of sea water resulting from the gravitational action of nearby bodies such as the Moon, and to a lesser extent the Sun. Tidal effects occur whenever a force between an extended body of mass m and another mass M, such as gravity GMm/r^2, varies with separation r so that the force on the side of the body facing the gravitating mass is greater than the force on the opposite side. In the classic model of the tides originally due to Newton, this leads to two raised "bulges" of water forming, centred on antipodal points on the Earth's surface and producing two tides per day as the Earth rotates beneath them. The relative motion of celestial bodies under gravity is predictable and on the face of it inexhaustible; however, in reality apart from the energy drawn off through power generation, there is also dissipation (and hence entropy production) due to friction wherever sea currents flow against the Earth's solid surface and also, deep beneath, by motion within the solid crust and mantle. The bulges act rather like a pair of brake pads, gradually slowing the Earth's rotation and ultimately leading to *tidal locking*,[6] when the Earth's rotation is brought in sync with the Moon's orbital motion so that it will appear to stand still in the sky, illuminating only one half of the Earth, and the tides cease.

What's behind the low entropy of sunlight? Its constituent photons have higher energy because the Sun is hot, and the ones reaching us move roughly in parallel because the Sun is concentrated in a sphere, subtending an angle of roughly half a degree when viewed from Earth. Both are a consequence of gravity; the mutual gravitational attraction of the material forming the Sun and all other stars causes it over time to contract and heat up, as gravitational potential energy is converted into the kinetic energy of thermal motion. At some point, the temperature is high enough for thermonuclear reactions to start up in the core, where the temperature and density are highest. The reactions in question are known as *nuclear fusion*, in which smaller nuclei fuse together to form a larger nucleus, which is more tightly bound by the *strong interaction*, thus releasing binding energy to heat the reaction products. Fusion requires the nuclei to come within $\sim 10^{-15}\,\mathrm{m}$ of each other, the range over which the strong interaction operates. High-speed collisions, and hence high temperatures, are needed in order to overcome the electrostatic repulsion between positively charged nuclei.

The predominant element in the Sun, and in the universe at large, is hydrogen formed shortly after the Big Bang, and the thermonuclear reactions convert the hydrogen found in the core of the Sun and more generally in the inner regions of all stars into helium. The detailed interaction pathways are complex and vary depending on the star's mass but are conveniently summarized by

$$2e^- + 4\mathrm{H}^+ \rightarrow \mathrm{He}^{2+} + 2\nu_e + 26.73\,\mathrm{MeV}, \tag{15.4}$$

i.e., four hydrogen nuclei (aka protons) fuse together to form a helium nucleus (aka α-particle), and two electrons e^- thereby converted into massless neutrinos ν_e which interact so weakly with everything else that they immediately stream away from the star and into space at light speed. From our perspective, the important number is the energy release of 26.73 MeV, originating in the mass difference between four bare protons and four nucleons bound inside a He nucleus, and determined by $E = mc^2$. This translates into 2.58×10^6 MJ mol^{-1}, which utterly dwarfs the 2.8 MJ mol^{-1} obtained from metabolizing glucose we learned about in Chapter 13 and is responsible for one of the more memorable astronomical statistics, namely that the Sun's luminosity of 3.8×10^{26} W equates to a mass loss of some four million tonnes per second!

While the characteristic energy scales between nuclear and chemical reactions differ by a factor of a million, the basic thermodynamic principles developed in Chapter 13 are just as valid. Reaction (15.4) proceeds in a direction dictated by the Law of Mass Action, governed by the relative concentrations of the nuclear and particle species on either side of the equation. In this case, however, gravity plays a key organizational role by acting principally on the more massive participants, namely the nuclei, but allowing neutrinos to flee the scene. Just as importantly, any massless photons produced in associated interactions are also able to escape, albeit on a much longer timescale,[7] ultimately resulting in the star's radiation. It is gravity that's responsible for maintaining the low-entropy conditions within star cores that permit them to act as primary power generators.

Finally, what of nuclear fission reactions which provide a significant contribution to our current and potentially future needs for carbon-neutral power? Fission occurs when a large[8] nucleus, such as $^{235}_{92}\mathrm{U}$, i.e., "uranium-235" containing 92 protons and $235-92=143$ neutrons splits into two smaller nuclei, such as rubidium $^{93}_{37}\mathrm{Rb}$ and caesium $^{141}_{55}\mathrm{Cs}$, together with a neutron and 181 MeV of released energy. In nuclear reactors, fission is induced by the nucleus first absorbing a slowly moving neutron from its environment. Tuning the supply of such "thermal neutrons" makes it possible to control the reaction rate. Fissile nuclei such as $^{235}\mathrm{U}$ are another example of low-entropy fuel and indeed are a scarce resource in the universe.[9]

While all nuclei are composed of both protons and neutrons, larger nuclei need a relatively larger fraction of neutrons in order to be stable. Their formation does not result from the relatively peaceful fusion processes prevalent in stars during their regular evolution. Rather, they require neutron fluxes some $10^9\times-10^{10}\times$ greater so that the absorption of neutrons by nuclei to form more massive nuclei can proceed rapidly via the so-called *r*-process along an isotopic sequence,[10] without intermediate highly radioactive (i.e., short half-life) isotopes having a chance to decay. Once the nucleus so formed is sufficiently neutron-rich, it may transform into a different element via β-radioactive decay, in which a neutron is transformed into a proton. Both *r*-process and β-decay make up the pathway to the heavier elements. A candidate for such extreme conditions are the rare cataclysmic events known as supernova explosions. These happen when a massive star[11] exhausts its readily processed thermonuclear fuel in the core so that pressure resulting from heat generation in the core can no longer sustain the weight of the star's outer layers, making the whole star unstable to gravitational collapse. The collapse has two large-scale consequences: firstly, the outer material of the star is subject to the extreme temperatures and neutron fluxes required for synthesis of the heavy elements and the elements so formed explosively redistributed within the surrounding interstellar medium. Secondly, an ultradense remnant known as a *neutron star* is formed, with a central density roughly $10^{15}\times$ that of water. Collisions between binary pairs of neutron stars have recently been observed via the emitted gravitational wave radiation,[12] and it is believed that such extraordinary mergers may actually be essential to the *r*-process synthesis of some heavy elements.

We have reached the conclusion that everything we need to keep the lights on and make a living is based on the transformation of low-entropy fuels into high-entropy exhaust and waste products. The Second Law states that this trend is irreversible; the "energy crisis" looming over us since the 1970s is actually an entropy crisis! Another theme which has emerged is the key role played by gravity in preparing the conditions in which low-entropy fuels are forged. We are left to contemplate the universe and everything that's in it marching steadily into the future, with its entropy growing inexorably and irreversibly, just as Clausius said.

15.3 THE ARROW OF TIME

It's tempting to view the Second Law in this context as some kind of cosmic principle underlying an "Arrow of Time", dictating a fundamental distinction between what's gone before and what is to come. This of course exactly matches our lived experience, but it's hard to justify in terms of known microscopic physical laws. Newton's Second Law of Motion provides the most immediate example; a body's acceleration is proportional to the implied force, or in equation form

$$F = ma. \tag{15.5}$$

The point here is that acceleration a is the second time derivative of the body's position $x(t)$; if the velocity is the change in position Δx in a short time interval Δt, then

$$a = \frac{d^2x}{dt^2} \equiv \lim_{\Delta t\to 0} \frac{\Delta x(t+\Delta t) - \Delta x(t)}{(\Delta t)^2}. \tag{15.6}$$

Equation (15.5) is unchanged if we "run time backwards" by reversing the sign of the t coordinate $t \mapsto -t$, because acceleration only depends on the squared quantity $(\Delta t)^2$ which is invariant under this mapping. Newtonian dynamics runs just as well backwards in time. If we view a filmed sequence in close-up of a head-on collision between two snooker balls (which, to good approximation is not unduly affected by dissipative processes such as friction with the table cloth or air resistance),[13] then actually, it can be quite difficult to work out whether the film is being run forwards or backwards. In either case, we see a moving ball approach and then collide with a stationary one. Following the collision, the first ball comes to a standstill while the second moves away with a departure speed equal to the speed of approach. It is similarly hard to guess the direction of time from a glancing collision between two balls in motion. Only once we view a more complex interaction, such as the disruption of the regularly arranged triangular pack of 15 reds by the initial strike of the cue ball, do we get a strong sense of evolution from order to disorder—reversing the film, in this case, would disorient us and look suspicious. We don't ascribe this to some cosmic principle, however, but merely to the fact that the referee has set the balls up in a particularly ordered way at the outset of the frame. From this perspective, the entropy increase of the Second Law *is due to special initial conditions*; in other words, the universe (or in microcosm, the snooker table) began in a low entropy state.[14]

We touched on these issues at the start of Chapter 5 in our discussion of the Joule expansion of a gas. Recall that immediately following the withdrawal of the partition, the molecules are all located on one side, but as a consequence of their motion, they almost instantly move to occupy the entire vessel. There is nothing precluding the possibility that at some point in the future, the molecules will again instantaneously be distributed in just one half of the vessel. The assumptions underlying ideal gases don't discriminate against this, so it's equally likely as any other molecular arrangement. However, as we argued, when set against the post-astronomic number of alternative possibilities, most of which have roughly the same fraction of molecules in either half, the likelihood of this recurrence is so vanishingly small that it can be ignored for all practical purposes. The associated entropy increase in the Joule expansion is really due to having started in a low-entropy state, and the Second Law is a natural consequence of evolving with time-reversible dynamics from an ordered state characterized by a small statistical weight Ω to one with a vastly greater one. A universe well-suited for life requires an operative Second Law, as discussed both above and in Chapter 13. In order to account for this, we next need to review what things were like in earlier times. It's time for a little cosmology.[15]

15.4 COSMOLOGY IN A NUTSHELL

The most important observation made about the universe is that, on a sufficiently grand scale, everything we can see is receding from us. The objects in question are typically galaxies whose individual stars are too far away to be resolved. The greater the galaxy's distance d, the faster the speed of recession v, as described by Hubble's Law:

$$v = H_0 d. \tag{15.7}$$

H_0 is called the Hubble Constant; its value is difficult to measure accurately but is roughly $70\text{kms}^{-1}\text{Mpc}^{-1}$.[16] In other words, a galaxy at a distance 400Mpc is receding from us at $2.8 \times 10^7\,\text{ms}^{-1}$, some 9% of the speed of light c. This has a noticeable effect on the light we see emitted by the galaxy—its wavelength is stretched by a factor

$$1 + z = \left(\frac{1 + \frac{v}{c}}{1 - \frac{v}{c}} \right)^{\frac{1}{2}} > 1, \tag{15.8}$$

known as a *red shift*: because the light has a longer wavelength it appears redder than expected based on the expected chemical composition of the stars responsible for the galaxy's light. The orthodox interpretation of Hubble's Law is that the universe is expanding uniformly and isotropically (i.e., the same in all directions). If the separation between any pair of galaxies is expressed as a multiple ga of some reference distance scale a then Hubble's Law can be recast as

$$v = \frac{1}{a}\frac{da}{dt}d \equiv H(t)d. \tag{15.9}$$

We don't need to specify an actual value for a, because it will cancel out in the calculation of anything we can observe or measure, but the idea of a reference scale $a(t)$ evolving in time but otherwise having the same value throughout the universe perfectly captures the nature of the expansion. Equation (15.9) is more general than (15.7) because it allows for the possibility that the Hubble "constant" $H(t)$ actually varies in time, with H_0 and a_0 the values as of today. A useful relation with cosmological red shift is

$$1 + z = \frac{a_0}{a}. \tag{15.10}$$

The expansion rate varies because the mutual gravitational attraction between all the matter in the universe acts as a decelerator. The appropriate equations were derived from our best theory of gravitation, General Relativity, in 1922 by Friedmann:

$$\frac{1}{a}\frac{d^2a}{dt^2} = -\frac{4\pi}{3}G\left(\rho + 3\frac{p}{c^2}\right); \tag{15.11}$$

$$\frac{1}{a^2}\left(\frac{da}{dt}\right)^2 = \frac{8\pi G}{3}\rho, \tag{15.12}$$

where ρ is the average density of matter in the universe, p its pressure, and the gravitational constant $G \simeq 6.674 \times 10^{-11}\,\mathrm{Nm^2kg^{-2}}$. Solving (15.12) at the present epoch and using (15.9) gives the current critical density corresponding to a flat universe[17]

$$\rho_{0c} = \frac{3H_0^2}{8\pi G} \approx 10^{-26}\,\mathrm{kgm^{-3}}. \tag{15.13}$$

Now, in the standard "Λ-CDM" cosmological model currently favoured by observation, visible matter accounts for only 5% of this critical density, with the other components believed to be invisible "dark matter" contributing 26% and a mysterious "dark energy" contributing 69%. Non-gravitational interactions between visible matter and the dark sector are currently not understood and remain too weak to be detected. To proceed, we'll assume the visible component of the universe to be very simple, consisting of visible matter with density $\rho_b = 0.05\rho_c$[18] in the form of an ideal gas of hydrogen atoms with mass $m_p \simeq 1\mathrm{u}$, and radiation in the form of a gas of photons. In this simple state we'll assume the expansion is adiabatic so that entropy is constant, and following $TV^{\gamma-1} = \text{constant}$ (3.31,37), with index $\gamma = \frac{5}{3}$ appropriate for a monatomic gas, we predict $T_bV^{\frac{2}{3}} \propto T_ba^2 = \text{constant}$. For radiation, we use the blackbody formula for the density (cf. Equation (15.1))

$$\rho_\gamma = \frac{\pi^2 k_B^4}{15\hbar^3 c^5}T_\gamma^4 \equiv \frac{\sigma_\gamma}{c^2}T_\gamma^4. \tag{15.14}$$

with the constant $\sigma_\gamma \simeq 7.56\times10^{-16}$ J m^{-3} K^{-4}. Since the energy per photon is inversely proportional to its wavelength λ, which increases (i.e., redshifts) as the universe expands, then

$$\rho_\gamma \propto \frac{\text{number of photons}}{a\times a^3} \tag{15.15}$$

and hence (15.14,15) imply $T_\gamma a = \text{constant}$.

Note that we distinguish between the baryon temperature T_b and the radiation temperature T_γ. In order for them to be equal, according to the Zeroth Law matter and radiation must be in thermal equilibrium, implying frequent collisions between atoms and photons. Adapting the arguments of Chapter 3, the average time between collisions is

$$\tau_{\text{col}} \approx \frac{\lambda_{\text{mfp}}}{c} = \frac{m_p}{\rho_b\sigma_H c}, \tag{15.16}$$

where $\sigma_H \sim 10^{-35}\,\text{m}^2$ is the collision cross-section. The timescale $\tau_{\text{col}} \gg \tau_{\text{Hubble}}$, the characteristic timescale defined by the universe expansion

$$\tau_{\text{Hubble}} \equiv a\left(\frac{da}{dt}\right)^{-1} = H^{-1}. \tag{15.17}$$

This means that collisions in our current epoch are extremely unlikely, and matter and radiation are *decoupled.* Decoupling is thought to have occurred when $a/a_0 \approx 1/300$. In subsequent evolution of this simple universe $T_\gamma/T_b \propto a$.

In our current epoch $T_{0\gamma} = 2.73$ K, the temperature of the CMB encountered at the beginning of the chapter. The radiation contribution $\rho_{0\gamma}$ (15.14) to the density ρ entering the Friedmann equation (15.11,12) is some 10^4 times smaller than that of visible matter ρ_{0b}. By contrast, the pressure is radiation dominated[19]:

$$p_0 = \rho_{0b}\frac{k_BT_{0b}}{m_p} + \frac{1}{3}\rho_{0\gamma}c^2 \simeq \frac{1}{3}\rho_{0\gamma}c^2 \ll \rho_{0b}c^2; \tag{15.18}$$

even so, the last inequality implies pressure gives a negligible contribution to (15.11). It is, however, interesting to compare the number density of hydrogen atoms

$$n_{0b} = \frac{\rho_{0b}}{m_p} \simeq 0.05\frac{\rho_{0c}}{m_p} \approx 0.3\text{m}^{-3} \tag{15.19}$$

with that of photons (using Wien's displacement law to estimate the energy per photon E_γ).

$$n_{0\gamma} = \frac{\rho_{0\gamma}c^2}{E_\gamma} \simeq \frac{\sigma_\gamma T_{0\gamma}^3}{4.965k_B} \approx 2.2\times10^8\,\text{m}^{-3}. \tag{15.20}$$

Now, for any ideal gas entropy density s is related to energy density and pressure via[20]

$$s = \frac{u+p}{T}. \tag{15.21}$$

For radiation $\rho_\gamma = \frac{\sigma_\gamma T^4}{c^2}$ (15.14) then yields

$$s_\gamma = \frac{4}{3}\sigma_r T_\gamma^3 \tag{15.22}$$

for the radiation entropy density, while using $\rho_{0bc} = 0.05 \times 3H_0^2/8\pi G$ (15.13) for an atomic gas in adiabatic expansion the entropy density[21]

$$s_b \approx 57 n_b k_B. \tag{15.23}$$

Note that because the expansion is assumed adiabatic following decoupling, relations (15.22,23) don't need to have a timescale specified; if we plug in today's value we find $s_\gamma/s_b \simeq 10^9$, i.e., the universe's entropy is overwhelmingly dominated by radiation, and roughly equal to the ratio n_γ/n_b.

At this point, let's pause and, Janus-like, look forward and backward in time. Professional cosmologists devote themselves mostly to looking backwards. Either relation $T_b a^2 = C_b$ or $T_\gamma a = C_\gamma$ with C_b, C_γ constants implies the universe was hotter at early times. Before decoupling, when $a/a_0 \approx 1/1500$, $T \approx 3500\,\mathrm{K}$, collisions between photons and H atoms were so frequent and energetic that the atoms would have become ionised, stripped of their attendant electrons. Under these circumstances, the mean free path of the photons becomes so short due to their continual scattering off electrons and ions that the material universe is no longer transparent but rather resembles the interior of a fluorescent light bulb. The glow from this medium, coming from all directions and red-shifted due to subsequent expansion, gives rise to the CMB we indirectly detect on cold clear nights. Because the universe was opaque before this *recombination*,[22] no direct observations of what was happening before this time are possible. To understand what conditions prevailed, it is necessary to extrapolate backwards using knowledge of nuclear and high-energy particle physics.[23] We need to understand the behaviour of fundamental constituents of matter because high temperatures imply high-energy collisions, which in turn imply that short-distance properties of matter are probed via the quantum nature of their interactions.[24] As we extrapolate backwards towards the initial singularity at $a = 0$, usually referred to as the *Big Bang*, more and more exotic particle species are thermally excited to become part of the story of early times: mesons, other baryons, heavier leptons such as muons, eventually quarks, Higgs bosons, and maybe even particles still awaiting discovery in terrestrial particle accelerators. Once $k_B T > Mc^2$, with M the particle's mass, then it must be *relativistic*, i.e., moving close to the speed of light and effectively an additional form of radiation. Relativistic matter consists of both particles and antiparticles.[25] For sufficiently extreme temperatures, it's even conceivable that the universe existed in a different *phase*,[26] in which particle/antiparticle properties are different from those observed under normal conditions, and the universal expansion governed by the Friedmann equation (15.11) is radically altered, with epochs of *exponential inflation* predicted having profound cosmological consequences.

15.5 WHY IS THERE SOMETHING RATHER THAN NOTHING?

We can't aspire to do justice to the full scope of contemporary cosmology here, but there is one fascinating issue we should discuss in passing. Following decoupling, we've deduced that there are roughly a billion photons in the universe for every baryon. Let's also assume, consistent with observations, that there are essentially no anti-baryons present in the universe today. Pre-recombination, in an era when all matter was relativistic and tightly coupled, this implies there must have been a billion and one baryons for every billion anti-baryons. As temperatures cooled below $m_p c^2/k_B$ the anti-baryons all annihilated in combination with an equal number of baryons, generating other forms of radiation and leaving behind a lone baryon to contribute to today's visible matter. To achieve this outcome, it seems as if a precision of at least 1 in 10^9 must have been needed to "set the Universe running": perhaps this is a specific instance of a low-entropy initial condition needed for the Second Law to manifest itself? Most physicists prefer the idea that the imbalance between matter and anti-matter has arisen as a consequence of natural processes inherent in fundamental physics. In 1967, Andrei Sakharov[27] set out three conditions for this to be possible. First, there must be fundamental processes which can change the net baryon number, i.e. the number of baryons minus the number of anti baryons.[28] Secondly, certain symmetries relating matter and anti-matter, as well as matter and its mirror image, must be violated, so that the particle interactions are endowed with a preferred direction.[29] Thirdly, the epoch in which baryon production occurred must have been out of thermal equilibrium,[30] possibly as a consequence of a phase transition. To visualize this, think of a boiling kettle—the water is heated so quickly that the transition from liquid to vapour doesn't happen in a spatially homogeneous way, but rather as a seething mass of bubbles, implying that the local density is varying from point to point in space. If something comparable happened as the universe in its first instants cooled, the first and second Sakharov conditions could act to create local concentrations of baryons, which then have no chance to vanish through reequilibration due to the rapid expansion. Of course, in such an irreversible process there is necessarily also significant entropy production. While the precise mechanism for *baryogenesis* is by no means a settled issue, it is fascinating that thermodynamic concepts lie at the heart of such a fundamental question.

15.6 HEAT DEATH...

Now, let's use our other Janus face to look forwards. Our over-simplified cosmological model has left us a universe populated with an ideal gas of hydrogen, comprising roughly 5% of the visible density of the universe, bathed in a relativistic gas of photons which dominates the entropy. The emergence of a recognizable universe requires organization driven by mutual gravitational attraction, which magnifies small inhomogeneities in the matter distribution so that stars, galaxies and clusters of galaxies can coalesce, heat up as they contract, and thereby initiate the thermonuclear reactions responsible for both the low-entropy energy supply fuelling all the things that make life interesting, and the generation of large and complex nuclei described earlier in this chapter needed to build planets, rocks, and people. Again, to try to be more specific in this simple account would be foolish; details of what happened in which order are complicated, requiring extensive calculation and computational modelling incorporating complex astrophysical processes. In addition, they may depend on the properties of the as-yet-unknown Dark Sector, thought to comprise the invisible 95% of the total density.

Let's return to energy supply, a perennial human concern, which in this chapter we have rephrased as an issue of low-entropy fuel supply. According to the Second Law, the entropy of the universe constantly increases, so such low-entropy sources may ultimately be in short supply. In the inexorable march towards universal thermal equilibrium, it will become ever harder to scratch out a living by exploiting temperature differences to run effective heat engines or maintaining homeostasis and reproduction through the consumption of low-entropy foodstuffs. At first sight, such fears naturally focus on the exhaustion of fossil fuel reserves, but in the wider cosmic perspective adopted in this chapter, we've seen that the question can be rolled back to the finite lifetime of stars such as the Sun,[31] and the potential for further stars to be born from the material available, which remember is constantly becoming ever more dilute due to Hubble expansion, which we have assumed[32] to be largely hydrogen gas. At some point, all available hydrogen will have been converted through thermonuclear fusion in star cores into heavier nuclei; in a sufficiently massive star, the sequence goes through helium ^{4}He, carbon ^{12}C, oxygen ^{16}O, neon ^{20}Ne, silicon ^{28}Si, and ultimately iron ^{56}Fe. Beyond this point, there's no further potential for power generation through fusion: stars will cease to shine brightly, and the low-entropy beacons populating our skies will fade.

The implications of Clausius' statement of the Second Law were not lost on his scientific contemporary Kelvin, who coined the term *Heat Death* to describe a Victorian nightmare of this grim and apparently inevitable outcome once the free energy of the universe is exhausted. The idea underlies a memorable excerpt from H.G. Wells' 1895 novel *The Time Machine,* in which the time-travelling protagonist describes his ultimate journey some 30 million years[33] into the future, finding himself on a lonely beach witnessing a total eclipse of the Sun in some future stage of advanced evolution:

> The darkness grew apace; a cold wind began to blow in freshening gusts from the east, and the showering white flakes in the air increased in number. From the edge of the sea came a ripple and whisper. Beyond these lifeless sounds the world was silent. Silent? It would be hard to convey the stillness of it. All the sounds of man, the bleating of sheep, the cries of birds, the hum of insects, the stir that makes the background of our lives – all that was over. As the darkness thickened, the eddying flakes grew more abundant, dancing before my eyes; and the cold of the air more intense. At last, one by one, swiftly, one after the other, the white peaks of the distant hills vanished into blackness. The breeze rose to a moaning wind. I saw the black central shadow of the eclipse sweeping towards me. In another moment the pale stars alone were visible. All else was rayless obscurity. The sky was absolutely black.[34]

Recall these words were written prior to a modern understanding of thermonuclear reactions powering the Sun; perhaps we can also forgive Wells the references to weather and starlight. The horror and desolation remain palpable.

15.7 ... AND RESURRECTION?

In a series of lectures given in New York in 1978, the theoretical physicist Freeman Dyson[35] took another look into the far future, going considerably further than Wells,[36] but reaching more optimistic conclusions. His starting point is that the basis of consciousness in living organisms is not necessarily a property of a specific biomolecular realization but rather is inherent in the way the molecules are organized. In other words, it should in principle be possible to make a copy of the brain using different material components that would behave cognitively in essentially the same way. Clearly anticipating much subsequent thought and discussion on the potential for artificial intelligence (AI), his key hypothesis was that the resulting intelligent entity could in effect choose its own operating temperature θ lower than the current 300 K we're used to and hence continue to operate in the challenging chills of the far future. The payoff is that the intelligence would experience a *subjective time* $\tilde{t}(t)$, measured with units *s*, running at a slower rate than physical time *t* by a factor θ/θ_0 where $\theta_0 = 300\,\mathrm{K}$.[37] Using language familiar from the management of computer systems, we might say that *t* is determined by the clock speed, and that survival depends on "clocking down". Dyson's subsequent argument is rooted in Information Theory; any intelligence may be characterized by a complexity $Q = dS/d\tilde{t}$ defined by its entropy production in unit subjective time. For instance, a single human being dissipating 200 W at room temperature 300 K to maintain homeostasis has complexity

$$Q = \frac{200}{300 \times k_B \ln 2} \approx 10^{23}\,\text{bits}\,\tilde{s}^{-1}. \tag{15.24}$$

Human civilization as a whole has $Q \approx 10^{33}$ bits$\tilde{s}^{-1}$. The assumption is that human civilization could conduct itself at a temperature $\theta < \theta_0$ provided the physical time needed to, say prove a theorem or compose a sonnet is stretched by a corresponding factor θ_0/θ. From the cognitive perspective of the population, however, the creative experiences involved would be indistinguishable from those at present.

There are, however, physical constraints on how low we can dial θ. In order to operate an organism or civilization has to dissipate energy at a rate

$$P = \theta \frac{dS}{dt} = \frac{\mathrm{k_B} \ln 2}{\theta_0} Q\theta^2, \tag{15.25}$$

where one factor of θ relates entropy to energy and the second is due to the time stretching factor when converting from S to Q. First, it is crucial that $\theta > T_\gamma(t)$, the temperature of the cosmic background radiation, so that the sky can be used as a heat sink as described at the start of the chapter; without this possibility it would be impossible to run a heat engine with even minuscule efficiency. Secondly, there are material constraints on the power that can be radiated by a device containing N electrons at temperature θ; using (15.24) and some radiation physics Dyson derives the limit

$$\theta > \frac{Q}{N} \times 10^{-12}\,\text{K}. \tag{15.26}$$

For any society with given Q/N there is thus a minimum feasible operating temperature, which for human civilization Dyson estimates to be 10^{-21} K. The initial, melancholy conclusion is that for a civilization located in any finite region of space,[38] because the amount of available energy is finite, there will inevitably come a point in the far distant future where there is not enough fuel left to maintain (15.26), and life will come to a halt.

However, there is a way out, through hibernation. Suppose the civilization devotes just a fraction g of its time to conscious activity, and spends the rest in dreamless sleep, consuming negligible power. The expressions for subjective time and energy consumption are modified accordingly:

$$\tilde{t}(t) = \frac{g(t)\theta(t)}{\theta_0}; \quad P(t) = \frac{k_B Q \ln 2\, g(t)}{\theta_0}\theta^2(t), \tag{15.27}$$

where we have made time dependence explicit. The constraint (15.25) is replaced by

$$\theta(t) > 10^{-12}\frac{Q}{N}g(t). \tag{15.28}$$

Dyson advocates a strategy with $g(t) = \theta(t)/\theta_0 = (t/t_0)^{-\alpha}$, where t_0 is the current age of the universe and the power $\alpha > 0$ is chosen so that the operating temperature θ decreases and the relative fraction of hibernation to activity g increases over time. The energy consumption is then

$$E(t) = \int_{t_0}^{t} P(t)dt = \frac{k_B Q \ln 2}{\theta_0}\int_{t_0}^{t} g(t)\theta^2(t)dt = k_B\theta_0 Q\ln 2\int_{t_0}^{t}\left(\frac{t}{t_0}\right)^{-3\alpha}dt$$

$$E(t) = \frac{k_B\theta_0 t_0 Q\ln 2}{3\alpha - 1}\left(1 - \left(\frac{t}{t_0}\right)^{1-3\alpha}\right), \tag{15.29}$$

where we have used (MA.15a). Provided $\alpha > \frac{1}{3}$, $E(t)$ tends to a finite value $Qk_B\theta_0 t_0 \ln 2/(3\alpha - 1)$ as the age of the universe $t \to \infty$. Dyson's estimate of $O(10^{30}\,\text{J})$ for the total requirement of a human-type civilization is surprisingly modest, less than the total energy radiated by the Sun in a day. For $t \gg t_0$ the subjective time the civilization lives through is

$$A(t) = \int_{t_0}^{t}\tilde{t}(t)dt = \int_{t_0}^{t} g(t)\frac{\theta(t)}{\theta_0}dt = \int_{t_0}^{t}\left(\frac{t}{t_0}\right)^{-2\alpha}dt \approx \frac{t_0^{2\alpha}}{1 - 2\alpha}t^{1-2\alpha}. \tag{15.30}$$

Provided $\alpha < \frac{1}{2}$,[39] $A(t)$increases without limit as $t \to \infty$; the civilization is effectively immortal. O death, where is thy sting?

Dyson's plan for cheating Heat Death has not yet featured in the ongoing debate about the merits and perils of developing AI. Perhaps it's just a bit premature. A common theme (dare we say trope?) of accounts of astronomy and cosmology is the emphasis on the comparative insignificance and mundaneness of our place in the universe. It all began with Copernicus, whose model displaced the Earth from the centre of the cosmos to be just one of several planets orbiting the Sun, and can be traced onwards: through the work of Galileo and William Herschel, who demonstrated the Sun is one of many myriads of stars populating the Milky Way, and a fairly small and relatively dim one one at that; Henrietta Leavitt, whose work showed that the Milky Way is just one "island universe" separated by vast distances from other galaxies; and Hubble himself, who first observed the cosmic expansion. With improved surveying techniques available in more recent times we've learned that galaxies themselves form clusters and even superclusters whose size can exceed 1000 Mpc. Each stage of this hierarchy seems further to underscore our insignificance.

However, from the thermodynamic perspective, these are still early days. Our best estimate of the time since Big Bang, based on observational estimates of Hubble flow and detailed properties of the CMB, is currently about 14 billion years. Our best estimate of the age of the Earth, based on radiometric dating, which requires a painstaking assay of the relative isotope fractions of radioactive elements found in rocks, is roughly 4.5 billion years. Remarkably, we're in almost from the start, well-placed to appreciate and prosper from the universe in its youth. There's plenty more entropy left to generate! Perhaps there is, after all, something special about us, not so much in *where* we live, but rather *when*. Enjoy it while you can.

NOTES

1 Derived as Equation (9.108) in Chapter 9.

2 Perhaps fortunately, radiative cooling is mitigated by the Earth's atmosphere which absorbs and reradiates infrared radiation, particularly once water vapour is present.

3 1 nanometre = 1 nm = 10^{-9} m.

4 Max Planck (1858–1947) was a theoretical physicist whose struggles in the early 20th-century to reconcile thermodynamics with electromagnetism fired the starting pistol for the emergence of quantum theory as the dominant physical paradigm. Together with Einstein's analysis of the photoelectric effect, his work forced a reinterpretation of all electromagnetic radiation, including light, as being composed of discrete quanta called photons, each moving with speed c and bearing energy hc/λ.

5 It is hardly necessary to note here that small changes in atmospheric composition due to anthropogenic fossil fuel burning may result in drastic changes to climate, essentially by slightly tweaking the efficiency of the greenhouse effect.

6 If we simplistically assume that the Earth is a sphere of uniform density throughout, that the Moon orbits the Earth in 27 days, that the ratio Moon distance: Earth radius is 55, and the mass ratio Earth: Moon is 81, then conservation of angular momentum and Kepler's Third Law of planetary motion predict that the Moon's distance ultimately increases by a factor of roughly 1.65 and the new "day/month" will be 57 current Earth days. 95% of the energy dissipated in this transition is due to the Earth's rotational slowing.

7 The number of photons is not conserved in nuclear reactions, so photons are not governed by the Law of Mass Action. In fact, photons travel from the star's core to its surface in a diffusive process by "random walking" between interactions with charged nuclei, with energy gradually becoming shared among more numerous but lower energy quanta along the way. Travel time estimates are very sensitive to details of the solar model, but typical estimates are $O(10^4)$ years.

8 The fact that smaller nuclei prefer to fuse together to form larger, while large nuclei prefer to fission to form smaller, is an interesting and complicated consequence of the physics of nuclear structure, which we needn't discuss here. The break-even point between large and small is ^{56}Fe, i.e. iron is the nucleus within which nucleons are most tightly bound. If our nuclei were in thermodynamic equilibrium with each other, then we'd clank when we walked. We have chemistry, via the Pauli exclusion principle and electrostatic repulsion, which keep our nuclei well-separated, to thank that we don't...

9 Easily mined ^{235}U resources may only meet the world's current energy demands for a few hundred years, but more efficient "fast breeder" reactors under development may mitigate this using the more abundant ^{238}U or ^{232}Th isotopes.

10 Isotopes are families of nuclei containing the same number of protons (and hence near-identical chemical properties) but differing numbers of neutrons (and hence masses). Different isotopes can have varying stability properties, leading to radically different half-lives.
11 A core collapse or "Type II" supernova can only occur for stars at least eight times more massive than the Sun.
12 B.P. Abbot et al., Multi-Messenger Observations of a Binary Neutron Star Merger, *Astrophysical Journal Letters*, **848**: L12 (2017).
13 We ignore the application of spin or "action" on the cue ball, which is of course precisely the factor that makes snooker such a skillful game.
14 These issues were discussed more technically in Chapter 10, where we saw the Second Law emerge from phase space volumes of dynamical systems apparently growing under time-reversible evolution as an inevitable consequence of our experimental inability to distinguish among the many microstates corresponding to the same observed macrostate.
15 Cosmology is the study of the structure and evolution of the Universe as a whole. It is a fascinating and rapidly developing subject mainly as a consequence of continual improvements in the scope, resolution and sensitivity of observational data. Our treatment here makes little attempt to keep up-to-date with the latest thinking impacting the earliest timescales; a useful survey at the level of this book is found in *Quarks, Leptons and the Big Bang*, J. Allday (CRC Press, Third Edition 2017).
16 Cosmic distances are conveniently measured in megaparsecs: $1\,\mathrm{Mpc} \simeq 3.09 \times 10^{22}\,\mathrm{m}$. The nearest galaxy beyond our Milky Way is roughly 0.76 Mpc distant.
17 Friedmann's equations take a more general form for "curved" Universes, which can be either closed (i.e. eventually recollapsing) or open (i.e. expanding without limit). Observations favour a universe remarkably close to the critical "flat" case described by Equation (15.11,12). In a flat universe, the volume of a sphere is $\frac{4}{3}\pi r^3$ and its surface area $4\pi r^2$ regardless of the magnitude of its radius r.
18 The b subscript denotes *baryon*, i.e. a particle such as a proton or a neutron composed of three quarks. Since baryons are much more massive than electrons, the baryon density is a useful proxy for the density of visible matter. Dark matter is believed to be non-baryonic in nature and remains a mystery for now.
19 For a relativistic ideal gas, i.e. one consisting of massless particles such as photons, we can use the result $u - 3p = 0$.
20 For a system in contact with a heatbath $dF = -SdT - pdV$ implying $p = -\partial F/\partial V = -f = -u + Ts$.
21 In obtaining the numerical factor in (15.22), we used the Sackur–Tetrode result (7.51), viz. $S/N = \frac{5}{2}k_B + \frac{3}{2}k_B \ln\left(2\pi m_p^{\frac{5}{3}} k_B T / h^2 \rho_b^{\frac{2}{3}}\right)$, but regardless of its precise value s_b is dwarfed by s_γ.
22 Strictly combination, yes, we know ….
23 It is anticipated that gravitational wave radiation emitted before recombination will eventually be detected.
24 The Heisenberg Uncertainty Principle $\Delta x \Delta p \sim \hbar$ implies that energetic interactions with large momentum exchanges Δp are necessarily probing small distance scales Δx.
25 Antiparticles have the identical mass but opposite electrical charge to particles, and particles and anti-particles are both created and destroyed pairwise in high-energy collisions. The best-known anti-particle is the anti-electron or *positron*, which was discovered in 1932 by Carl Anderson. Care is needed for neutral particles; both neutrons and neutrinos are distinct from their antiparticles, but the photon is *self-conjugate*, i.e. its own anti-particle.
26 Matter can exist in different *phases* with very different bulk properties, depending on external conditions such as temperature, concentration or applied magnetic field. The best-known example are the gaseous, liquid and solid phases exhibited by H_2O, all readily available in most kitchens.
27 Andrei Sakharov (1921–1989) made key contributions to the Soviet nuclear weapons programme but also faced years of persecution and internal exile for his championing of human rights causes. He was awarded the Nobel Peace Prize in 1975.
28 Initially, this condition appeared to require Grand Unified Theories, which predict that the proton is ultimately unstable, but we now believe there are also rare baryon-number violating events even in the Standard Model under extreme conditions; the jury is still out over the details.
29 The symmetries in question are known as C-symmetry and CP-symmetry; both are weakly violated in the Standard Model.
30 Since baryons and anti-baryons have identical masses, their abundance in thermal equilibrium $\propto \exp\left(-m_b c^2 / k_B T\right)$ would be the same.
31 Estimated to be of order 10^{10} years.
32 Rather inaccurately, it turns out. Approximately 25% of the baryonic mass a few minutes after the Big Bang was actually in the form of ^{4}He.
33 The novel was written before knowledge of nuclei and thermonuclear fusion; the most informed estimate of the Sun's age at the time, based on a theory of solar radiation arising from continual gravitational contraction due to Kelvin, was between 20 and 60 million years.
34 *The Time Machine*, H.G. Wells (J.M. Dent New Centennial Edition, 1995).
35 Freeman Dyson (1923–2020) contributed to several areas of science in often unconventional ways, as well as being a writer of compelling lucidity. He is best known for the development of a systematic framework for relativistic quantum

field theory calculations for particle physics in a readily implementable way following pioneering work by Feynman, Schwinger, and Tomonaga.

36 The timescales set out in Dyson's accompanying paper. Rev. Mod. Phys. **51** (1979) 447 include stars cooling off after 10^{14} years, the dissolution of galaxies after 10^{19} years, all matter decaying to iron after 10^{1500} years, and finally, the collapse of all stars to neutron stars after $10^{10^{76}}$ years!

37 The fundamental equation of quantum theory $i\hbar\partial\Psi/\partial t = E\Psi$ implies that it is possible, in principle, to reproduce arbitrarily complex processes evolving more slowly in time by working at a lower energy scale; this scale is directly proportional to the operating temperature θ.

38 Dyson proposed as an example complex intelligence something like the "Black Cloud", a vast diffuse assembly of electrically charged dust grains, with internal communication via electromagnetic impulses over scales of an astronomical unit, in the eponymous 1957 sci-fi novel by astronomer Fred Hoyle (1915–2001). Hoyle's research straddled many of the themes explored in this chapter, including seminal contributions to stellar nucleosynthesis—he also helped develop a radical alternative to Big Bang cosmology, invoking the continuous, if unobservably rare, spontaneous creation of matter and entropy in order to maintain the expanding Universe in a *Steady State*. It is difficult to reconcile Steady State with the CMB.

39 This upper bound on α also ensures the constraint $\theta > T_\gamma \propto a^{-1}$ can be satisfied, since $a \propto t^{\frac{2}{3}}$ for a flat matter-dominated Universe. For a Λ-CDM universe $a \propto (\sinh \frac{t}{Ct_0})^{\frac{2}{3}}$ and the constraint still easier to satisfy.

16 Black

Black Holes are regions of space–time where strong gravitational fields prevent matter and radiation from escaping. To understand them, we develop concepts from relativity theory, defining the interval in flat space–time and demonstrating that the equivalence of inertial observers leads to such characteristically relativistic phenomena as time dilation. In general the interval is a function of space–time coordinates dictated by local distributions of matter and energy, as encapsulated in Einstein's field equations of General Relativity. We discuss the spherically symmetric Schwarzschild solution and show that it predicts the existence of a horizon, a surface surrounding the black hole delineating the hidden region. The more general Kerr solution for rotating black holes demonstrates that the horizon area can never decrease as a result of interactions with external matter. The analogy with the Second Law leads us to deduce a version of the First Law in which the horizon area is proportional to entropy and consequently that black holes have a non-zero temperature. Quantum theory is essential for the consistency of this picture; thermal Hawking radiation from the black hole is interpreted in terms of the spontaneous creation of particle—antiparticle pairs near the horizon, which evade the quantum Uncertainty Principle once the negative energy partner crosses the horizon to go on-shell. This radiation implies that black holes must eventually evaporate on post-astronomic timescales, raising the question of the fate of the information originally present in the infalling matter. The discussion is framed in terms of entanglement entropy and a possible resolution sketched in terms of a prototype model of quantum gravity called String Theory. Finally, we argue that the overwhelming fraction of the universe's entropy resides in black holes lying at the heart of galaxies.

16.1 HIDDEN PLACES

There are places in the universe that are closed off from us and from which no information appears to emerge. Known with striking simplicity as *black holes,* they are regions where gravitational forces are so strong that nothing is able to escape. Even light, borne by massless photons travelling at speed $c = 299792458\,\mathrm{ms}^{-1}$,[1] remains trapped inside a black hole so that no signal can be sent revealing what's inside. Black holes have been theorized for well over a century, but by now, they are known to exist and have been identified either through their formation, as a natural endpoint of stellar evolution, whenever the pressure within a collapsing star is unable to withstand the inward gravitational force due to its own mass, or as supermassive black holes found at the centres of galaxies, with masses ranging from millions to billions of solar masses.[2] In both cases, the black hole's existence is revealed through observation of high energy radiation[3] from matter, which undergoes compression as it falls towards the hole, converting gravitational potential energy to heat and radiation in the process.[4]

 DOI: 10.1201/9781003121053-20

By this stage of our entropic journey, some bells may already be ringing. Black holes form and subsequently evolve as matter falls inside; even if we restrict our attention to stable particles, that matter can in principle be formed from one of over a hundred chemical elements and be assembled in many different ways—as gas, dust, rocks, space junk, people—all of which require a lot of information to describe them or, better still, specify their makeup. Once inside the hole, what do we have left? One way of thinking about a black hole, at least from the outside, is as a gravitational field shorn of all but the minimal information concerning the gravitating body within. A moment's thought suggests that at least the body's mass M is accessible to us, and further investigation shows that we can also in principle know its angular momentum $\boldsymbol{J}$,[5] and its charge Q. So, after the material falls into a black hole, all the manifold things we can know about it are absorbed into changes in value of just five numbers. In other words, an immense number of different microstates all effectively correspond to just one of a far, far smaller number of physically distinguishable macrostates. This is not the first time we've encountered this idea — remember Chapter 5? — might all this state-counting suggest a useful application of entropy to describe what's going on? We'll see in this chapter that the answer is in the affirmative and that black holes do indeed furnish a novel and fascinating aspect of thermodynamics. As you might anticipate, in order to tease this out, we'll need to learn more about black holes using our best theory of gravity: *General Relativity*. Less expected is that another crucial piece needed to complete the theoretical jigsaw has to come from our best theory of microscopic processes: *Quantum Theory*.

16.2 THE INTERVAL IN SPECIAL RELATIVITY

Relativity is a geometrical theory, concerned with the measurement of distance between points. Actually, points in relativity theory live in *space–time*[6] and are specified by not just spatial coordinates x, y, and z which tell where they are located, but also by a time of occurrence t; for this reason it's more customary to refer to them as *events*. The separation between two events is then given by four numbers:

$$\Delta x = x_2 - x_1;\ \Delta y = y_2 - y_1;\ \Delta z = z_2 - z_1;\ \Delta t = t_2 - t_1. \tag{16.1}$$

A central concept is the *interval* between the events defined by

$$\Delta s^2 = -c^2(\Delta t)^2 + (\Delta x)^2 + (\Delta y)^2 + (\Delta z)^2. \tag{16.2}$$

Because of the minus sign in front of the first term in (16.2), the interval can take either a positive or a negative value or even vanish. Let's consider events separated along just the x-axis so that $\Delta y = \Delta z = 0$, and solve for just such a *null interval* with $\Delta s^2 = 0$:

$$0 = -c^2(\Delta t)^2 + (\Delta x)^2 \Rightarrow \frac{\Delta x}{\Delta t} = c. \tag{16.3}$$

Equation (16.3) describes a moving particle whose velocity $v = \Delta x / \Delta t$ coincides with the speed of light; in other words, in relativity light rays move along *null trajectories*. It's now straightforward to characterize other possibilities. If $\Delta x/\Delta t < c$ then the interval $\Delta s^2 < 0$, and is known as *timelike*. Since all particles are constrained to move at less than the speed of light, the trajectory of a particle through space–time inevitably follows a timelike interval. By contrast, if $\Delta s^2 > 0$ then no physical particle or agency can traverse this interval, which is then called *spacelike*. Two events separated by a spacelike interval can exert no influence on each other[7]; we say they are *causally separated*.

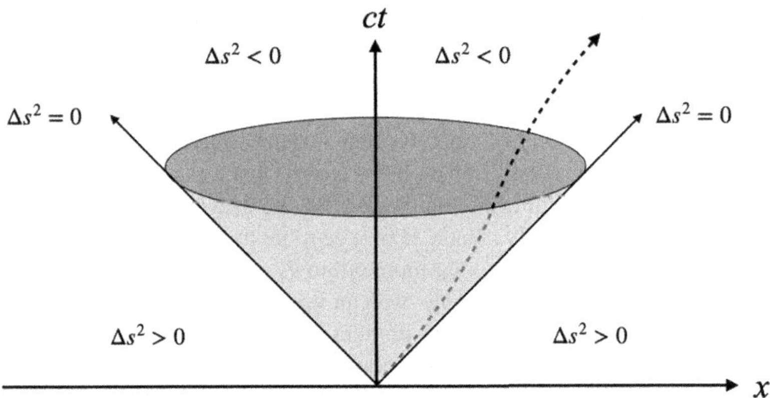

FIGURE 16.1 Space–time diagram showing the forward light cone and the world-line of a physical particle.

The diagram in Figure 16.1 summarizes the situation. We have chosen to plot ct versus x so that null intervals form lines at 45° to the vertical. Timelike intervals then have to live in between the two null trajectories. While only the x-axis is formally plotted, we can also imagine including a y-axis coming straight out of the page; in this case, the set of null trajectories sweeps out a cone, and all timelike trajectories must lie within this *forward light cone*.[8] Similarly, matter and information can only arrive at a point from within the *backward light cone*, which is what we call "the past" in relativity-speak! The dashed curve in Figure 16.1 shows one possible particle trajectory curving through space–time; in this context, the trajectory is often called a *world-line*.

A key relativity concept is that physical predictions shouldn't depend on the details of the coordinates we choose to describe them with. Suppose instead of (t, x, y, z) we chose to use a system (t', x', y', z') defined by

$$t' = \gamma\left(t - \frac{vx}{c^2}\right); \quad x' = \gamma(x - vt); \quad y' = y; \quad z' = z. \tag{16.4}$$

Here the factor $\gamma \equiv (1 - v^2/c^2)^{-\frac{1}{2}}$ is known as the *Lorentz factor*. It is defined for $v < c$ and always has a numerical value exceeding 1. It is a simple exercise to show that in these new primed coordinates, the interval is given by

$$\Delta s'^2 \equiv -c^2(\Delta t')^2 + (\Delta x')^2 + (\Delta y')^2 + (\Delta z')^2 = -c^2(\Delta t)^2 + (\Delta x)^2 + (\Delta y)^2 + (\Delta z)^2 \equiv \Delta s^2; \tag{16.5}$$

in other words, the interval (16.2) has the same value in either set of coordinates and is thus invariant under a *Lorentz transformation*[9] of the form (16.4). In particular, null trajectories are still drawn at 45° angles to the ct' and x' axes; the speed of light is unchanged. How can we physically interpret the Lorentz transformation? Consider two causally connected events: a space vessel leaves base with event coordinates [0,0,0,0] and travels in a straight line at speed v until it arrives at a spaceport located a distance ℓ along the x-axis. Since time elapsed = distance/speed, we deduce the coordinates of the arrival event to be $[\ell/v, \ell, 0, 0]$. Now using (16.4) to transform the arrival event to primed coordinates along with $y' = y = 0$, $z' = z = 0$, and assuming that v in the definition of the Lorentz factor γ is the same as the vessel's speed, we find

$$t' = \gamma\left(\frac{\ell}{v} - \frac{v\ell}{c^2}\right) = \gamma\left(1 - \frac{v^2}{c^2}\right)\frac{\ell}{v} = \frac{\ell}{\gamma v}; \quad x' = \gamma\left(\ell - \frac{\ell v}{v}\right) = 0. \tag{16.6}$$

So, in the primed system, the departure event still occurs (trivially) at [0,0,0,0], but arrival now takes place at $\left[\ell/(v\gamma),0,0,0\right]$. Because the x' coordinate has the same value 0 at both departure and arrival, we deduce that the primed system of coordinates is appropriate for someone travelling *aboard* the vessel who cannot perceive their own motion except in relation to their surroundings and for whom departure and arrival occur at the same location. The Lorentz transformation relates the frames of reference of observers moving at uniform speeds with respect to each other.

A more striking result concerns the elapsed time between the events. Since $\gamma > 1$:

$$\frac{\ell}{v\gamma} < \frac{\ell}{v} \quad \Rightarrow t' - 0 = \Delta t' < \Delta t. \tag{16.7}$$

Therefore, the moving observer aboard the vessel perceives a shorter elapsed time between the two events than someone at rest, perhaps based in either Departures or Arrivals. This phenomenon of *time dilation* is the first of several unexpected aspects of relativity; we're simply not accustomed to the idea that elapsed time depends on the details of the journey taken, principally because for speeds accessible to everyday human experience $v \ll c$ and hence γ effectively indistinguishable from unity. The time $\Delta\tau$ experienced by an observer at rest in their own frame of reference, as measured by a clock moving along the same world-line (i.e. the only contribution to the interval is proportional to $(\Delta t)^2$), is known as the *proper time*. It is related to the interval by $(\Delta\tau)^2 = -\Delta s^2 / c^2$.

16.3 THE SCHWARZSCHILD SOLUTION IN GENERAL RELATIVITY

So far, in considering intervals of the form (16.2), we have limited ourselves to observations made in frames of reference moving with a uniform velocity $v < c$, with the Lorentz transformation (16.4) specifying how to relate frames with different v. The inhabitants of such frames are known as *inertial observers*, and the restriction to inertial observers forms the domain of *Special Relativity,* first published by Einstein in 1905.[10] Its most well-known prediction is the mass–energy relation $E = mc^2$, perhaps the most famous equation in all physics. However, there are more general possibilities for the form of Δs^2; equation (16.2) describes a "flat" space–time in which, say, the angles of every triangle sum to 180°. There are curved geometries, such as the surface of a sphere,[11] where this is not the case. Einstein's greater achievement in 1915 was to relate such space–time curvature, as described by the interval, to the space–time distribution of matter and energy. The relevant equations (we won't write them here) are known as *Einstein's Field Equations of General Relativity*. While they represent a bravura exercise in geometry, the leap of genius was to interpret the resulting particle motion through curved space–time as equivalent to motion in a gravitational field: general relativity is thus a theory of gravity.

Einstein's field equations are challenging to solve, and modern applications frequently rely on high-performance computers to do the job. The earliest pencil-and-paper solution, which has proved supremely useful in astrophysics, is due to Schwarzschild,[12] who considered a single spherical non-rotating body of mass M. The solution, applicable in all regions of space–time exterior to the body, is given by[13]

$$\Delta s^2 = -\left(1 - \frac{2GM}{c^2 r}\right)(c\Delta t + \Delta r)^2 + 2\left(c\Delta t + \Delta r\right)\Delta r + r^2[(\Delta\theta)^2 + \sin^2\theta(\Delta\phi)^2]. \tag{16.8}$$

The interval is expressed in coordinates suitable for a spherical geometry: r is a radial coordinate for the distance from the centre, while θ and ϕ are angles akin to latitude and longitude. In subsequent work we will only consider radial trajectories, i.e. those aimed directly at the centre, so

$\Delta\theta = \Delta\phi = 0$. The constant G is the gravitational constant appearing in Newton's law of gravitation, i.e. that the attractive force between two bodies of mass m_1, m_2 and separation r is given by an inverse square relation of the form Gm_1m_2 / r^2. Note that each term on the right-hand side of (16.8) contains two powers of a Δ increment, but in contrast to (16.2) there are now mixed terms proportional to $\Delta t \Delta r$.

We will use (16.8) to solve for the trajectories of light beams moving along the radial direction, characterized by the null condition $\Delta s^2 = 0$:

$$-\left(1-\frac{2GM}{c^2 r}\right)(c\Delta t + \Delta r)^2 + 2(c\Delta t + \Delta r)\Delta r = 0. \tag{16.9}$$

An immediately obvious solution to (16.9) is

$$c\Delta t + \Delta r = 0, \tag{16.10}$$

implying that $\Delta r / \Delta t = -c$, ie. the beam is flying inwards towards the centre. To find other solutions, including possible outgoing beams, divide (16.9) by $(c\Delta t + \Delta r)$ to find

$$-\left(1-\frac{2GM}{c^2 r}\right)(c\Delta t + \Delta r) + 2\Delta r = 0 \;\Rightarrow\; ct + r = 2\int \frac{dr}{1-\frac{R_S}{r}} = 2\int dr\left(1+\frac{R_S}{r-R_S}\right),$$

where in the second step we have taken Δ to be infinitesimal, used integral calculus, and defined the *Schwarzschild radius*

$$R_S = 2GM / c^2. \tag{16.11}$$

The integral is a fairly standard adaptation of (MA.15b), but needs to be handled carefully according to whether r is greater or less than R_S. The result is

$$ct - r - 2R_S \ln\left|\frac{r}{R_S} - 1\right| = \text{constant}; \tag{16.12}$$

the vertical lines signify that we are to take the *absolute value* of the enclosed expression, so the argument of the logarithm is positive. Solutions for radial null trajectories are plotted as functions of r and t in Figure 16.2. The axis scales are chosen so that the ingoing rays (16.10) are tilted at 45°. At various points in the plane the light cone is shown spanning the angle between ingoing and would-be outgoing rays (16.12). For $r > R_S$ the light cone includes trajectories which eventually travel towards $r \to \infty$; in other words, it is possible for both light and material particles in this region to escape the black hole. For $r < R_S$ by contrast all light beams, and by extension any other material object, are doomed to travel to the origin $r = 0$. The thick vertical line precisely at $r = R_S$ denotes the worldline of stationary light, hovering at the *horizon* separating these two regions.

In three dimensions, the horizon of a Schwarzschild black hole is not a line but a closed spherical surface of area

$$A_S = 4\pi R_S^2 = 16\pi G^2 M^2 / c^4. \tag{16.13}$$

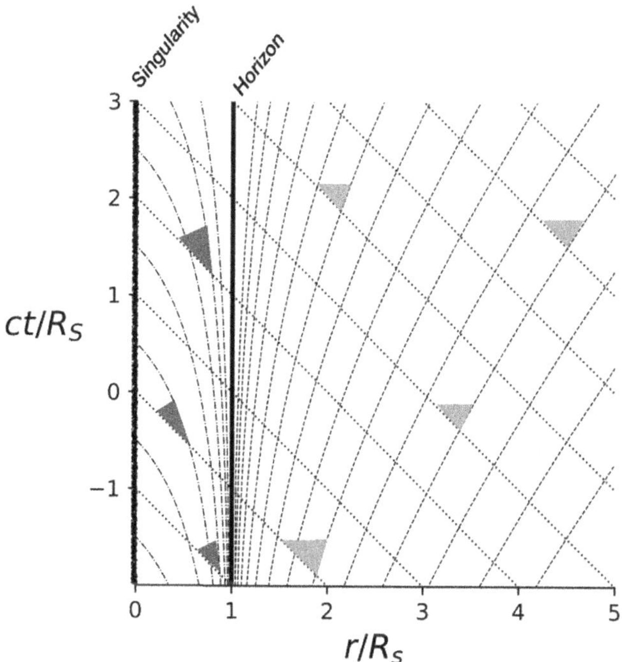

FIGURE 16.2 Radial light beam trajectories in the vicinity of a Schwarzschild black hole. Solutions (16.10) are shown as dotted lines, and (16.12) as dashed lines ($r > R_S$), and dash-dotted lines ($r < R_S$). The shaded triangles show forward light cones in profile at various space–time locations.

It is the horizon that closes off a region of space from which nothing can escape, and apparently nothing can be learned. More bad news: the vertical line at $r = 0$ in Figure 16.2 denotes a *singularity* where the first term in the interval Δs^2 in (16.8) diverges; at this point of infinite space–time curvature gravitational forces become infinitely strong, and everything that falls into the black hole will be compressed and crushed into an infinitesimal volume.

The Schwarzschild geometry described by (16.8) is valid for the exterior of every spherically symmetric gravitating body, regardless of whether a black hole forms. Only if all the matter comprising the body lies *within* the Schwarzschild radius does the black hole form, whereupon collapse to the singularity is inevitable and in realistic cases rather rapid. For a body of the Earth's mass, some 6×10^{24} kg, $R_S \simeq 9\,\text{mm}$, so we are in no imminent danger. For 5 solar masses ($5M_\odot \approx 10^{31}$ kg), $R_S \simeq 15\,\text{km}$ which is roughly the same size as a neutron star remnant formed after a core-collapse supernova. Depending on the mass of the progenitor star, the as-yet poorly determined equation of state of high-density neutron matter, and the details of the explosion, black hole formation from exhausted stars is a very real possibility. Finally, the black holes in galactic nuclei can weigh as heavy as $10^9 M_\odot$, leading to $R_S \approx 10^9$ km $\approx 1\,\text{AU}$.

While the horizon is a very special place, exactly what happens there depends on M. For a stellar-mass black hole, the gravitational fields vary so greatly over short spacelike intervals in the horizon's vicinity that huge tidal forces are induced, which would rip an unfortunate space traveller apart long before they get to traverse it; however, a freely falling observer crossing the horizon of a much larger black hole would observe no singular behaviour and conceivably might not even notice. Once inside, the travel time to reach and merge with the singularity is of order R_S / c: roughly 10^{-5} s for a stellar-mass black hole (so not much time for regrets…), but perhaps a couple of hours for a galactic core black hole (one exciting ride!).

Something strange happens to geometry inside the horizon. For $r > R_S$, surfaces of constant r, which are spherical in three dimensions but represented as vertical lines in Figure 16.2, always lie within the forward light cone of every point they intersect; in other words they are timelike. Within the horizon by contrast, such trajectories, which include the singularity at $r = 0$, are spacelike. Since no material particle can follow a spacelike trajectory, staying at fixed r, and hence clear of the singularity, is not an option. We could almost instead think of r in this region as a timelike coordinate, and the singularity as a particular instant in time. Another aspect to note is that the factor $-\left(1-\frac{R_S}{r}\right)$ multiplying $(c\Delta t)^2$ in (16.8), which is negative-valued in flat space so that trajectories of stationary objects are timelike, actually becomes positive for $r < R_S$.

16.4 THE KERR SOLUTION AND HORIZON AREA

The Schwarzschild solution is by no means the most general solution for a black hole; a more astrophysically realistic case is one in which the black hole has angular momentum $J > 0$.[14] Not only do most stars spin on their axes, but also subsequent matter injection as the black hole grows will typically be from an accretion disk in the equatorial plane, and continually spin up the system as it falls in. The pertinent solution of Einstein's equations is due to Kerr[15]; for a black hole with mass M, angular momentum J, and in the equatorial plane with $\theta = \frac{\pi}{2}$,[16] the interval is

$$\Delta s^2 = -\left(1-\frac{R_S}{r}\right)(c\Delta t)^2 - 2a_K\frac{R_S}{r}(c\Delta t)(\Delta\phi) + \frac{(\Delta r)^2}{1-\frac{R_S}{r}+\frac{a_K^2}{r^2}} + \left(r^2 + a_K^2\left(1+\frac{R_S}{r}\right)\right)(\Delta\phi)^2, \tag{16.14}$$

with R_S the Schwarzschild radius for mass M and the *Kerr length*

$$a_K = \frac{J}{cM}. \tag{16.15}$$

Our Sun has $R_S \sim 3000\,\text{m}$ and $a_K \sim 300\,\text{m}$. For $a_K = 0$ (16.14) reduces to the Schwarzschild solution (using Schwarzschild coordinates). For general a_K an analysis of null trajectories reveals the presence of a horizon with equatorial radius

$$r_+ = \frac{1}{2}\left(R_S + \sqrt{R_S^2 - 4a_K^2}\right). \tag{16.16}$$

We see that $r_+ < R_S$ in general, and the maximum allowed value of $a_K = R_S/2$, corresponding to angular momentum

$$J_{\max} = \frac{GM^2}{c}. \tag{16.17}$$

This solution is known as an *extreme* Kerr black hole. Consideration of the interval for general θ reveals that this time the horizon is not spherical, but rather is flattened at the poles. The horizon area is

$$A = 4\pi R_S r_+ = 8\pi\frac{G^2M^2}{c^4}\left(1+\sqrt{1-\frac{J^2c^2}{G^2M^4}}\right) = \frac{A_S}{2}\left(1+\sqrt{1-\frac{16\pi a_K^2}{A_S}}\right), \tag{16.18}$$

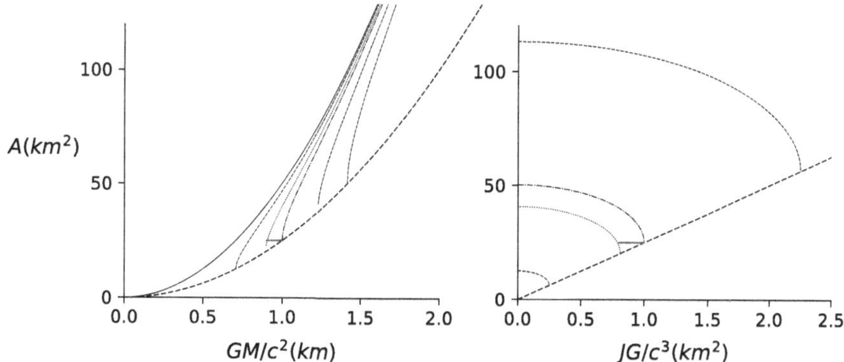

FIGURE 16.3 Horizon area A vs. M, for $JG/c^3 = 0$ (full line), 0.5,0.8, 1.0, 1.5, 2.0 km² (left panel). J, for GM / c^2 =0.5,0.9,1.0,1.5 km (right). The lower dashed curves show the extreme black hole bound. The horizontal lines show a Penrose process for an extreme black hole which results in $\Delta M / M = -0.1, \Delta J / J = -0.2$ but $\Delta A = 0$.

with A_S the Schwarzschild horizon area (16.13). For a given mass M, A lies between $A_S / 2$ and A_S. The behaviour of A as a function of both M and J is shown in Figure 16.3.

For Schwarzschild black holes, options for evolution are limited. As further material falls in along radial trajectories, the resulting increase of M must lead to increases in both R_S and A_S. For Kerr black holes, things are more interesting because infalling matter alters both M and J, and the linkage between the two can lead to some subtle effects. Careful study reveals the possibility of so-called *Penrose processes*,[17] in which a particle close to the horizon decays into two particles, one of which is able to escape from the region of the black hole while the other disappears below the horizon. Since angular momentum carries an orientation as well as a magnitude, it can be shown that there are processes which not only reduce J but also reduce M, thereby releasing some of the rotational energy of the black hole,[18] according to (with $\Delta M, \Delta J$ both negative quantities)

$$\Delta A \propto c^2 \Delta M - \omega_{BH} \Delta J \geq 0, \tag{16.19}$$

where

$$\omega_{BH} = \frac{c^2 J}{2GM^2 r_+} \tag{16.20}$$

can be thought of as the angular velocity of the black hole.[19] The horizon area A does *not* decrease. Remarkably, there is a theorem due to Hawking[20] stating that *any* interaction between a black hole and external matter can only ever increase A, or at best leave it unchanged.

16.5 A GEOMETRICAL DEFINITION OF ENTROPY

The similarity between the unidirectional horizon area increase and the Second Law of Thermodynamics stimulated Bekenstein[21] to pursue the analogy still further. In a famous 1973 paper, he started from the Kerr horizon area (16.18) in differential form:

$$dA = \frac{8\pi G}{c^2}\left(r_+ dM + M dr_+\right). \tag{16.21}$$

Using $r_+ = \frac{1}{2}\left(R_S + \sqrt{R_S^2 - 4a_K^2}\right)$ (16.16) to evaluate the differential dr_+,

$$r_+ = \frac{GM}{c^2} + \sqrt{\frac{G^2M^2}{c^4} - \frac{J^2}{c^2M^2}} \Rightarrow$$
$$dr_+ = \frac{G}{c^2}dM + \left(\frac{G^2M^2}{c^4} - \frac{J^2}{c^2M^2}\right)^{-\frac{1}{2}}\left[\left(\frac{G^2M}{c^4} + \frac{J^2}{c^2M^3}\right)dM - \frac{J}{c^2M^2}dJ\right],$$

and performing some simplifications along the way we arrive at[22]

$$c^2 dM = \Theta dA + \omega_{BH} dJ, \tag{16.22}$$

with

$$\Theta = \frac{c^4}{GA}\sqrt{\frac{G^2M^2}{c^4} - \frac{J^2}{c^2M^2}}. \tag{16.23}$$

Noting that each term has the dimensions of energy (i.e. is measured in Joules), and moreover that c^2dM can be identified with dU, the change in internal energy of a black hole, Bekenstein proposed (16.22) as an expression of the First Law of Thermodynamics for black holes. The term $\omega_{BH}dJ$ represents work done on the black hole by an external agency increasing the angular momentum, e.g. by throwing rocks into the hole along a non-radial trajectory, or by the Penrose process discussed above. By analogy with the Second Law Θ and A should then correspond with temperature T and entropy S, but how is this to be quantified so it can be related to other thermodynamic systems?

Using Information Theory and the Landauer limit introduced in Section 14.6, Bekenstein proposed to identify the smallest possible black hole entropy change $\delta S = k_B \ln 2$, namely the loss of one bit of information, with the minimum change in area $\delta A_{\min}$ whenever a single particle crosses the horizon. As indicated by the short horizontal lines in Figure 16.3, there are Penrose processes for extreme black holes with $J = J_{\max}$ for which $\Delta A = 0$, so classically Bekenstein's proposal looks set to fail. However, the classical calculation assumes a point-like particle with vanishing spatial extent. For an elementary particle it is more reasonable to apply a quantum description; we can develop a rough-and-ready estimate as follows. For any quantum particle there is an uncertainty δx in its position constrained by the Heisenberg Uncertainty Principle $\delta x \delta p \geq \hbar$ (6.30), where the reduced Planck's constant $\hbar = h/2\pi \simeq 1.055 \times 10^{-34}$ Js. For a relativistic particle of rest mass m we may take the momentum uncertainty to be $\delta p \approx mc$, whereupon

$$\delta x = \frac{\hbar}{mc} \equiv r_C. \tag{16.24}$$

The length scale r_C defined in (16.24) is known as the particle's *Compton wavelength*.[23] Accordingly it makes more sense to consider the particle to have a spatial size ~ δx: a detailed analysis of infalling trajectories then shows that the minimal area increase

$$\delta A_{\min} = \frac{8\pi G}{c^2} m \delta x = \frac{8\pi G\hbar}{c^3}, \tag{16.25}$$

clearly independent of M and J. Since by hypothesis $S \propto A$, we deduce the black hole entropy to be proportional to the total number of bits $A/\delta A_{\min}$[24];

$$S \approx k_B \ln 2\left(\frac{A}{\delta A_{\min}}\right) = k_B \frac{c^3 \ln 2}{8\pi G\hbar} A. \tag{16.26}$$

Following a different route, in 1974 Hawking used quantum field theory in the curved spacetime near the horizon to derive an expression for the temperature of a Schwarzschild black hole:

$$T = \frac{c^3\hbar}{8\pi GMk_B}. \tag{16.27}$$

We can now use the First Law (16.22) with $dJ = 0$ together with (16.27) to determine S:[25]

$$TdS = c^2 dM \Rightarrow \int MdM = \frac{c\hbar}{8\pi Gk_B}\int dS \Rightarrow$$

$$S = \frac{c^3 k_B}{4G\hbar} A, \tag{16.28}$$

where in the last step we used the Schwarzschild area $A_S = 16\pi G^2 M^2 / c^4$(16.13). Hawking's expression (16.28) for S,[26] now held to be the correct result for all types of black hole, gives a similar answer to Bekenstein's approximation (16.26) up to a numerical factor of roughly ten; by now we've learned to be comfortable with approximate methods yielding the correct answer up to some numerical factor.

Despite the cosy familiarity of classical thermodynamics, we should grasp that black holes are exotic beasts; for instance, they have a negative heat capacity

$$C_J = c^2 \left.\frac{\partial M}{\partial T}\right|_{J=0} = -\frac{c^5\hbar}{8\pi Gk_B T^2} < 0, \tag{16.29}$$

unlike anything we've encountered before. "Heating" a black hole, say by bathing it in radiation, actually causes it to cool down! Again, the primacy of entropy over heat as a classical concept, because it is a function of state, is underlined. Also note that while expression (16.28) for S is true for all black holes, in general the temperature T has a complicated dependence on M and J; indeed for extreme black holes $T \to 0$, as highlighted by the curves for $A = 4G\hbar / c^3 k_B \times S$ approaching the extreme limit as vertical lines in the left pane of Figure 16.3, using the First Law (16.22) in the form $T^{-1} = \partial S / c^2 \partial M\,|_J$. Spinning black holes are cooler.

16.6 HAWKING RADIATION

Hawking's work also gave physical insight into the mechanism of the black hole to have $T > 0$. In quantum theory, all possible states of a system which respect constraints such as charge and momentum conservation must be taken into account when computing possible outcomes. This aspect is particularly impactful when considering relativistic systems. Consider processes in which a particle–antiparticle pair is spontaneously created out of the vacuum. The pair's electric charges necessarily sum to zero, and their momenta and angular momenta also cancel when summed. In an important sense, the presence of the pair is indistinguishable from empty space, and the physical vacuum is best envisaged as a seething soup of particle–antiparticle pairs of all species. The particle–antiparticle pair is not directly detectable because energy and momentum conservation force at least one of them to be created "off-shell"[27] in a form in which they cannot propagate arbitrarily far; rather, they are produced in *virtual* states and survive only for a brief time δt before reannihilating back into the empty vacuum.[28] Their lifetime is constrained by the time-energy uncertainty relation $\delta E\,\delta t \geq \hbar$, where $\delta E \sim mc^2$ with m the particle rest mass, i.e. $\delta t \sim r_C / c$.

Now consider pair formation near a black hole, with one member of the pair remaining outside the horizon while its conjugate partner has sufficient time within uncertainty constraints to move just inside. Since space time is non-singular at the horizon, there is nothing forbidding this. Energy conservation implies that if the outside particle has energy $E > 0$ consistent with the properties of a physical particle, the inside one has $E < 0$ and is apparently unphysical. However, since within the horizon a particle's world-line is spacelike, quickly terminating on the singularity, it turns out it is possible to obey the on-shell condition even with $E < 0$, and hence evade any further constraint imposed by the uncertainty principle. The inside particle falls into the black hole and *decreases* the hole's mass by an amount E / c^2, while the outside particle with positive energy E can escape to be detected remotely. From the outside, it looks as if the black hole is radiating particles and decreasing mass to $M - E / c^2$ in so doing.

Suppose the pair initially forms a distance δr above the horizon; the proper time for the freely falling negative energy twin move inside the horizon and go on shell is

$$\delta\tau \approx \sqrt{\frac{2GM\,\delta r}{c^4}}. \tag{16.30}$$

From the uncertainty relation we deduce the energy of the particle set to escape:

$$\mathcal{E} = \frac{\hbar}{\delta\tau} = \frac{\hbar c^2}{\sqrt{2GM\,\delta r}}. \tag{16.31}$$

However, before it can be detected remotely the particle must climb out of the gravitational influence of the black hole, undergoing *gravitational red-shift*.[29] The final observed energy

$$E = \mathcal{E}\sqrt{\frac{\delta r c^2}{2GM}} \approx \frac{\hbar c^3}{2GM} \tag{16.32}$$

is independent of δr: the characteristic energy of the radiated particle just depends on the parameters of the black hole and is independent of the pair's point of origin. Hawking's more rigorous calculation showed the energy spectrum of radiated particles is *thermal*, i.e. obeying the Black Body relation (9.102,103), with the characteristic temperature given by (16.27). Black holes are not really black!

For a solar mass black hole with $M_\odot \approx 2\times10^{30}$ kg, the Hawking temperature (16.27) is $T_{H\odot} \approx 6\times10^{-8}$ K. In isolation, such a black hole would radiate energy at a rate $\sigma_\gamma c A T_{H\odot}^4 / 4$, where the constant σ_γ was introduced in (15.14).[30] Using the Schwarzschild results (16.13,27) we deduce the general rate of mass loss

$$\frac{dM}{dt} = -\frac{\hbar c^4}{15360\pi M^2 G^2} \tag{16.33}$$

i.e.

$$\int_{M(t)}^{0} M^2 dM = -\frac{M^3}{3} = -\frac{\hbar c^4}{15360\pi G^2}\int_t^{t_{evap}} dt$$

with solution

$$M(t) = \left[\frac{\hbar c^4}{5120\pi G^2}\left(t_{\text{evap}} - t\right)\right]^{\frac{1}{3}}; \tag{16.34}$$

in other words, the black hole radiates mass at progressively faster pace, becoming ever more luminous until eventually it evaporates after a time

$$t_{\text{evap}} = \frac{5120\pi G^2 M^3}{\hbar c^4}. \tag{16.35}$$

There are a couple of other aspects of Hawking radiation worth highlighting. First, we can use Wien's displacement law (9.108), to predict the wavelength dominating the radiated energy. Using $T = c^3\hbar / 8\pi GMk_B$ (16.27) and recalling $R_S = 2GM / c^2$:

$$\lambda = \frac{2\pi\hbar c}{4.965 k_B T} = \frac{8\pi^2}{4.965} R_S; \tag{16.36}$$

in other words, the ratio of the typical wavelength of Hawking radiation to the Schwarzschild radius is about 16. This unexpectedly large value runs counter to a naive visualization of a black hole as something akin to a glowing hot cannonball. Indeed, a much more interesting comparison is with light emission from an atom as it deexcites by an electron transitioning from an outer to an inner shell. The shortest wavelength λ_{Hmin} emitted by a hydrogen atom lies in the ultra-violet region of the spectrum, in the so-called *Lyman limit:* its ratio to the characteristic size of the atom specified by the Bohr radius $r_B = 4\pi\varepsilon_0\hbar^2 / m_e c^2$ is given (through solution of the Schrödinger equation for the hydrogen atom) by

$$\frac{\lambda_{\text{Hmin}}}{r_B} = \frac{4\pi\hbar^3 c(4\pi\varepsilon_0)^2}{m_e e^4} \times \frac{m_e e^2}{4\pi\varepsilon_0\hbar^2} = 4\pi\left(\frac{4\pi\varepsilon_0\hbar c}{e^2}\right) = \frac{4\pi}{\alpha}. \tag{16.37}$$

Here $e \simeq 1.6\times10^{-19}$ C is the fundamental quantum of electric charge carried by a single proton, and m_e the electron mass. The dimensionless combination of constants α between the brackets after the second = sign is known as the *fine structure constant* of electromagnetism, well known by generations of physics students to have a numerical value close to 1/137. Thus $\lambda_{\text{Hmin}} / r_B \approx 1722$. In some sense, then, Hawking radiation resembles the result of a coherent quantum process,[31] although there are also profound differences: while the Hawking spectrum is continuous, atoms emit radiation at discrete wavelengths, yielding a characteristic line spectrum. Secondly, we can use (16.33) and (16.36), together with $d(Mc^2)/dt = -(hc/\lambda)dN_\gamma/dt$ to estimate the rate of photon emission:

$$\frac{dN_\gamma}{dt} \approx \frac{c}{4.965\times960\times R_S}. \tag{16.38}$$

For a solar-mass black hole the result is about 20 photons per second: Hawking radiation in this case has a much more granular nature than the glowing cannonball picture would suggest.[32]

For $M = 1M_\odot$ the black hole evaporation timescale (16.35) is immense: $t_{\text{evap}} \sim 2\times10^{67}\,y$. This exceeds the current age of the universe by some 57 orders of magnitude, and it would be remarkable (and possibly a testament to Dyson's foresight) if human civilization were still around to observe such an event. Indeed, in the current epoch, stellar-mass black holes are actually still growing through absorption of the much hotter CMB radiation with $T_{0\gamma} = 2.73K$. For Hawking radiation to be the process dominating black hole evolution we need to wait for the universe expansion to reach $a/a_0 = T_{0\gamma}/T_{H\odot}$, which assuming a matter-dominated expansion will only happen after

$$t = \left(\frac{2.73}{6\times10^{-8}}\right)^{\frac{3}{2}} \times 14\times10^{9} \approx 4\times10^{21}\,y. \tag{16.39}$$

The feeling that we are very new around here keeps getting stronger… For Hawking radiation to contribute to making the lives of contemporary astronomers more interesting by inducing black holes to explosively evaporate round about now, the black holes must have formed just after the formation of the universe and have $M \sim 10^{11}$ kg, roughly equal to the mass of all the buildings on Manhattan Island. It's conceivable that such objects may have grown through gravitational collapse of some early density fluctuation in the universe, but so far there is no observational evidence for such *primordial black holes*.

16.7 WHERE DOES ALL THE INFORMATION GO?

Perhaps the most interesting aspect of Hawking's prediction that black holes radiate and ultimately evaporate is the challenge it poses to fundamental physics. As discussed above, the black hole forms from matter belonging previously, say, to a star interior and consisting of a variety of nuclei and more elementary particles. The information about this initial state is at first somehow hidden behind the horizon. The thermal nature of the output radiation, however, is entirely determined by the parameters M and $\boldsymbol{J}$ (and Q, if charged); in other words, the information resides in just five numbers. Once evaporation proceeds to completion, there is no horizon left to hide behind, and on the face of it, we are left with nothing. Over the course of its lifetime, the black hole has seemingly removed information from the universe. Yet, assuming the black hole's evolution is governed by some unitary quantum theory in which, as we saw in Chapter 10, the Gibbs entropy is strictly unchanged, and information loss is forbidden. Where, then, has the information gone?

The "unitary quantum theory" referred to above, which must inevitably describe gravity in addition to all the other relevant forces and particles, is as yet unknown. Perhaps resolving the black hole information problem might point us in the right direction. In 1993, Don Page set the stage for tackling these issues using the concept of entanglement entropy, which was introduced in Chapter 12. Suppose the black hole and the emitted radiation are each considered to be components of a combined system which starts in a pure quantum mechanical state, described by a wavefunction $\Psi(\alpha,\beta;t)$, where α denotes observables related to the black hole and β the radiation. The appearance of t in the list of arguments shows that Ψ evolves in time, but from now on we'll take this as read and leave it out. Subsequently Ψ evolves unitarily so that $\Sigma_{\alpha\beta}\Psi^*(\alpha,\beta)\Psi(\alpha,\beta)=1$, and remains in a pure state throughout. Now define density matrices $D_{\mathrm{BH}}, D_{\mathrm{rad}}$ governing all measurements pertinent to each subsystem (for economy we now consider α,β as matrix indices):

$$(D_{\mathrm{BH}})_{\alpha\alpha'} = \sum_{\beta}\Psi^*(\alpha,\beta)\Psi(\alpha',\beta);\quad (D_{\mathrm{rad}})_{\beta\beta'} = \sum_{\alpha}\Psi^*(\alpha,\beta)\Psi(\alpha,\beta'). \tag{16.40}$$

Recall from Chapter 11 that D is Hermitian (i.e. $D_{\alpha\alpha'} = D^*_{\alpha'\alpha}$), with eigenvalues which are positive real numbers $0 \le \lambda \le 1$, and that $\mathrm{Tr}[D] \equiv \Sigma_\alpha D_{\alpha\alpha} = 1$. If one of the eigenvalues equals one, therefore, all the others vanish, and we can say the subsystem is also in a pure state. This occurs when Ψ can be written as a product of factors $\Psi(\alpha,\beta) = \Phi_{\mathrm{BH}}(\alpha)X_{\mathrm{rad}}(\beta)$. Recall from Section 11.5.5 that if the combined state $\Psi(\alpha,\beta)$ is pure then the density matrices $D_{\mathrm{BH}}, D_{\mathrm{rad}}$ share the same spectrum of eigenvalues, i.e. there are states $\phi^{(\lambda)}(\alpha), \chi^{(\lambda)}(\beta)$ such that

$$(D_{\rm BH})_{\alpha\alpha'}\phi_{\alpha'}^{(\lambda)}=\lambda\phi_{\alpha}^{(\lambda)}\Leftrightarrow(D_{\rm rad})_{\beta\beta'}\chi_{\beta'}^{(\lambda)}=\lambda\chi_{\beta}^{(\lambda)}. \tag{16.41}$$

Since on the face of it we don't know anything about the internal states of the black hole, it makes sense to focus for now on $D_{\rm rad}$. The identical spectra of $D_{\rm BH}, D_{\rm rad}$ imply they have the same *entanglement entropy,* which for convenience we define in dimensionless units[33]:

$$\mathcal{S}=\mathcal{S}_{\rm BH}=-{\rm Tr}\left[D_{\rm BH}\ln D_{\rm BH}\right]=-{\rm Tr}\left[D_{\rm rad}\ln D_{\rm rad}\right]=\mathcal{S}_{\rm rad}. \tag{16.42}$$

Note the combined state, with a single non-zero unit eigenvalue, has vanishing entanglement entropy, i.e. $\mathcal{S}_{\rm BH+rad}=0\neq\mathcal{S}_{\rm BH}+\mathcal{S}_{\rm rad}$. Entanglement entropy is not in general an extensive quantity.[34]

Consider for a moment a large system in a pure state Ξ composed of many smaller mutually interacting subsystems ξ_i labelled by $i=1,\ldots,N$. If we repeat the above analysis for sufficiently large N, then we expect the density matrix of each subsystem to be thermal:

$$D_{\xi_i}=\frac{e^{-\hat{H}_{\xi_i}/k_BT}}{Z_{\xi_i}}, \tag{16.43}$$

where H is the Hamiltonian operator acting within the subsystem. The *coarse-grained* or *thermodynamic* entropy S is then defined to be the sum of the entanglement entropies of all the subsystems:

$$S=k_B\sum_i\mathcal{S}_{\xi_i}. \tag{16.44}$$

We describe S as coarse-grained because each of the individual terms in the sum (16.44) has information about the underlying pure state removed as a consequence of the trace over unobserved degrees of freedom. This S is the same as the entropy considered throughout the rest of the book, satisfying the Second Law; to stress this we've restored the conventional units with a factor k_B.

Suppose initially Ξ is a product state over pure subsystem states ϕ_{ξ_i}, i.e. $\Psi_\Xi=\prod_i\phi_{\xi_i}$: in this case all the D_{ξ_i} have a unit eigenvalue implying all entanglement entropies vanish, i.e. $\mathcal{S}_\Xi=\mathcal{S}_{\xi_i}=0$. Over time the subsystems interact and become entangled, so that in general $\mathcal{S}_{\xi_i}>0$, and therefore $S>0$; however, the combined system remains in a pure state so $\mathcal{S}_\Xi$ remains zero.

Now return to the bipartite black hole/radiation system. Bearing in mind that contributions to $\mathcal{S}$ are not in general additive, whereas we expect S to be extensive, we deduce the inequality $S_{\rm rad}/k_B>\mathcal{S}_{\rm rad}\equiv\mathcal{S}$. Following (12.43) we now define *mutual information I*:

$$I=S_{\rm rad}/k_B-\mathcal{S}. \tag{16.45}$$

Information quantifies the difference between the course-grained thermodynamic entropy obeying the Second Law and the fine-grained entanglement entropy which necessarily vanishes for a pure state obeying unitary time evolution.

Page[35] used a simple model of the black hole/radiation system consisting of a total of N quantum states described by an $N\times N$ matrix, with a number m of the states associated with the output radiation[36] and $n=N/m$ with the as yet unknown quantum states of the black hole.

Initially $m=1$ and $n=N$, but as the black hole radiates over the course of time, as quantified by the ever-increasing thermal radiation entropy $S_{\rm rad}/k_B=\ln m$, m increases and n decreases such that N remains constant. He was able to deduce the behaviour of I and $\mathcal{S}$ shown in Figure 16.4. The main features are that $\mathcal{S}$ begins from zero and ends in zero; in both limits the entire system in a pure state is being described. In fact, $\mathcal{S}$ is symmetric about the midpoint of the plot, growing steadily while $m\le\sqrt{N}$ (i.e. $\ln m\le\frac{1}{2}\ln N$) and then decreasing thereafter. In the first stage of the process $\mathcal{S}\approx S_{\rm rad}/k_B$

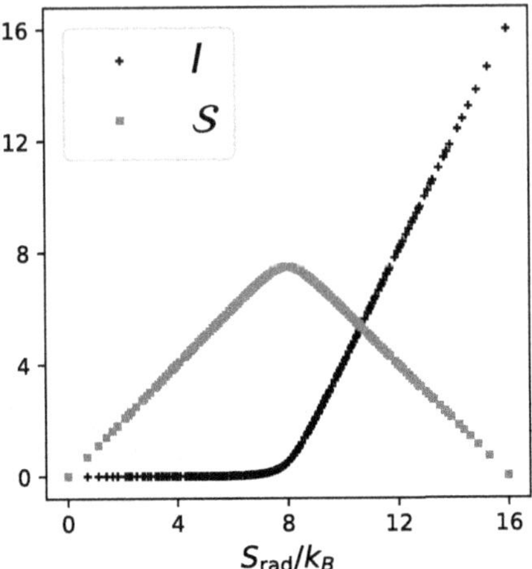

FIGURE 16.4 Predictions for entanglement entropy S and information I as a function of the radiation entropy S_{rad} for a Page model with $N = 8748000$.

and no information emerges. Almost all the information I emerges from the black hole horizon[37] only in the latter stage once $m > \sqrt{N}$. Since the endpoint is a pure quantum state with vanishing $\mathcal{S}$, information about the internal state of the black hole could in principle be reconstructed by examining correlations between radiation quanta emitted across the entire sky over the whole history of the black hole; such correlations are ignored in Hawking's semiclassical calculation predicting a thermal spectrum, described by five numbers. For all practical purposes, of course, this is impossible, and we can continue to regard black holes as places where information leaves the universe.

16.8 A LITTLE STRING THEORY

While the Page curve shown in Figure 16.4 provides a template for the resolution of the black hole information paradox, we can't be satisfied until we have a microscopic theory of the black hole microstates, which requires a full-blown theory of quantum gravity, valid at all length scales so that the physics all the way to evaporation is faithfully captured. Such a theory is not currently available, and we can only speculate about future directions. One promising route is offered by *String Theory*, a description of nature which views all matter as composed of elementary one-dimensional entities called *strings* existing in a variety of different states of excitation. There's no room[38] here to do the theory full justice, but the following basic notions will serve for now. A string is an object of length $\mathcal{L}$ and mass M. It has a constant tension κ, measured in units J m^{-1}; hence $M = \kappa\mathcal{L} / c^2 \propto \mathcal{L}$. The reason we're not aware of stringiness in everyday life is their supposed physical scale: $\kappa \sim O\left(c^4 / G\right) \simeq 10^{44} N$; $\mathcal{L} \sim \sqrt{G\hbar / c^3} \simeq 10^{-35}$ m. Strings exist both in open-ended form and as closed loops; the latter form is particularly apt to describe gravity since it has excitations which share the quantum numbers of the putative *graviton* thought to mediate low-energy gravitational interactions between massive bodies. Strings interact with each other either by splitting or fusing at the ends in the case of open strings or for closed strings by a Y-shaped process visualized in Figure 16.5[39]; in either case we can specify the likelihood of such an event by a dimensionless number g known as the *string coupling*.

Now, through analysis of the force between massive strings the following relation holds in string theory:

$$G \sim \frac{g^2 c^4}{\kappa} = \frac{g^2 c^3 \ell_s^2}{\hbar}, \tag{16.46}$$

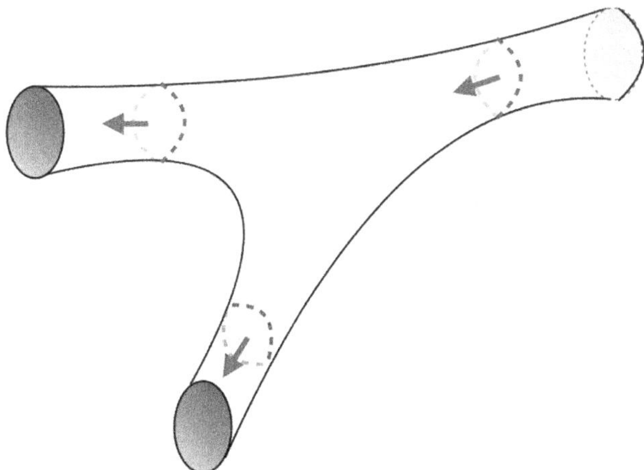

FIGURE 16.5 Visualization of an interaction between closed strings, as one closed string splits into two. String worldsheets sweep out a tubelike structure in space–time; strings themselves trace out cross-sections as indicated by the dashed lines and arrows.

where the second equality defines the *string bit length* ℓ_s, which can be thought of as the length of an elementary unit out of which longer strings can be built, as if by gluing matchsticks together end to end. The ~ symbol in Equation (16.46) indicates that from here on we are playing fast-and-loose, setting all purely numerical factors to unity in an effort to get to where we're heading with as few bumps as possible. In this spirit we can rewrite expressions for the black hole radius and entropy as

$$R_0 \sim \frac{GM_0}{c^2} \sim g_0^2 \frac{c\ell_s^2 M_0}{\hbar}; \quad \frac{S}{k_B} \sim \frac{GM_0^2}{\hbar c} \sim g_0^2 \frac{c^2 \ell_s^2 M_0^2}{\hbar^2}. \tag{16.47}$$

Check that these expressions are dimensionally consistent; in particular S / k_B is dimensionless, as must be the case. The 0 subscripts in (16.47) denote the fact that they are derived with g_0, R_0, and M_0 sufficiently large that the results of classical general relativity are valid; in other words it is safe to ignore string theory corrections to the horizon properties derived earlier in the chapter. This in particular implies $R_0 \gg \ell_s$.

Now let's perform a "theoretical experiment" in which the coupling g is permitted to smoothly decrease from its starting value g_0, so that gravitational interactions parametrized by G also decrease in strength while ℓ_s, an intrinsic property of the string, remains constant. R and M will each also evolve in response. If the decrease is gradual enough then the change must be adiabatic so $S \sim k_B c^2 \ell_s^2 / \hbar^2 \times (Mg)^2$stays constant, implying that $M \propto g^{-1}$, $R \propto g^2 M \propto g$. Eventually by some point $g = g_* \ll g_0$, R has decreased sufficiently to be comparable with the string bit length scale ℓ_s: now we can no longer trust general relativity, but deduce from the constancy of S that

$$\frac{R}{\ell_s} \sim 1 \Rightarrow M_* \sim \frac{\hbar}{g_*^2 \ell_s c}; \quad \frac{S}{k_B} \sim g_*^2 \frac{c^2 \ell_s^2 M_*^2}{\hbar^2} \sim \frac{1}{g_*^2}. \tag{16.48}$$

In this weakly coupled regime the appropriate physical picture is a single weakly interacting string formed from n string bits, with mass $M_* = \kappa \mathcal{L} / c^2 \sim \hbar \mathcal{L} / \left(\ell_s^2 c\right)$, and length $\mathcal{L}_* \sim n\ell_s$ implying $M_* \ell_s \sim n\hbar / c$. The dimensionless number g_* characterizing this crossover from a general relativistic to a stringy description is in principle calculable but requires theoretical control over the regime where string interactions parametrized by $g \sim g_0$ are too large to be neglected, which presents a severe technical challenge.

The entropy of a single weakly interacting string can be calculated using string theory, but the following rough-and-ready argument is just as good in the current context. Imagine building an open string from n string bit elements joined end to end, aligned along the links of a three-dimensional cubic lattice (think of it as scaffolding to help in gluing the matchsticks together!). If we glue the bits in consecutive order, and allow back-tracking, then at each step there are six choices for how to place the next bit, corresponding to the six links emerging from each lattice point. The total number of possible string states is thus 6^n, so $S / k_B = n\ln 6 \sim n$. We conclude

$$\frac{S}{k_B} \sim n \sim \frac{cM_* \ell_s}{\hbar} \sim \frac{1}{g_*^2}; \tag{16.49}$$

in other words, string theory predicts the same form (16.48) for S as the Bekenstein–Hawking result from general relativity, and is thus a candidate theory on which to build a statistical mechanics description of black holes based on identifiable microstates. The adiabatic process of the previous paragraph starts with a black hole across whose horizon is stretched a strongly interacting string spaghetti to a depth ℓ_s: as the string coupling g decreases the horizon shrinks so that its area A is a constant multiple of G,[40] until by the time $g \to g_*, R \to \ell_s$ we are left with a single crumpled string. It's a promising start, but at the time of writing the dynamical principles underlying string theory are not yet fully elucidated.

16.9 WHERE IS ALL THE ENTROPY?

Before closing, let's return to stellar mass black holes. In 2016, an exciting new era of astronomy commenced with the discovery of a merger between two black holes through the detection of gravitational radiation (another remarkable prediction of general relativity) emitted during the final few milliseconds of this event by the LIGO gravitational wave detector based in the United States.[41] In the first such event to be observed, labelled GW150914, a merger between black holes with masses $36M_\odot$ and $29M_\odot$ some 400 Mpc from Earth resulted in final state black hole of mass $62M_\odot$, the $3M_\odot$ deficit being emitted as energy in the form of gravitational radiation.

Gravitational waves are detected as they pass via their effect on the measured distance L between two stationary (i.e. having constant x, y, and z coordinates) objects; the first observations, shown in Figure 16.6, were made using an interferometer consisting of two arms oriented at right angles of physical length 4 km, whose effective length is boosted to 1200 km through the use of multiple mirror reflections; changes δL in this effective length can be measured using laser interferometry to a precision of one part in 10^{21}. In fact, LIGO consists of two near-identical observatories, one located at Hanford WA and one at Livingston LA, separated by 3000 km. In order for an event to be judged genuine, a coincident signal at both observatories is required to eliminate spurious signals due to local noise or seismic activity. The sharp increase[42] in both amplitude and frequency towards the close of the signal is characteristic of a black hole pair merging through a rapid "inspiralling" as modelled by numerical general relativity calculations. Most of the energy is radiated in the final 20 ms; the power of 3×10^{49} W exceeds the luminosity of all the stars in the observable universe.

For simplicity let's assume all three black holes are Schwarzschild, to calculate the entropy increase:

$$\Delta S = \frac{4\pi G k_B}{\hbar c} M_\odot^2 \left(62^2 - 29^2 - 36^2\right) = 1707 \times 1.47 \times 10^{54} \simeq 2.5 \times 10^{57}\,\mathrm{J\,K^{-1}} \tag{16.50}$$

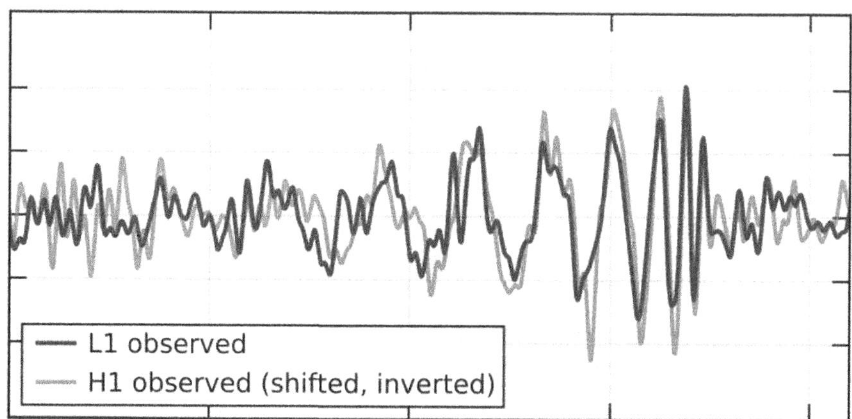

FIGURE 16.6 Gravitational wave signal observed for GW150914; the darker colour (L1) is the signal observed at the Livingston observatory and the lighter one (H1), shifted by a time-delay of 6.9 ms, at Hanford.[a] Each gradation on the horizontal axis corresponds to 0.05 s, and on the vertical axis to a strain $\delta L / L$ of 0.5×10^{-21}.

a This figure is adapted from one presented in the discovery paper, Observation of Gravitational Waves from a Binary Black Hole Merger, B.P. Abbott et al., *Phys. Rev. Lett.* **116** (2016) 061102, DOI 10.1103/PhysRevLett.116.061102.

After all the algebra it's good to calculate an actual number! In roughly a quarter of second, a single cataclysmic event generates some 10^{46} times more entropy than the entirety of humanity as estimated in the previous chapter. Perhaps we're just not trying hard enough! Moreover, the entropy per unit mass of a black hole scales with M. In Chapter 15 we claimed the entropy per baryon of the CMB $s_\gamma / k_B \sim 10^9$ as the biggest entropy reservoir in the universe. Step forward the supermassive black holes at the heart of active galactic nuclei. Suppose the universe is populated by $10^9 M_\odot$ galaxies, each with a $10^6 M_\odot$ black hole at its centre. The number of baryons in each galaxy $N_b = 10^9 M_\odot / m_p \approx 10^{66}$. The entropy per baryon of the black hole at the core,

$$\frac{S_{BH}}{k_B N_b} = \frac{4\pi G (10^6 M_\odot)^2}{10^{66} \times \hbar c} \approx 8 \times 10^{22}, \tag{16.51}$$

exceeds that originating from the CMB by over 13 orders of magnitude. This rather conservative estimate demonstrates that black holes inevitably dominate the entropy budget of the universe.

In summary, the remarkable insights of Bekenstein and Hawking have opened up a new chapter in thermodynamics, which shows that no account of physical processes at cosmic scales is complete without a discussion of black holes. Along the way, the conceptual challenges provided by the riddle of where all the information goes is providing fruitful stimulus for theoretical efforts to reconcile general relativity with quantum theory, the two dominant physical paradigms bequeathed us by 20th century physics. Entropy, the mysterious quantity with a meaning so often hard to grasp, is central to these developments. Should we be surprised? It is worth recalling the words of Max Planck, initiator of the quantum revolution in physics with his celebrated treatment of black body radiation in 1900, obtained following several years of intellectual struggle; he later wrote[43] "*This approach was opened to me by maintaining the two laws of thermodynamics. The two laws, it seems to me, must be upheld under all circumstances. For the rest, I was ready to sacrifice every one of my previous convictions about physical laws.*"

Plus ça change…

NOTES

1 The speed of light has this exact value by definition, and forms the basis for the definition of the SI length unit, namely the metre. Our favourite illustration of its huge magnitude (by human terms) is that in one second a light beam travels a distance roughly seven and a half times the length of the Earth's equator.
2 The current record for a supermassive black hole is ~30 billion $M_\odot$, inferred from its gravitational lensing influence on views of a more distant galaxy. See https://arxiv.org/abs/2303.15514
3 Stellar mass black holes are revealed by X-rays generated by matter accreting from a closely orbiting binary companion star; supermassive black holes are thought responsible for Active Galactic Nuclei or quasars, small regions at the core which may be many thousands of times more luminous than all the galaxy's stars put together.
4 This process is actually more efficient at releasing energy from matter than the thermonuclear reactions discussed in Chapter 15.
5 Measured in units $\text{kg m}^2\ \text{s}^{-1}$: angular momentum is a vector quantifying a spinning body's tendency to remain spinning in the absence of external influences. For a system of particles each having position $\boldsymbol{r}_i$ and linear momentum $\boldsymbol{p}_i$, angular momentum is given by $\boldsymbol{J} = \Sigma_i \boldsymbol{r}_i \times \boldsymbol{p}_i$.
6 An introductory account of relativity can be found in *Space-time: an Introduction to Einstein's Theory of Gravity* by J. Allday (CRC Press, 2019).
7 Indeed, it makes no sense even to specify which event precedes the other.
8 Of course, if all three spatial dimensions are included, the shape accommodating all timelike trajectories is much harder to draw, but it's still referred to as a *cone*.
9 Hendrik Lorentz (1865–1928) was a physicist who shared the 1902 Nobel Prize for his study of the effect of magnetic fields on atomic spectra.
10 That year, in the same journal, *Annalen der Physik* Einstein, aged 26, published four showstopper papers: one on special relativity; one on the mass–energy equivalence; one explaining Brownian motion in terms of molecular collisions mentioned in Chapter 5; and one explaining the photoelectric effect by describing light as a stream of discrete photons. 1905 was a very good year…
11 Imagine drawing three great circles on the Earth's surface: starting from the North Pole, draw two meridians at 0° and 90°W longitude, respectively, until each meets the equator. Each angle of the resulting triangle is a right angle, and the sum is 270°.
12 Karl Schwarzschild (1873–1916) was a physicist and astronomer who found the solution while serving in the German army during World War I. He subsequently died from disease while serving at the Russian front.
13 Experts may not immediately recognize this expression since it uses so-called Eddington–Finkelstein coordinates. The more familiar expression in Schwarzschild coordinates, i.e., $ds^2 = -c^2\,(1 - R_s/r)(\Delta t)^2 + (1 - R_S/r)^{-1}(\Delta r)^2 + r^2[(\Delta\theta)^2 + \sin^2\theta(\Delta\phi)^2]$, contains a spurious singularity at the Schwarzschild radius $r = R_S = 2GM/c^2$ which obscures the discussion. The two time coordinates are related by $t_{EF} = t_S + R_S / c \ln\left|r / R_S - 1\right|$.
14 For simplicity we'll treat J as a positive scalar quantity from here on, but remember it's really a component of a vector.
15 Roy Kerr (1934–) developed the theory of spinning black holes in 1963 while working at the University of Texas.
16 In differential expressions such as (16.14) it is convenient to measure angles in radians. A full circle = 360° = 2π radians, so $\frac{\pi}{2}$ is a right angle.
17 Roger Penrose (1931–) is a mathematician and physicist whose work on black holes was recognized by the award of the 2020 Nobel Prize. He is also a compelling science writer, and has proposed that human consciousness cannot be explained algorithmically but rather requires a theory of quantum gravity.
18 The argument is subtle and relies on the existence of a region *outside* the horizon known as the *ergosphere* where trajectories which would be timelike in flat space become spacelike; for a Schwarzschild black hole this only happens *inside* the horizon; the resulting reduction in M will be explored more fully when we discuss Hawking radiation. Requiring that corotating observers within the ergosphere don't observe unphysical particles with negative energy, whose existence could be reported back to base, leads to a bound on the accompanying decrease in J ensuring that the area A given by (16.18) can only increase.
19 Measured in radians s^{-1}. Physically this corresponds to the rotation rate of photons moving along null trajectories *within* the horizon. For a $(M_\odot, J_\odot)$ black hole $\omega_{BH} \sim 1.12 \times 10^4\ \text{s}^{-1}$, i.e., some 1800 rotations per second.
20 Stephen Hawking (1942–2018) was a theoretical physicist and cosmologist who made seminal contributions to study of black holes. His motor neurone disease confined him to a wheelchair communicating using a robot voice for most of his adult life. His achievements in both fundamental science and science popularization made him an inspirational figure with influence transcending traditional scientific circles, and enabled him to make several guest appearances on *The Simpsons*.
21 Jacob Bekenstein (1947–2015) initiated the study of black hole thermodynamics while a graduate student working at Princeton University; reading his paper *Black Holes and Entropy*, published as *Phys. Rev.* **D7** (1973) 2333, was one of the inspirations for writing this book.

22 In case you were wondering, if the black hole carries a charge Q measured in Coulombs, then the First Law reads $c^2dM = \Theta dA + \omega_{BH} dJ + \Phi dQ$, with the electrostatic potential $\Phi = Qr_+ / A\varepsilon_0$, $r_+ = GM/c^2 + \sqrt{G^2m^2/c^4 - J^2/c^2M^2 - GQ^2/4\pi\varepsilon_0 c^4}$, and the constant $\varepsilon_0 = 8.854 \times 10^{-12}$ F m^{-1}.

23 The Uncertainty Principle is a mathematical expression of wave-particle duality, according to which matter has both particle and wave-like aspects. Quantum field theory describes particles as *quanta,* namely propagating groups of waves of some underlying *quantum field* which pervades space-time. However, the field aspect can also be identified in certain limits with classical force fields such as those well-known from electromagnetism or gravity. The Compton wavelength is an expression of the range over which the resulting force acts. Long-ranged forces such as electromagnetism and gravity correspond to $m \to 0$ in (16.24)—the associated quanta, namely the photon and the graviton, are massless.

24 In order to convert from A to S we need to find a combination of constants with units J K^{-1} m^{-2}. The factor of k_B provides the J K^{-1}, but there is no combination of G and c, the two constants appearing in classical general relativity, with dimensions m^{-2}. The introduction of $\hbar$ from quantum physics permits the combination $c^3/\hbar G$.

25 We assume that the entropy vanishes for $M \to 0$.

26 There's something very special about this equation: it contains no fewer than four fundamental physical constants: $c, \hbar, G$, and k_B.

27 In flat space-time an "on-shell" particle satisfies $E = +\sqrt{p^2c^2 + m^2c^4}$.

28 The "vacuum polarization" resulting from the presence of virtual particle—antiparticle pairs produces small corrections to the details of atomic spectra calculated using quantum mechanics. Improvements in spectroscopic measurements of e.g., the *Lamb shift* between $S_{1/2}$ and $P_{1/2}$ states of hydrogen in the 1940s inspired a flourishing in quantum field theory calculation culminating in the award of a Nobel Prize to Feynman, Schwinger and Tomonaga in 1965. More recently the vacuum polarization correction to the magnetic moment of a fundamental particle called the muon has been the focus of intense theoretical scrutiny as precision experiments probe the validity of the Standard Model.

29 Gravitational red-shift, namely the change in a photon's frequency as a consequence of moving in a gravitational field, was detected in terrestrial experiments with a gamma-ray source located at the top of a 20 m tower by Pound and Rebka in 1960.

30 At such a low temperature the content of the radiation would be exclusively photons; hotter black holes also radiate more massive particles.

31 The limit $\lambda / r \to \infty$ corresponds to the *dipole approximation* in quantum transition calculations; a typical application is to the emission of a photon having the same direction and wavelength as one already present. Such *stimulated emission* processes underpin the operation of a laser.

32 To paraphrase a famous misquotation from *Star Trek's* Mr. Spock: "it's Thermodynamics, Jim, but not as we know it."

33 Indeed, $\mathcal{S} / \ln 2$ is measured in bits.

34 Since $\mathcal{S}$ is not extensive and doesn't obey the Second Law, we find it helpful to give it an alternative symbol to S. In our humble opinion, it would be more helpful still to give it a different name…

35 D.N. Page, Information in Black Hole Radiation, *Phys. Rev. Lett.* **71** (1993) 3743.

36 If you're worried about how to count m it's possible to define the whole system in a large cavity so that discrete radiation modes are well-defined.

37 It is perfectly possible to use Page's model to plot the information in the black hole I_{BH} decreasing with time, to obtain the mirror-image of I in Figure 16.4. Without a microscopic quantum theory of the black hole we have no idea how to measure this, however.

38 Nor adequate authorial expertise….

39 Often referred to as the "pants diagram". Take a look in your laundry basket….

40 Alternatively, A is a constant multiple of ℓ_P^2, where the *Planck length* $\ell_P = \sqrt{G\hbar/c^3} = 1.61 \times 10^{-35}$ m.

41 LIGO scientists Rainer Weiss, Barry Barish and Kip Thorne were awarded the 2017 Nobel Prize for this epoch-making discovery.

42 Sometimes known as a *chirp*.

43 In a letter to physicist R.W. Wood dated 1931.

17 Afterthoughts

This chapter takes the form of a "conversation" between the two authors, with occasional interruptions of varying degrees of politeness.

One voice is in italics, and the other is in normal font.

17.1 FINAL REFLECTIONS

Entropy isn't what it used to be...

SH : Not long after the book was conceived, I discovered this *graffito* in the bathroom at my place of work,[1] and logged it as useful material. Rather like C.P. Snow, we consider Second Law gags somewhat underrepresented in popular culture, (*agreed!*) and are delighted with this addition to the canon. However, the quip permits a second interpretation. Entropy is a difficult concept to grasp at first or even second exposure (*we can both attest to that*); in this book we've presented and tried to explain it from several perspectives, taking a roughly historical path pretty much corresponding to the order in which we first encountered them ourselves. It has turned out there are several entropy variants, often associated with the originator's name : Clausius, Boltzmann, Gibbs, Shannon, von Neumann, Hawking-Bekenstein. It seems natural to expect evolution in the mathematical formalism used to describe entropy, but one observation I think worth making is that the very language, and those features which are stressed, appear to evolve according to the prevailing scientific and technological concerns of the time.

***JA:** Evidence would certainly suggest that you are right, and this does not help when you try and go back to the source material. Perhaps when teaching the subject, we should be more open about such things...*

We began with classical thermodynamics, where we learned that a precise, quantitative, and useful treatment is phrased not in terms of the everyday concepts of *work* and *heat* but rather in terms of the more abstract state functions, *energy* and *entropy*. These are properties that anyone and everyone can experience, either through physical contact or remote thought, and each is accessible to both measurement and calculation.

The ideas were articulated and developed during the Industrial Revolution, when the new-found capacity to generate wealth using force-multiplying machines such as steam-driven pumps, locomotives, power looms, and nautical engines, all powered by the burning of fossil fuels, had a profound impact on life.

DOI: 10.1201/9781003121053-21

In the first half of the twenty-first century, the concept of energy is by now commonplace—most of us have a working understanding of what it means, why it's important, increasingly expensive, and how it can be transformed into different forms. It has become a crucial element in our conceptual toolbox for understanding the world.

Yet here again, I am not sure we ever convey a clear idea of what it means. We talk about what it can do 'the capacity to do work', whatever that is, but we avoid a clear and meaningful fundamental definition. When pressed in class, I always used to say that a definition was not possible as that would require an understanding in terms of something more fundamental. Perhaps we need to sort this one out next...

In theoretical physics, energy is a well-defined conserved quantity associated with any dynamical system whose basic rules don't change over time. Entropy, by contrast, remains harder to grasp. In Chapter 4, we exerted no little mental effort in following Clausius, outlining how entropy changes in a body can be calculated via the reversible heat exchange with a reservoir of known temperature, and illustrating how engine performance can be analysed in terms of a cyclic path in (S,T) space. All this close, abstract reasoning gives little idea of what entropy actually *is*.

I always felt slightly cheated by these important considerations, as if the blend of the abstract and the practical were being used to pull something over my eyes.

For me, these powerful, economic, and austere arguments bear a similar relation to the rest of physics as do Latin & Greek to the Arts and Humanities. However, the standout results from thermodynamics, as encapsulated in the first two laws, *has* had a wider impact; while energy is exactly conserved when processes are analysed with care, the total entropy in a closed system does not keep its value, but can only increase in value. That sounds portentous, but what does it actually mean?

I have never been convinced by the 'arrow of time' – the notion that temporal progression is entropy increase; that time derives from entropy. Partly, this is because I think that time is a much more important and slippery concept than that, but also as I think entropy results from the limits of our approach to the world, not 'in the world' itself.

The great conceptual leap was made by Boltzmann, working in an atomistic paradigm permitting a view of the world in which things can be counted as well as weighed, and measured. As we have seen, Boltzmann was active when atomism was gaining traction and would shortly go on to revolutionize and dominate 20th-century science, but at the time things were far from fixed. His famous formula identifies entropy with the logarithm of the number of distinguishable microstates of a system composed of many discrete components, all consistent with some particular macroscopic state specified by just a few parameters such as temperature, volume, and pressure. The logarithm ensures the key property of extensivity. The counting aspect yields a more tangible aspect of entropy—a system with large entropy is in some sense more disordered (*hmmm...*), and its microscopic state less predictable (*yes!, or...less easily tracked?*).

Quantum theory helps define the accessible microstates of a system in a precise and countable way (*as developed in Part II of the book*). A lazy reading of the Second Law then suggests that the disorder in the world is bound to increase with time, promoting the idea that things are "running down" as we slide inexorably towards universal equilibrium and that we're all doomed. At the very least, these forebodings need to be quantified and relevant timescales understood, as well as what we mean by "system" and what by "environment". We've tried to illustrate this point by reviewing the basic thermodynamics underlying living processes, demonstrating that entropy production is not a bug but rather a feature of how we live.

Looking back on it now, I think that I never found the idea of disorder that helpful. I know that it's used a lot in popular accounts, and it has become part of the Zeitgeist with respect to entropy. For me the notion of predictability is much more useful. Perhaps it's because I live in a disordered world and can't imagine it getting any more disordered, but I find the term vague. It's all very well talking about the order of cards in a deck, but what does order mean for a macroscopic system in the scientific sense? If we say that the universe gets less predictable over time, that means something in general. We can't pin down the initial microstate, so our predictions are limited and over time those limitations get worse as the system gets lost to us in its phase space. We don't have a phase space GPS to keep track of exactly where things are. If we did, arguably there would be no entropy increase.

By changing the focus from isolated systems to those in contact with a heat reservoir characterized by a specified temperature,[2] Gibbs reformulated Boltzmann's entropy into an expression phrased in terms of the *probability* of specific microstates.

For me a crucial moment came when I realized that the Gibbs formula was a measure or ranking of probability distributions. In a sense, it's not an 'entropy' itself until it is maximised. At least it's not an entropy in any way that can be compared with thermodynamic entropy. Here I think language is important. I believe that Shannon has used the probability measure idea, but it was perhaps unwise to take von Neumann's advice. Having said that, having an information theory perspective on thermodynamic entropy has been very helpful. That has been the fundamental transformation in my thinking as a result of writing this book.

From a practical point of view, this change of emphasis (*to probability*) ushers in the concept of *free energy* $F = U - TS$, quantifying the amount of energy available to do useful work which, as a consequence of the minus sign, clearly diminishes as entropy increases, and which has become a key guiding principle, especially in chemical and biological thermodynamics. Gibbs' work set up the statistical mechanical machinery still routinely used across many scientific disciplines to analyse systems far more complicated than the ideal gases we've focussed on in this book. Formulae for physical observables are expressed as integrations over probability distribution functions, whose value can be estimated using computer programmes via a sampling technique which makes extensive use of a random number generator. As we saw in Chapter 10, however, the resulting theoretical edifice is not straightforward to reconcile with the Second Law under either classical or quantum time evolution. Entropy increase only follows once we take into account the immense practical difficulties in keeping track of microstates; the pragmatic solution of coarse-graining our description, thereby blurring the distinction between microstates, restores the Second Law at the cost of diminished experimental resolution. Unlike quantum uncertainty, which presents a fundamental constraint on what we can know about a physical system as articulated through the Heisenberg Uncertainty Principle, the uncertainty introduced by coarse-graining is *instrumental* in nature. In principle, there is nothing to prevent us from dreaming about the perfect unitary evolution of a complicated quantum state in which all conceivable information about a system is preserved; the Second Law is merely the emergent outcome expected by weary pragmatists.

Nicely put! I would only add weary authors as well…

There is an important point here about how science works. Too often science is portrayed as describing the world, whereas it is somewhat more subtle than that. Although that is our ultimate aim, we can't avoid the fact that science describes the world as revealed by our experimentation. Quantum theory makes that point decisively. The choice of what we intend to measure has physical consequences. This is overblown in some 'New Age' quarters, but it is important never-the-less. The more I have come to see entropy as intimately connected to our ability to control the preparation and influence the progression of systems, the less I see it as a factor 'out there' in the world. I do wonder if 'information' will come to play a part in physics, however.

Ha! For me, the switch to probabilistic language simply means we can set up a theoretical "washing machine": load up your problem, programme the partition function, shut up and calculate, rinse and repeat ....

Pragmatist...

However, this is very important. That approach can be carried over into different fields, as you are about to explain. However, I am not sure it helps to drag the word 'entropy' along as well. The Gibbs measure of a probability distribution is flexible and useful. It's only a thermodynamic entropy when maximised in a physical situation...

The switch from state counting to probability meant that entropy could in principle be used as an abstract quantity to characterize any phenomenon depending on a probability distribution. Gibb's result found an unexpected echo in the 1940s work of Shannon, from a period when there was increased incentive, both economic and geopolitical, to communicate rapidly over large distances. Shannon found an essentially identical formula to that of Gibbs in his account of Information Theory, devised as a means to quantify the information content of a message and hence optimize the speed and accuracy of its transmission. As we saw, while the forum of application is completely different to that of thermodynamics, some of the methodology, including the use of the thermodynamic limit and ergodicity in proving key results, carries straight across. Entropy in information theory is essentially a dimensionless quantity measured in *bits*. We can associate entropy with the information lost when we focus on a macroscopic description of a system rather than attempting to keep track of its particular microstate. This way of thinking influences the arguments of Bennett, Dyson and Bekenstein, reviewed in Part III.

This is exactly what I meant by the useful information theory perspective feeding back into the physics.

In those later chapters, we found ourselves acknowledging this shift of emphasis by increasingly working with the dimensionless combination S/k_B. It's amusing to contemplate the consequences had Shannon's work predated Clausius and Gibbs—we would surely measure temperature in Joules per bit! *That's another point worth making in the classroom!*

Von Neumann set out to come to grips with the measurement problem in fundamental quantum theory. He extended the description of the state by using density matrices and applied them to thought experiments designed to tease out any connection between measurement and entropy. Here, reversibility was a key idea for him. Another striking concept for me was the von Neumann entropy being zero for a pure state. Hence it is our information about the state, not the information in the state that matters. Entanglement entropy comes from an application of von Neumann's formula by tracing over aspects that lie outside of experimental control in that context. Wiping out information generates entropy. The idea extends as a measure of correlations between systems that are entangled, but in this context perhaps a different term ought to be used.

Agreed! Entanglement entropy is a measure of the extent to which different subsystems are correlated, mirroring a similar Information Theory concept known as *associated entropy*; in many cases, the dominant contribution scales as the *area* of the boundary separating the subsystems. In the application explored in Chapter 16, entanglement entropy is an important diagnostic tool that may ultimately yield insight into foundational questions such as what happens to the information carried by material that falls through a black hole horizon. Entanglement entropy also informs the study of complex quantum systems, which is growing in importance as technological advances promise new quantum devices such as analogue simulators and quantum computers. However, as the specific example of Chapter 16 shows, entanglement entropy is neither in general extensive (since it scales as the volume of the boundary, not the bulk) nor does it obey the Second Law.

I think these ought to be decisive points to suggest that it is not an 'entropy' in these sense we are after in this book.

The final development covered in Chapter 16 is the Hawking–Bekenstein entropy associated with the horizon area of a black hole. In general relativity, this is precisely defined via equation (16.28); moreover, unlike the other variants we've covered (with the exception of von Neumann's), this version of entropy *only* makes sense in a quantum context—there is no classical counterpart.

That is so cool ...

It predicts a new phenomenon, Hawking radiation, whose main impact is more fundamental than astrophysical. The fact that black holes will evaporate in some post-astronomically distant future raises the question of what happens to the information contained in the infalling matter over the course of the black hole's lifetime. This is not merely assessing a pinhead for its suitability for angelic terpsichorean activity (*useful though that may be*); any resolution promises to shed light on the unification of gravity and quantum theory, a major piece of unfinished business in theoretical physics. We also learned that black holes are probably already the dominant locus of entropy in the universe.

This may be a pivot point in understanding what entropy is. I've come to think of it, along with von Neumann and others, in 'subjective' terms. It is a result of our perspective on the world. Given the scale at which we exist, microscopic control is not possible. Yet microscopic events determine what we will observe macroscopically. Experiments will drift 'out of our control', their systems exploring volumes of phase space deterministically, but untraceably. Black holes absorb matter and, whatever the fate of the corresponding information, we lose the ability to track what happens. Hence, the subjectivist would say, the entropy increases as a result.

Well, everybody agrees on what entropy is in the thermodynamic equilibrium limit.

Despite what is sometimes said by famous physicists, there are still very important issues to resolve in our theoretical knowledge. From my perspective, the measurement problem remains annoyingly intractable. I am interested in physical approaches, such as Penrose's proposals to do with gravity and I wonder how this will impact on entropy, given von Neumann's analysis. I suspect that the random element of the quantum probabilities will remain, and so the subjectivist approach will still be the best one.

Please enlarge on these ideas, which haven't been discussed up to now. And what exactly is a subjectivist? Do any live near me?

Well, we live within driving distance of each other... I don't think that 'subjectivist' is the best term, but it seems to be around in the literature. What I mean by it is that entropy is to do with our lack of knowledge or control about or over the system. It's a feature of our ability to experiment on nature, and also to a degree a function of the variables we choose to manipulate and use to describe the system. The fact that everything works out in the perspective vanishing point of the thermodynamic limit it what prevents it being entirely subjective in the common sense of the word and ensures that physics is universal.

Instinct, such as I have, tells me that the resolution of the measurement problem will come from a new version of quantum theory which takes a modified form of the Schrödinger equation and adds some stochastic element. The random, probabilistic nature of state collapse will remain, but instigated by a new form of physical process. Penrose suggests that this will be a gravitational 'instability' between possible outcomes. As the process retains a random element, state collapse will still convert a pure state ensemble into a mixed state. Hence it will continue to blur information for us.

At the outset, I expected to find foundational discussions such as those concerning Maxwell's Demon to be the most interesting aspects of entropy and the Second Law to research and think about; in the event, they left me cold. What I've learned to value more is the astonishing power and universality of the classical thermodynamic arguments reviewed in Part I—in retrospect, it's almost criminal that this material should be covered relatively early in undergraduate syllabuses. To me, the concepts involved demand a far more mature outlook to be appreciated. But when teaching, we have to hurry on to quarks and lasers...

Like you, I always had this hazy idea that entropy had something to do with "disorder"—yet in classical thermodynamics it's a thing! Once you start to talk about systems away from equilibrium and the thermodynamic limit, then entropy is a more slippery concept, and I like the idea that, in general, it's more of an emergent phenomenon related to instrumental resolution, as set out in Chapter 10. What makes it special, and useful, is the unique form taken for macroscopic equilibrium systems. Finally, I do find black hole entropy fascinating—the issues have been around since we were at high school, and yet only now have I begun to absorb the implications relating to the information paradox. It would be nice to be around to see its resolution—I only hope we don't have to rely on Dyson...

Not a bad person to rely on in a physics corner...

17.2 EPILOGUE

The fundamental aim of this book has been an attempt at digging up the conceptual roots of entropy. This has meant having to take some difficult decisions about excluding some material from consideration. However, we ought to justify those decisions to the reader, which we try to do here.

Firstly, despite our best intentions, we haven't discussed one of the most spectacular aspects of thermodynamics, namely *phase transitions*, the sudden changes in bulk properties of a substance as a control parameter, such as temperature or pressure, is varied infinitesimally. As already mentioned, well-known examples of this can be explored using a kettle or a freezer compartment of a fridge, but phase transitions crop up across many areas of physics. Some of the most interesting applications concern the state of the entire universe at early times and the impact this has had on what's observed today. Phase transitions generally occur in systems where particles interact with each other. These are typically passed over in elementary treatments which focus on the ideal gas.[3] Transitions between ordered and disordered states of a substance (such as ice to water), which are induced as a consequence of varying the temperature, can often be analysed in terms of a competition between energy and entropy.

Knowledgeable readers may feel cheated by our omitting explicit discussion of the Third Law, which asserts that entropy vanishes in the limit $T \rightarrow 0$, or equivalently, that the ground state of any finite system is non-degenerate (so in the absence of thermal excitations there is only one microstate to count).[4] Vanishing entropy means it grows harder and harder to devise a thermodynamic cycle to remove heat from a system via work as $T \to 0$, so an alternative formulation of the Third Law asserts the impossibility of reaching absolute zero.

The Third Law is challenged to some extent by the existence of amorphous substances such as glasses which exist for long times in metastable states, and which may never succeed in relaxing to a true ground state. From the perspective of this book, extreme black holes, whose Hawking temperature vanishes even though $S(A) > 0$ as shown in Figure 16.3, present a clearer challenge—might this be a further paradox that needs a more complete quantum description to be resolved?

Finally, and most importantly, we have neglected any serious discussion of processes out of equilibrium. To paraphrase the Talking Heads, thermal equilibrium is a place where nothing ever happens.[5] As we have seen, one characteristic of equilibrium is *detailed balance*: microscopic processes that change the system in opposite directions occur with equal probability, so that on average over time the system remains in the same macrostate. Equilibrium thermodynamics, beautiful and powerful as it is, often isn't the right tool to say what happens next. Small departures may induce an orderly flow of some conserved quantity such as energy, electric charge or molecular species in a direction tending to bring things back to equilibrium; this can be brought relatively straightforwardly into a thermodynamic framework via the introduction of *transport coefficients* such as thermal or electrical conductivity, diffusion coefficient, or fluid viscosity.

If we attempted to model a more cataclysmic event, such as a bubble bursting, a supersonic shock wave forming, or two nuclei colliding at relativistic speeds, we would of course assume without question that it is possible to monitor the energy summed over all the component parts of the system and that this energy should be constant, at least to within some acceptable tolerance. Non-equilibrium thermodynamics asks the same question of entropy: if we accept the Clausius definition of entropy as a state function, then this ought to be possible, but how? What variables would entropy depend on, and how would it depend on instrumental resolution? Remember, this time, we don't have a global conservation condition to help; indeed, the Second Law alerts us to expect entropy increase in a time-evolving process.

Over the course of the book, we've learned about many facets of entropy and seen it evolve and perhaps even mutate as a concept, following insight and innovation from different directions. Perhaps this is the hallmark of a truly important idea. Whatever happens next, we are confident of one thing: entropy is only going to get bigger ....

NOTES

1 OK, it was in a university physics department.
2 The technical term for this shift of view, or change of variables, is the *Legendre transformation*.
3 An important exception is the Bose–Einstein condensation occurring in cold atomic systems.
4 The Third is the only thermodynamic law to be recognized by the award of a Nobel Prize, in Chemistry, to Walther Nernst in 1920.
5 In quantum field theory, thermal equilibrium is modelled by evolution through *imaginary* rather than real time.

Appendix MA
Mathematical Preliminaries

This appendix reviews some of the mathematical machinery we're going to be reliant on. Much of this may already be familiar; if that's the case then so much the better, but a brief refresher is never bad. It's particularly important to read on if you've not encountered a logarithm before.

It is very common in science to represent the value of a physical quantity not by giving a numerical value but by writing it as a *variable* using a symbol such as x or y Mathematical shorthands have been developed to concisely express related quantities, e.g., $3x$ has three times the value of x, and $2x+5$ is the result of doubling the quantity and then adding 5. Very often, we represent constant numbers whose value is assumed not to change throughout the calculation by another letter a conventionally (but not necessarily) chosen from near the beginning of the alphabet, leading to expressions like $5y-b$. One of the most powerful shorthands uses a superscript: $x^2 = x \times x$ is the result of multiplying x by itself, and we read it aloud as "x squared". In general, we can write x^p as the result of multiplying x with itself p times; p is called the *power* of x.

Now consider the action of multiplying together two quantities expressed as powers: $x^2 \times x^3 = (x \times x) \times (x \times x \times x) = x^5 = x^{2+3}$.

Multiplying two quantities based on the same x results in the powers adding up. In general

$$x^p x^q = x^{p+q}. \quad \text{(MA.1)}$$

By the same token, dividing one quantity by another results in the subtraction of one power from another:

$$\frac{x^p}{x^q} = x^{p-q}. \quad \text{(MA.2)}$$

Since nobody said whether p is greater than q or not, we need to consider negative powers:

$$x^{-p} = \frac{1}{x^p}. \quad \text{(MA.3)}$$

The special case of (MA.2) with $p = q$ also yields an important result:

$$x^0 = 1. \quad \text{(MA.4)}$$

Finally, consider the relation $y = x^2$, which we can rewrite in several ways: $x^2 = x \times x = y = y^{1/2} \times y^{1/2} \Rightarrow x = y^{1/2} = \sqrt{y}$.

In other words, fractional powers are also allowed and correspond to taking the pth root:

$$x^{\frac{1}{p}} = \sqrt[p]{x}. \quad \text{(MA.5)}$$

Suppose we always stick to the same base variable x. Could we work directly just with powers? That way, using (MA.1, MA.2), we could replace multiplication/division with addition/subtraction, which many of us find easier. There is indeed a suitable mathematical operation called a *logarithm*. If $y = x^p$, $z = x^q$ then

$$p = \log_x y,\ q = \log_x z \text{ so that } \log_x yz = \log_x y + \log_x z.$$

The little subscript x is there to remind us that we must always specify a base for logarithms. An obvious and commonly used choice is $x = 10$, so that $\log 10 = 1$, $\log 100 = 2$, $\log 1000 = 3$, etc.; for such *common logarithms*,[1] it's fine to omit the subscript. The results derived above using powers can then be recast:

$$\log xy = \log x + \log y; \tag{MA.6a}$$

$$\log \frac{x}{y} = \log x - \log y; \tag{MA.6b}$$

$$\log x^p = p\log x; \tag{MA.6c}$$

$$\log \sqrt[p]{x} = \frac{1}{p}\log x; \tag{MA.6d}$$

$$\log 1 = 0. \tag{MA.6e}$$

While these results have been written using common logarithms, they apply to *any* choice of base. How can we convert between logarithms defined using different bases? Suppose $y = \log_a x$ and $z = \log_b x$. It follows that $x = a^y = b^z$ so that $y = \log_a b^z = z\log_a b$, i.e.,

$$\log_b x = \frac{\log_a x}{\log_a b}. \tag{MA.7}$$

Suppose we had a means to easily convert between a variable and its logarithm and back again. In that case, one can get by with addition and subtraction (reserving multiplication and division for evaluating roots and powers). Both authors started high school before cheap electronic calculators were available and were taught to do complicated problems involving products and quotients of several seemingly arbitrary, ugly numbers using printed "log tables" bound in a little booklet. As pictured in Figure MA.1, slide rules are a much more elegant embodiment of the same principles.

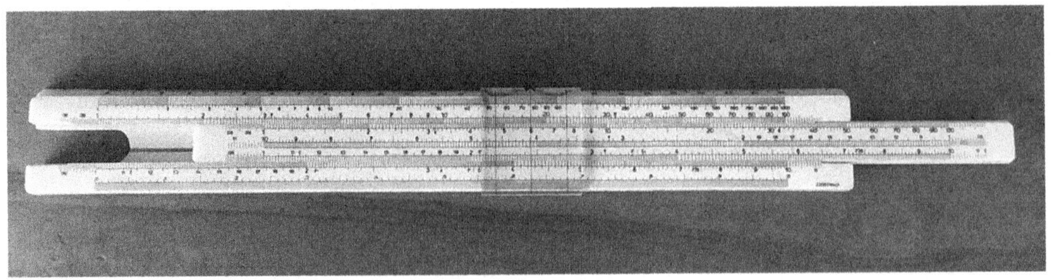

FIGURE MA.1 A slide rule configured to multiply numbers by 3.

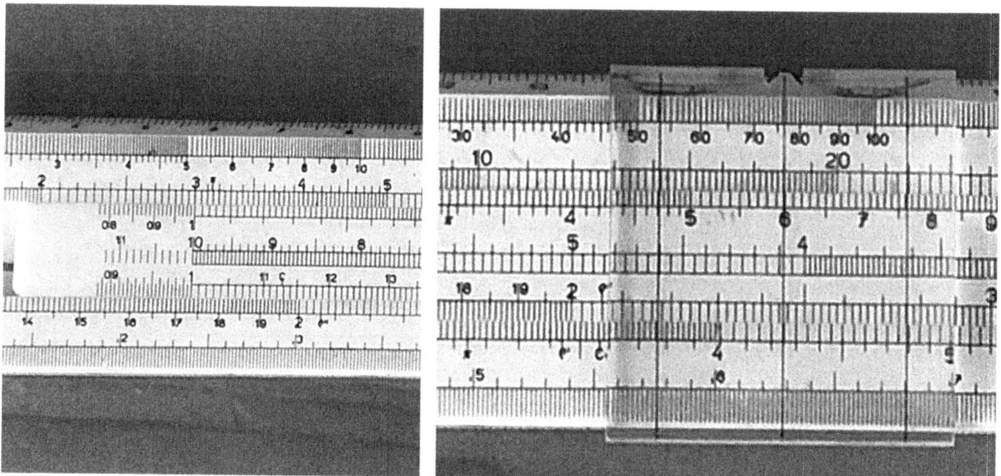

FIGURE MA.2 Detail showing (L) how 3 is set as multiplier, (R) cursor showing $3 \times 6 = 18$.

At its heart, the slide rule has two logarithmic scales (notice the gradations get narrower the further you look to the right) accurately engraved on adjacent sections, which slide freely and smoothly against each other. To multiply two numbers, slide the middle section until the 1 on the scale sits exactly beneath the first number to be multiplied (3, as pictured) on the top scale, shown in Figure MA.2(L). Next, move the transparent cursor until it's centred over the second number (6) on the sliding scale (Figure MA.2(R)).

The cursor then indicates the product on the upper scale. If you didn't fancy using log tables, examination boards permitted slide rules in exams — provided you wrote "SR" in the margin each time!

Logarithms as a method for doing arithmetic were originally proposed by John Napier in 1614; but Napier's logarithms used a different base. The Naperian[2] or *natural* logarithm is based on the irrational[3] constant $e = 2.71828182846\ldots$ and is usually distinguished by writing $\ln x = \log_e x$. What's so "natural" about this definition? It will become clear soon; for now, note one interesting route to the numerical value for e:[4]

$$e = \lim_{n \to \infty} \left(\frac{n}{n-1} \right)^n . \tag{MA.8}$$

Try it on a calculator!

Next, let's briefly survey techniques to quantify how variables change. Suppose a variable y depends on another variable x. We call y a *function* of x, written formally as $y(x)$. If x changes its value in a smooth, continuous fashion, how does y respond? The most straightforward possibility, sketched in Figure MA.3, is called a *linear response*, i.e., if we graph y against x the result is a straight line.

The vertical dashed line signifies the change Δy resulting from a change Δx. The ratio $\Delta y/\Delta x$, known as the *gradient* of the line, is a measure of how rapidly y is changing as x changes. For this linear case, the evolution is uniform, and $\Delta y/\Delta x$ is constant along the line. A simple example is a moving object: if y represents its position along a particular direction, measured in metres and x the time measured in seconds, then $\Delta y/\Delta x$ is the object's speed measured in units ms^{-1}. Note that $\Delta y/\Delta x$ can also be negative; in this case, the object would be moving backwards, and the line would slope the other way.

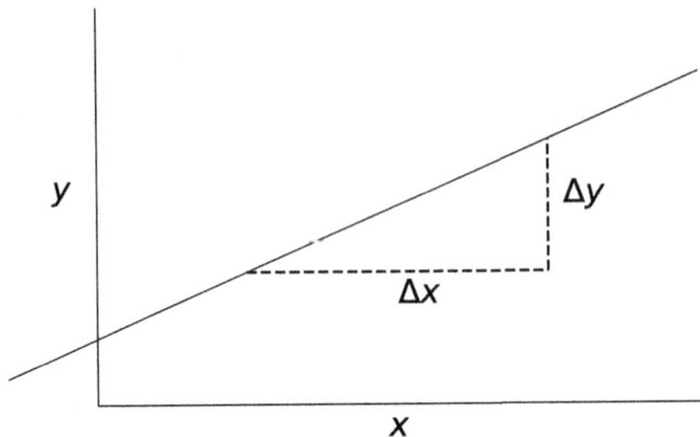

FIGURE MA.3 Linear variation of y as a function of x.

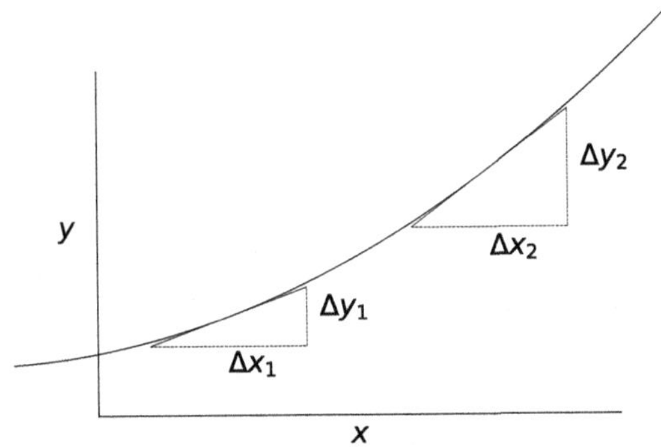

FIGURE MA.4 Variation of a general y as a function of x: note $\Delta y_2/\Delta x_2 \neq \Delta y_1/\Delta x_1$.

In general, things are more complicated because y may be a *non-linear* function of x, represented by a curved line as in Figure MA.4.

In this case, we have to generalize the definition of the curve's gradient to be that of a line just kissing the curve at a single point of the curve: such a line is called the *tangent*. Figure MA.4 shows that now the gradient takes different values depending on where we are along the curve, i.e., it depends on x. The mathematical technique for calculating the gradient is called *differential calculus*,[5] and since it now varies from point to point, the notation used is dy/dx. An alternative name for dy/dx is the *derivative*. There are several helpful rules for calculating dy/dx given a function $y(x)$; usually, the first to be taught is

$$y = x^p \Rightarrow \frac{dy}{dx} = px^{p-1}. \tag{MA.9}$$

Sometimes, when a function $f(x, y)$ depends on more than one variable, we specify which kind of variation is being examined using a slightly different notation $\partial f/\partial x$, $\partial f/\partial y$, etc.; such expressions are known as *partial derivatives*.

Now we can reveal what's special about using e as a base in the definition of natural logarithms. If $x = \ln y$, then $y = e^x$. Then

$$\frac{dy}{dx} = \frac{d}{dx}e^x = e^x; \tag{MA.10}$$

In other words, the curve $y = e^x$ has the property that at every point, the gradient has the same value as the curve itself. It turns out this is enormously useful in physics applications. What about the natural logarithm? If $y = \ln x$ then

$$\frac{dy}{dx} = \frac{d}{dx}\ln x = \frac{1}{x}. \tag{MA.11}$$

It's helpful to generalize results (MA.9– MA.11) to include a constant multiplier a:

$$\frac{d}{dx}ax^p = apx^{p-1}; \tag{MA.12a}$$

$$\frac{d}{dx}e^{ax} = ae^{ax}; \tag{MA.12b}$$

$$\frac{d}{dx}\ln ax = \frac{1}{x}. \tag{MA.12c}$$

Sometimes, the power to which e is raised can be quite a complicated expression, in which case the alternative notation $e^x = \exp(x)$ comes in handy. We describe this function as the *exponential* of x, which is the inverse function of the natural logarithm, viz. $\exp(\ln x) = \ln(e^x) = x$. Note that we also regard the logarithm as a function with its *argument* put between brackets; both forms are used in this book.

Another important application of calculus is to evaluate the area lying beneath a curve $y(x)$, i.e., between the curve and the x-axis. In Figure MA.5, this area A is approximated by a sequence of tall, thin rectangles whose height is adjusted to coincide with $y(x)$ at their midpoints. Imagine a vertical line sweeping from left to right across the graph uniformly. At any given point, the rate at which the area is being swept out is approximately given by the height of the rectangle "under the cursor". In the limit that the rectangles are allowed to become infinitesimally thin and simultaneously infinite in number, we obtain

$$\frac{dA}{dx} = y(x). \tag{MA.13}$$

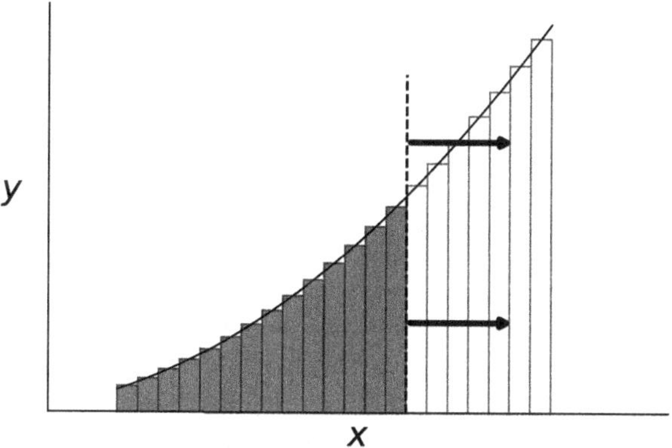

FIGURE MA.5 Area under the curve approximated by a sequence of rectangular elements.

To evaluate A, we therefore need the inverse of the derivative. This is sometimes called the antiderivative, but more usually the *integral*, written as[6]

$$A = \int y dx. \tag{MA.14}$$

Here are some useful integrals:

$$\int ax^p dx = \frac{a}{p+1} x^{p+1};\ p \neq -1 \tag{MA.15a}$$

$$\int \frac{a}{x} = a\ln x; \tag{MA.15b}$$

$$\int e^{ax} dx = \frac{1}{a} e^{ax}; \tag{MA.15c}$$

$$\int \ln ax dx = \frac{1}{x} + x\ln a. \tag{MA.15d}$$

When we want to evaluate the area under the curve lying between two positions a, b on the x-axis, the procedure is to subtract the value of the integral $\int y(x)dx$ evaluated at $x = a$ from its value evaluated at $x = b$. This is known as a *definite integral* between limits a and b, and is notated, e.g.,

$$\int_a^b x^2 dx = \left.\frac{x^3}{3}\right|_a^b \equiv \frac{b^3}{3} - \frac{a^3}{3}. \tag{MA.16}$$

Sometimes, it is useful to have approximations which hold when a variable x is very small, in the sense that $x \ll 1$. Here are some useful examples:

$$(1+x)^p = 1 + px + \frac{p(p-1)}{2} x^2 + \frac{p(p-1)(p-2)}{6} x^3 + \cdots \approx 1 + px; \tag{MA.17a}$$

$$e^x = 1 + x + \frac{x^2}{2} + \frac{x^3}{6} + \frac{x^4}{24} + \cdots \approx 1 + x; \tag{MA.17b}$$

$$\ln(1+x) = x - \frac{x^2}{2} + \frac{x^3}{3} - \cdots \approx x \tag{MA.17c}$$

In each case, the function can be expressed as a *power series* expansion; in principle, the series is indefinite (i.e., goes on forever),[7] but successive terms diminish in magnitude since $x \ll 1 \Rightarrow x^2, x^3, \ldots, x^n, \ldots \ll x$.

An important point is that when a variable x is associated with a physical quantity, as in the example of speed in Figure MA.3, then it is endowed with units of measurement (e.g., ms^{-1}) which need to be kept track of. When multiplying variables in mathematical expressions together, we also need to multiply the units. For example, if x is a length measured in m, then x^2 is an area measured in m^2 and x^3 a volume measured in m^3. Another rule is that it's only possible to add or subtract variables measured *in the same units.* The power series expressions in equation (MA.17) add together different powers of x, which is only possible if x is a *dimensionless* variable, i.e., a pure number not requiring units. *In particular, we can only exponentiate or take logarithms of dimensionless expressions.* This surprisingly under-appreciated principle is very helpful when checking calculations!

We close with a couple of notational shortcuts, which are useful when working with a lot of numerical data. Suppose we have a collection of N particles distributed along a line, each of which is a distance x_i from a fixed point on the line. Here the i subscript is known as an index and takes numerical values $i = 1,2,3,\ldots,N$. The average position $\overline{x}$ of the particles is given by

$$\overline{x} = \frac{1}{N}\left(x_1 + x_2 + x_3 + \cdots + x_N\right). \tag{MA.18}$$

A much slicker way to write this is

$$\overline{x} = \frac{1}{N}\sum_{i=1}^{N} x_i = \frac{1}{N}\sum_{i} x_i \tag{MA.19}$$

which employs the summation symbol Σ, the upper-case Greek letter sigma, standing for sum. The values over which the index ranges can be shown explicitly, as in the expression immediately following the first = sign, or are understood based on context, as in the second example. So, for instance, it's possible to reexpress the last relation of (MA.17) more concisely as

$$\ln\left(1+x\right) = \sum_{n=1}^{\infty}(-1)^{n-1}\frac{x^n}{n}. \tag{MA.20}$$

Similarly, it's possible to express a product over many similar-looking factors via

$$y_1 y_2 y_3 \cdots y_N = \prod_{i=1}^{N} y_i \tag{MA.21}$$

where the product symbol $\prod$ is the upper-case Greek letter pi.

NOTES

1 Common logarithms are frequently employed as the basis of measurement scales in science: decibels (acoustics); the Richter scale for earthquakes (geophysics); pH for acidity/alkalinity (chemistry); and magnitude for star brightness (astronomy).

2 Strictly, Napier's logarithms used base $1/e$.

3 This means it can't be expressed as the ratio of two integers m,n, i.e., $e \neq \frac{m}{n}$. Indeed, e is *transcendental*, meaning it is not the solution of a polynomial equation with integer-valued coefficients.

4 e is known as Euler's number after Leonhard Euler (1707–1783).

5 Independently formulated by Isaac Newton (1642–1727) and Gottfried Wilhelm Leibniz (1646–1716).

6 Strictly in (MA.14) A is defined only up to the addition of an overall constant since curves can be shifted up or down by an arbitrary amount without changing dA/dx.

7 The series expansion for $(1+x)^p$ eventually terminates if p is integer.

Appendix PA1 Physics Appendix

PA1.1 MULTIPARTICLE PARTITION FUNCTION

The Boltzmann distribution gives us the probability of finding a system with energy ε_i from a collection of possible energy states$\{\varepsilon_j\}$:

$$p(\varepsilon_i) = \frac{1}{Z} e^{-\varepsilon_i / k_B T} \tag{PA 1.1}$$

Here Z is the partition function:

$$Z = \sum_{i=1}^{r} e^{-\varepsilon_i / k_B T} \tag{PA 1.2}$$

Now, if our system is one particle, then this is the single particle partition function.

However, with a collection of N particles making up the system of interest, the situation needs a little more careful thought.

If the particles are freely able to occupy the various energy levels without hindrance and without getting in each other's way (i.e., they are non-interacting), then we can specify the overall state of the system by listing the state of each particle:

$$\lambda = (\mathbb{S}_1, \mathbb{S}_2, \ldots, \mathbb{S}_N) \tag{PA 1.3}$$

Here λ is indexing the state of the system, not the energy level (which is i) and this notion shows that particle 1 is in state $\mathbb{S}_1$, particle 2 in state $\mathbb{S}_2$, … and particle N in state $\mathbb{S}_N$. We will use j as our index for the particles, so that particle j is in state $\mathbb{S}_j$.

The total energy of the system is then:

$$E_\lambda = \varepsilon_{\mathbb{S}_1} + \varepsilon_{\mathbb{S}_2} + \ldots \varepsilon_{\mathbb{S}_N} \tag{PA 1.4}$$

Using $\varepsilon_{\mathbb{S}_1}$ to indicate the energy of particle 1 which happens to be in state $\mathbb{S}_1$, etc.

The probability of finding the system with this energy flows from the Boltzmann distribution, as before, PA 1.1

$$p(E_\lambda) = \frac{1}{\mathcal{Z}} e^{-E_\lambda / k_B T} \qquad \mathcal{Z} = \sum_\lambda e^{-E_\lambda / k_B T}$$

We need to break this sum down from being indexed over the states of the *system* to being indexed over the states of the *individual particles*:

$$\mathcal{Z} = \sum_i e^{-E_\lambda/k_BT} = \sum_{\mathbb{S}_1,\mathbb{S}_2,\ldots,\mathbb{S}_N} \exp\left(-\frac{1}{k_BT}\left(\varepsilon_{\mathbb{S}_1} + \varepsilon_{\mathbb{S}_2} + \ldots \varepsilon_{\mathbb{S}_N}\right)\right) \tag{PA 1.5}$$

If we assume that the particles are distinguishable as well as non-interacting, then summing over the combination $\mathbb{S}_1, \mathbb{S}_2, \ldots, \mathbb{S}_N$ is the same as summing over each one independently. In other words:

$$\mathcal{Z} = \sum_{\mathbb{S}_1}\sum_{\mathbb{S}_2}\cdots\sum_{\mathbb{S}_N} \exp\left(-\frac{1}{k_BT}\left(\varepsilon_{\mathbb{S}_1} + \varepsilon_{\mathbb{S}_2} + \ldots \varepsilon_{\mathbb{S}_N}\right)\right)$$

$$= \sum_{\mathbb{S}_1}\sum_{\mathbb{S}_2}\cdots\sum_{\mathbb{S}_N} \exp\left(-\frac{\varepsilon_{\mathbb{S}_1}}{k_BT}\right)\exp\left(-\frac{\varepsilon_{\mathbb{S}_2}}{k_BT}\right)\ldots\exp\left(-\frac{\varepsilon_{\mathbb{S}_N}}{k_BT}\right)$$

$$= \left(\sum_{\mathbb{S}_1}\exp\left(-\frac{\varepsilon_{\mathbb{S}_1}}{k_BT}\right)\right)\left(\sum_{\mathbb{S}_2}\exp\left(-\frac{\varepsilon_{\mathbb{S}_2}}{k_BT}\right)\right)\cdots\left(\sum_{\mathbb{S}_N}\exp\left(-\frac{\varepsilon_{\mathbb{S}_N}}{k_BT}\right)\right) \tag{PA 1.6}$$

Now, when we are summing over the various $\mathbb{S}_j$, we are summing over the energy level that particle j happens to find itself in. This is the same as simply summing over i. Hence, our partition function is:

$$\mathcal{Z} = \prod_{j=1}^{N}\left(\sum_{i=1}^{r}\exp\left(-\frac{\varepsilon_i}{k_BT}\right)\right) \tag{PA 1.7}$$

However, the term in ( ) is the single particle partition function from earlier. So, $\mathcal{Z} = Z^N$, and hence:

$$p(E_\lambda) = \frac{1}{Z^N} e^{-E_\lambda/k_BT} \tag{PA 1.8}$$

PA1.2 VARIATIONS ON A THEME

There is another way of looking at our multi-particle partition function, $\mathcal{Z}$. After all, it is very likely that multiple particles will end up in the same energy level. For example, we might find both particle 3 and particle 7 in the same state, so that $\mathbb{S}_3 = \mathbb{S}_7$.

If we sweep up all the particles that are in state $i = 1$, and find n_1 of them, and then continue in the same fashion for $i = 2 \to r$, then $\mathcal{Z}$ can also be written as:

$$\mathcal{Z} = \sum_{\mathbb{S}_1,\mathbb{S}_2,\ldots,\mathbb{S}_N} \exp\left(-\frac{1}{k_BT}\left(\varepsilon_{\mathbb{S}_1} + \varepsilon_{\mathbb{S}_2} + \ldots \varepsilon_{\mathbb{S}_N}\right)\right)$$

$$= \sum_{n_1,n_2,\ldots n_N} \exp\left(-\frac{1}{k_BT}\left(n_1\varepsilon_1 + n_2\varepsilon_2 + \ldots n_N\varepsilon_N\right)\right) \tag{PA 1.9}$$

We are now summing over all the various occupations, n_i, within each level, being careful to respect the constraint:

$$\sum_{i=1}^{r} n_i = N \tag{PA 1.10}$$

This would be a very tricky sum to do in practice, but we don't need to carry it out as we already know that the answer must be Z^N. However, this form of the partition function is going to be very useful in the next calculation.

PA1.3 AVERAGE STATE OCCUPATION

It is very reasonable to enquire after the number of particles that we would expect to find in any one of the states. This is tantamount to calculating the average number of particles in that state across all possible permutations, which is done using the probability distribution. So:

$$\langle n_I \rangle = \frac{1}{\mathcal{Z}} \sum_{n_1, n_2, \ldots n_N} n_I \exp\left(-\frac{1}{k_B T}\left(n_1 \varepsilon_1 + n_2 \varepsilon_2 + \ldots n_N \varepsilon_N\right)\right) \tag{PA 1.11}$$

Here we have picked the specific state I from $\{i\}$ to be the focus of our interest.

Intriguingly, we note that the sum we see here can be obtained from the partition function in its latest form, by differentiating with respect to ε_I:

$$\sum_{n_1, n_2, \ldots n_N} n_I \exp\left(-\frac{1}{k_B T}\left(n_1 \varepsilon_1 + n_2 \varepsilon_2 + \ldots n_N \varepsilon_N\right)\right)$$

$$= -k_B T \frac{\partial}{\partial \varepsilon_I}\left(\sum_{n_1, n_2, \ldots n_N} \exp\left(-\frac{1}{k_B T}\left(n_1 \varepsilon_1 + n_2 \varepsilon_2 + \ldots n_N \varepsilon_N\right)\right)\right)$$

$$= -k_B T \frac{\partial \mathcal{Z}}{\partial \varepsilon_I} \tag{PA 1.12}$$

Using this result tidies things somewhat:

$$\langle n_I \rangle = -\frac{1}{\mathcal{Z}} k_B T \frac{\partial \mathcal{Z}}{\partial \varepsilon_I} \tag{PA 1.13}$$

Now we deploy $\mathcal{Z} = Z^N$:

$$\langle n_I \rangle = -\frac{1}{Z^N} k_B T \frac{\partial Z^N}{\partial \varepsilon_I}$$

$$= -\frac{1}{Z^N} k_B T N Z^{N-1} \frac{\partial Z}{\partial \varepsilon_I}$$

$$= -\frac{k_B TN}{Z}\frac{\partial Z}{\partial \varepsilon_I}$$

$$= -\frac{1}{Z}k_B TN\frac{\partial}{\partial \varepsilon_I}\left(\sum_{i=1}^{r} e^{-\varepsilon_i/k_B T}\right)$$

$$\langle n_I \rangle = \frac{1}{Z}Ne^{-\varepsilon_I/k_B T} \qquad \text{(PA 1.14)}$$

Admittedly, this is exactly the result that we might have expected from a naïve approach using $p_I \sim n_I / N$. The difference here is that we have calculated the average n_I, which approaches the value $n_I = p_I N$ in the limit large N and clarified any issues that may have arisen in the jump from a single particle to a multi-particle approach.

This argument can also be generalized to cater for degenerate energy levels without altering the result.

PA1.4 INDISTINGUISHABLE MULTI-PARTICLE PARTITION FUNCTION

One limitation of the calculation as carried out is the assumption that the particles were distinguishable. Even without the extra considerations coming from boson or fermion statistics, quantum particles of the same species are fundamentally indistinguishable. This means that when we transition from PA 1.9:

$$\mathcal{Z} = \sum_{\mathbb{S}_1,\mathbb{S}_2,\ldots,\mathbb{S}_N} \exp\left(-\frac{1}{k_B T}\left(\varepsilon_{\mathbb{S}_1} + \varepsilon_{\mathbb{S}_2} + \cdots \varepsilon_{\mathbb{S}_N}\right)\right)$$

to:

$$\mathcal{Z} = Z^N$$

we are being too hasty. The crucial step which is at risk is the point at which we factorized the sum into a product of identical sums:

$$\mathcal{Z} = \left(\sum_{\mathbb{S}_1}\exp\left(-\frac{\varepsilon_{\mathbb{S}_1}}{k_B T}\right)\right)\left(\sum_{\mathbb{S}_2}\exp\left(-\frac{\varepsilon_{\mathbb{S}_2}}{k_B T}\right)\right)\cdots\left(\sum_{\mathbb{S}_N}\exp\left(-\frac{\varepsilon_{\mathbb{S}_N}}{k_B T}\right)\right) \qquad \text{(PA 1.15)}$$

which will result in us over-counting the possibilities.

To see this more clearly, consider the case of just two particles. The partition function is then:

$$\mathcal{Z} = Z^2 = \left(\sum_{\mathbb{S}_1} e^{-\varepsilon_{\mathbb{S}_1}/k_B T}\right)\left(\sum_{\mathbb{S}_2} e^{-\varepsilon_{\mathbb{S}_2}/k_B T}\right)$$

$$= \sum_{\mathbb{S}_1,\mathbb{S}_2} e^{-\left(\varepsilon_{\mathbb{S}_1}+\varepsilon_{\mathbb{S}_2}\right)/k_B T}$$

$$= \sum_{\mathbb{S}_j} e^{-2\varepsilon_{\mathbb{S}_j}/k_B T} + \sum_{\mathbb{S}_1}\sum_{\mathbb{S}_2,\mathbb{S}_2\neq\mathbb{S}_1} e^{-\left(\varepsilon_{\mathbb{S}_1}+\varepsilon_{\mathbb{S}_2}\right)/k_B T} \qquad \text{(PA 1.16)}$$

In the final line, the sum has been split over two terms. The first deals with the parts of the overall sum where the particles are in the same state, $\mathbb{S}_1 = \mathbb{S}_2 = \mathbb{S}_j$. The second term runs over all values of $\mathbb{S}_1, \mathbb{S}_2$, where they are not equal, i.e., $\mathbb{S}_1 \neq \mathbb{S}_2$ and the particles are in different states. It is this latter term that needs a little more work. As the sum runs over all possible (different) values of $\mathbb{S}_1, \mathbb{S}_2$, it overcounts the number of cases when the particles are identical. With identical particles, the set $\{\mathbb{S}_1 = 3, \mathbb{S}_2 = 4\}$ can't be distinguished from $\{\mathbb{S}_1 = 4, \mathbb{S}_2 = 3\}$, but the summation will contain both. In this case, the overcounting is easily dealt with—we just need an extra factor of $1/2$ in front of the summation:

$$\mathcal{Z} = \sum_{\mathbb{S}_j} e^{-2\varepsilon_{\mathbb{S}_j}/k_BT} + \frac{1}{2}\sum_{\mathbb{S}_1}\sum_{\mathbb{S}_2,\mathbb{S}_2\neq\mathbb{S}_1} e^{-\left(\varepsilon_{\mathbb{S}_1}+\varepsilon_{\mathbb{S}_2}\right)/k_BT} \quad \text{(PA 1.17)}$$

This factor of $1/2$ effectively comes from the standard way in which we deal with over counting, referenced in Section 7.6.1, which is to divide by $N!$, which in this case is $2! = 2$.

Things get to be considerably trickier as we move to the N particle case. Our sum now has multiple terms along the following lines:

$$\mathcal{Z} = \sum_{\mathbb{S}_j} e^{-N\varepsilon_{\mathbb{S}_j}/k_BT} + \left\{\frac{1}{2!}\sum_{\mathbb{S}_1=\mathbb{S}_2}\cdots\sum_{\mathbb{S}_j}\cdots\sum_{\mathbb{S}_N} e^{-\left(2\varepsilon_{\mathbb{S}_1}+\cdots+\varepsilon_{\mathbb{S}_N}\right)/k_BT}\right\}_{\text{first two same}}$$

$$\ldots+\{\ \}+\ldots+\left[\frac{1}{N!}\sum_{\mathbb{S}_1}\cdots\sum_{\mathbb{S}_N} e^{-\left(2\varepsilon_{\mathbb{S}_1}+\cdots+\varepsilon_{\mathbb{S}_N}\right)/k_BT}\right]_{\text{all different}} \quad \text{(PA 1.18)}$$

The first term is a sum over states given that all the particles are in the same state. The second term, cast in $\{\ \}$, is a summation where the first two particles are in the same state, $\mathbb{S}_1 = \mathbb{S}_2$, but the others are in different states. This has a factor of $1/2!$ To avoid the inherent over-counting. Explicitly omitted from this expression but implicit within the $\ldots\{\ \}\ldots$ are all other sums where two particles from the set are in the same state. However, this is not all. We must also account for summations where three particles (pick any three you want from the N) are in the same state, four in the same state, etc. Each of these summations must have the necessary factor to remove overcounting. You can see that it's a bit of a mess.

The final term which is shown has all the particles in different states, with an overcounting factor of $N!$.

In the classical regime, where the quantum numbers are astronomically high, the number of available states is much greater than the number of particles. Consequently, the probability that a state will have more than one particle resident becomes vanishingly small. In other words, to a very good approximation, we can discount all the terms in the partition function aside from the last one, where all the particles are in different states.

$$\mathcal{Z} \cong \left[\frac{1}{N!}\sum_{\mathbb{S}_1}\cdots\sum_{\mathbb{S}_N} e^{-\left(2\varepsilon_{\mathbb{S}_1}+\ldots+\varepsilon_{\mathbb{S}_N}\right)/k_BT}\right]_{\text{all different}}$$

$$\mathcal{Z} \cong \frac{1}{N!}Z^N \quad \text{(PA 1.19)}$$

which is the typically quoted result.

PA1.5 FERMION AND BOSON ENTROPY CALCULATIONS

In Section 9.5.1, we got to the fermion entropy in the form (9.110):

$$\mathcal{S}_{GF} = k_B \sum_j \left\{ \ln\left(1 + e^{(\mu-\varepsilon_j)/k_BT}\right) + \left(\frac{1}{k_BT}\right)\frac{(\varepsilon_j - \mu)}{e^{(\varepsilon_j-\mu)/k_BT}+1} \right\}$$

At the time, we moved on to another form of this expression without going into mathematical detail regarding the various manoeuvres needed. Here, we explore that calculation.

A convenient starting point to move on is the second term in the {} brackets. If we go back to the average occupation number for fermions (9.52):

$$\langle \mathrm{N}_j \rangle = \frac{1}{1+e^{(\varepsilon_j-\mu)/k_BT}} = \frac{e^{(\mu-\varepsilon_j)/k_BT}}{e^{(\mu-\varepsilon_j)/k_BT}+1}$$

we can calculate, apparently on a whim....

$$1-\langle \mathrm{N}_j \rangle = 1 - \frac{e^{(\mu-\varepsilon_j)/k_BT}}{e^{(\mu-\varepsilon_j)/k_BT}+1} = \frac{e^{(\mu-\varepsilon_j)/k_BT}+1-e^{(\varepsilon_j-\mu)/k_BT}}{e^{(\mu-\varepsilon_j)/k_BT}+1} = \frac{1}{e^{(\mu-\varepsilon_j)/k_BT}+1}$$

$$= \frac{e^{(\varepsilon_j-\mu)/k_BT}}{e^{(\varepsilon_j-\mu)/k_BT}+1} \tag{PA 1.21}$$

Now our cunning becomes apparent, since if we take logs:

$$\ln\left(1-\langle \mathrm{N}_j \rangle\right) = \left(\frac{\varepsilon_j - \mu}{k_BT}\right) - \ln\left(e^{(\varepsilon_j-\mu)/k_BT}+1\right) \tag{PA 1.22}$$

we obtain:

$$\left(\frac{\varepsilon_j - \mu}{k_BT}\right) = \ln\left(1-\langle \mathrm{N}_j \rangle\right) + \ln\left(e^{(\varepsilon_j-\mu)/k_BT}+1\right) \tag{PA 1.23}$$

which we can paste into the entropy formula:

$$\mathcal{S}_{GF} = k_B \sum_j \left\{ \ln\left(1 + e^{(\mu-\varepsilon_j)/k_BT}\right) + \left(\frac{1}{k_BT}\right)\frac{(\varepsilon_j - \mu)}{e^{(\varepsilon_j-\mu)/k_BT}+1} \right\}$$

$$= k_B \sum_j \left\{ -\ln\left(1-\langle \mathrm{N}_j \rangle\right) + \frac{\ln\left(1-\langle \mathrm{N}_j \rangle\right) + \ln\left(e^{(\varepsilon_j-\mu)/k_BT}+1\right)}{e^{(\varepsilon_j-\mu)/k_BT}+1} \right\}$$

$$= k_B \sum_j \left\{ -\ln\left(1-\langle \mathrm{N}_j \rangle\right) + \langle \mathrm{N}_j \rangle \ln\left(1-\langle \mathrm{N}_j \rangle\right) + \langle \mathrm{N}_j \rangle \ln\left(e^{(\varepsilon_j-\mu)/k_BT}+1\right) \right\} \tag{PA 1.24}$$

This tidies to:

$$\mathcal{S}_{GF} = -k_B \sum_j \left\{ \left(1 - \langle \mathrm{N}_j \rangle\right) \ln\left(1 - \langle \mathrm{N}_j \rangle\right) + \langle \mathrm{N}_j \rangle \ln\left(\langle \mathrm{N}_j \rangle\right) \right\} \qquad \text{(PA 1.25)}$$

which is the result quoted in 9.111.

Boson Entropy
Starting from 9.116:

$$\mathcal{S}_{GB} = -k_B \sum_j \left\{ \ln\left(1 - e^{(\mu-\varepsilon_j)/k_BT}\right) + \left(\frac{1}{k_BT}\right) \frac{(\varepsilon_j - \mu)}{e^{(\varepsilon_j-\mu)/k_BT} - 1} \right\}$$

we carry out a shell-game, as we did for the fermions, noting that for bosons (9.59):

$$\langle \mathrm{N}_j \rangle = \frac{1}{e^{(\varepsilon_j-\mu)/k_BT} - 1} = \frac{e^{(\mu-\varepsilon_j)/k_BT}}{1 - e^{(\mu-\varepsilon_j)/k_BT}}$$

so that:

$$\ln\left(\langle \mathrm{N}_j \rangle\right) = \ln\left(e^{(\mu-\varepsilon_j)/k_BT}\right) - \ln\left(1 - e^{(\mu-\varepsilon_j)/k_BT}\right) \qquad \text{(PA 1.27)}$$

or:

$$\frac{\mu - \varepsilon_j}{k_BT} = \ln\left(\langle \mathrm{N}_j \rangle\right) + \ln\left(1 - e^{(\mu-\varepsilon_j)/k_BT}\right) \qquad \text{(PA 1.28)}$$

Inserting this into the entropy:

$$\mathcal{S}_{GB} = -k_B \sum_j \left\{ \ln\left(1 - e^{(\mu-\varepsilon_j)/k_BT}\right) + \frac{\ln\left(\langle \mathrm{N}_j \rangle\right) + \ln\left(1 - e^{(\mu-\varepsilon_j)/k_BT}\right)}{e^{(\varepsilon_j-\mu)/k_BT} - 1} \right\} \qquad \text{(PA 1.29)}$$

Flipping the exponents in the second term:

$$= -k_B \sum_j \left\{ \ln\left(1 - e^{(\mu-\varepsilon_j)/k_BT}\right) + \frac{e^{(\mu-\varepsilon_j)/k_BT} \ln\left(\langle \mathrm{N}_j \rangle\right) + e^{(\mu-\varepsilon_j)/k_BT} \ln\left(1 - e^{(\mu-\varepsilon_j)/k_BT}\right)}{1 - e^{(\mu-\varepsilon_j)/k_BT}} \right\}$$

$$= -k_B \sum_j \left\{ \ln\left(1 - e^{(\mu-\varepsilon_j)/k_BT}\right) + \langle \mathrm{N}_j \rangle \ln\left(\langle \mathrm{N}_j \rangle\right) + \frac{e^{(\mu-\varepsilon_j)/k_BT} \ln\left(1 - e^{(\mu-\varepsilon_j)/k_BT}\right)}{1 - e^{(\mu-\varepsilon_j)/k_BT}} \right\} \qquad \text{(PA 1.30)}$$

and then gathering terms:

$$= -k_B \sum_j \left\{ \frac{\left(1-e^{(\mu-\varepsilon_j)/k_BT}\right)\ln\left(1-e^{(\mu-\varepsilon_j)/k_BT}\right)+e^{(\mu-\varepsilon_j)/k_BT}\ln\left(1-e^{(\mu-\varepsilon_j)/k_BT}\right)}{1-e^{(\mu-\varepsilon_j)/k_BT}} + \langle \mathbb{N}_j \rangle \ln\left(\langle \mathbb{N}_j \rangle\right) \right\}$$

$$\mathcal{S}_{GB} = -k_B \sum_j \left\{ \frac{\ln\left(1-e^{(\mu-\varepsilon_j)/k_BT}\right)}{1-e^{(\mu-\varepsilon_j)/k_BT}} + \langle \mathbb{N}_j \rangle \ln\left(\langle \mathbb{N}_j \rangle\right) \right\} \quad \text{(PA 1.31)}$$

As:

$$1+\langle \mathbb{N}_j \rangle = 1+\frac{e^{(\mu-\varepsilon_j)/k_BT}}{1-e^{(\mu-\varepsilon_j)/k_BT}} = \frac{1-e^{(\mu-\varepsilon_j)/k_BT}+e^{(\mu-\varepsilon_j)/k_BT}}{1-e^{(\mu-\varepsilon_j)/k_BT}} = \frac{1}{1-e^{(\mu-\varepsilon_j)/k_BT}} \quad \text{(PA 1.32)}$$

we finally get:

$$= -k_B \sum_j \left\{ \left(1+\langle \mathbb{N}_j \rangle\right)\ln\left(1-e^{(\mu-\varepsilon_j)/k_BT}\right) + \langle \mathbb{N}_j \rangle \ln\left(\langle \mathbb{N}_j \rangle\right) \right\} \quad \text{(PA 1.33)}$$

Hence, the boson entropy relationship is:

$$\mathcal{S}_{GB} = k_B \sum_j \left\{ \left(1+\langle \mathbb{N}_j \rangle\right)\ln\left(1+\langle \mathbb{N}_j \rangle\right) - \mathbb{N}_j \ln\left(\langle \mathbb{N}_j \rangle\right) \right\} \quad \text{(PA 1.34)}$$

or:

$$\mathcal{S}_{GB} = k_B \sum_j \left(1+\langle \mathbb{N}_j \rangle\right)\ln\left(1+\langle \mathbb{N}_j \rangle\right) - k_B \sum_j \langle \mathbb{N}_j \rangle \ln\left(\langle \mathbb{N}_j \rangle\right) \quad \text{(PA 1.35)}$$

Boson entropy and probability

We recall the relationship for bosons (9.82):

$$\langle \mathbb{N}_j \rangle = \frac{e^{(\mu-\varepsilon_j)/k_BT}}{1-e^{(\mu-\varepsilon_j)/k_BT}} = \frac{1-p_j(0)}{p_j(0)}$$

This allows us to construct:

$$1+\langle \mathbb{N}_j \rangle = 1+\frac{1-p_j(0)}{p_j(0)} = \frac{p_j(0)+1-p_j(0)}{p_j(0)} = \frac{1}{p_j(0)} \quad \text{(PA 1.36)}$$

which we can make use of to convert the previous expression:

$$\mathcal{S}_{GB} = k_B \sum_j \left\{ \left(1+\langle \mathbb{N}_j \rangle\right)\ln\left(1+\langle \mathbb{N}_j \rangle\right) - \mathbb{N}_j \ln\left(\langle \mathbb{N}_j \rangle\right) \right\}$$

$$= k_B \sum_j \left(\frac{1}{p_j(0)} \right) \left\{ (p_j(0) - 1) \ln \left(\frac{1 - p_j(0)}{p_j(0)} \right) - \ln(p_j(0)) \right\} \quad \text{(PA 1.37)}$$

This, after a couple of lines of work, becomes:

$$\mathcal{S}_{GB} = -k_B \sum_j \left(\frac{1}{p_j(0)} \right) \left\{ (1 - p_j(0)) \ln(1 - p_j(0)) + p_j(0) \ln(p_j(0)) \right\} \quad \text{(PA 1.38)}$$

As this is a slightly unusual relationship, it is worth confirming the result from a different direction. The raw Gibbs entropy is (5.46):

$$\mathcal{S}_G = -k_B \sum_{n=1}^{\infty} p_n \ln(p_n)$$

Where the n sums over the number of particles and the p_n is the probability of finding that many particles in the system with energy E_n. Things are different here, as our probability distributions apply within an orbital, hence we generalize the relationship to:

$$\mathcal{S}'_G = -k_B \sum_j \sum_{\mathbb{N}=0}^{\infty} p_j(\mathbb{N}) \ln(p_j(\mathbb{N})) \quad \text{(PA 1.39)}$$

Now we rummage back in our toolkit and extract (9.81):

$$p_j(\mathbb{N}) = (1 - p_j(0))^{\mathbb{N}} p_j(0)$$

for bosons. Using this in our entropy expression:

$$\mathcal{S}'_{GB} = -k_B \sum_j \sum_{\mathbb{N}=0}^{\infty} (1 - p_j(0))^{\mathbb{N}} p_j(0) \ln\left((1 - p_j(0))^{\mathbb{N}} p_j(0) \right)$$

$$= -k_B \sum_j \sum_{\mathbb{N}=0}^{\infty} (1 - p_j(0))^{\mathbb{N}} p_j(0) \left\{ \mathbb{N} \ln(1 - p_j(0)) + \ln(p_j(0)) \right\}$$

$$= -k_B \sum_j \sum_{\mathbb{N}=0}^{\infty} \left\{ \mathbb{N} p_j(0) (1 - p_j(0))^{\mathbb{N}} \ln(1 - p_j(0)) + p_j(0) (1 - p_j(0))^{\mathbb{N}} \ln(p_j(0)) \right\} \quad \text{(PA 1.40)}$$

Applying the results:

$$\sum_{n=0}^{\infty} n x (1 - x)^n = \frac{1 - x}{x} \qquad \sum_{n=0}^{\infty} x (1 - x)^n = 1 \quad \text{(PA 1.41)}$$

we can remove the sum over ℕ to get:

$$\mathcal{S}'_{GB} = -k_B \sum_j \left\{ \left(\frac{1-p_j(0)}{p_j(0)} \right) \ln\left(1-p_j(0)\right) + \ln\left(p_j(0)\right) \right\}$$

$$= -k_B \sum_j \left(\frac{1}{p_j(0)} \right) \left\{ \left(1-p_j(0)\right) \ln\left(1-p_j(0)\right) + p_j(0) \ln\left(p_j(0)\right) \right\} \quad \text{(PA 1.42)}$$

as we obtained before.

PA1.6 THE REDUCED DENSITY MATRIX

Starting from the density matrix in Section 11.5.4 (11.142):

$$D = \frac{1}{2} \begin{pmatrix} 0 & 0 & 0 & 0 \\ 0 & 1 & -1 & 0 \\ 0 & -1 & 1 & 0 \\ 0 & 0 & 0 & 0 \end{pmatrix}$$

which we calculated for the state:

$$D = |\Phi\rangle\langle\Phi| = \frac{1}{2}\left(|R\rangle|L\rangle - |L\rangle|R\rangle \right)\left(\langle R|\langle L| - \langle L|\langle R| \right) \quad \text{(PA 1.43)}$$

we wish to calculate the partial trace over photon 2, which is the second photon in the construction:

$$D = |\Phi\rangle\langle\Phi| = \frac{1}{2}\left(|R\rangle_1|L\rangle_2 - |L\rangle_1|R\rangle_2 \right)\left(\langle R|_1\langle L|_2 - \langle L|_1\langle R|_2 \right) \quad \text{(PA 1.44)}$$

The definition of the partial trace is (11.134):

$$\mathrm{Tr}_2[D] = \sum_k \langle k|D|k\rangle = \langle R|_2 D|R\rangle_2 \langle R|_2 D|R\rangle_2 + \langle L|_2 D|L\rangle_2$$

Or more properly, to ensure that we don't mess with photon 1:

$$\mathrm{Tr}_2[D] = \left(I_1 \otimes \langle R|_2\right) D \left(I_1 \otimes |R\rangle_2\right) + \left(I_1 \otimes \langle L|_2\right) D \left(I_1 \otimes |L\rangle_2\right) \quad \text{(PA 1.45)}$$

Now we set this up in matrix terms by using:

$$I_1 = \begin{pmatrix} 1 & 0 \\ 0 & 1 \end{pmatrix} \quad |R\rangle_2 = \begin{pmatrix} 1 \\ 0 \end{pmatrix} \quad \langle L_2| = \begin{pmatrix} 0 \\ 1 \end{pmatrix} \tag{PA 1.46}$$

The first step is to calculate the items we are going to need in the right-hand terms:

$$I_1 \otimes |R\rangle_2 = \begin{pmatrix} 1 & 0 \\ 0 & 1 \end{pmatrix} \otimes \begin{pmatrix} 1 \\ 0 \end{pmatrix} = \begin{pmatrix} 1 & 0 \\ 0 & 0 \\ 0 & 1 \\ 0 & 0 \end{pmatrix}$$

$$I_1 \otimes |L\rangle_2 = \begin{pmatrix} 1 & 0 \\ 0 & 1 \end{pmatrix} \otimes \begin{pmatrix} 0 \\ 1 \end{pmatrix} = \begin{pmatrix} 0 & 0 \\ 1 & 0 \\ 0 & 0 \\ 0 & 1 \end{pmatrix} \tag{PA 1.47}$$

And now the left-hand side:

$$I_1 \otimes \langle R|_2 = \begin{pmatrix} 1 & 0 \\ 0 & 1 \end{pmatrix} \otimes \begin{pmatrix} 1 & 0 \end{pmatrix} = \begin{pmatrix} 1 & 0 & 0 & 0 \\ 0 & 0 & 1 & 0 \end{pmatrix}$$

$$I_1 \otimes \langle L|_2 = \begin{pmatrix} 1 & 0 \\ 0 & 1 \end{pmatrix} \otimes \begin{pmatrix} 0 & 1 \end{pmatrix} = \begin{pmatrix} 0 & 1 & 0 & 0 \\ 0 & 0 & 0 & 1 \end{pmatrix} \tag{PA 1.48}$$

We use these to build the first parts of the two terms:

$$D\left(I_1 \otimes |R\rangle_2\right) = \frac{1}{2}\begin{pmatrix} 0 & 0 & 0 & 0 \\ 0 & 1 & -1 & 0 \\ 0 & -1 & 1 & 0 \\ 0 & 0 & 0 & 0 \end{pmatrix}\begin{pmatrix} 1 & 0 \\ 0 & 0 \\ 0 & 1 \\ 0 & 0 \end{pmatrix} = \frac{1}{2}\begin{pmatrix} 0 & 0 \\ 0 & -1 \\ 0 & 1 \\ 0 & 0 \end{pmatrix}$$

$$D\left(I_1 \otimes |L\rangle_2\right) = \frac{1}{2}\begin{pmatrix} 0 & 0 & 0 & 0 \\ 0 & 1 & -1 & 0 \\ 0 & -1 & 1 & 0 \\ 0 & 0 & 0 & 0 \end{pmatrix}\begin{pmatrix} 0 & 0 \\ 1 & 0 \\ 0 & 0 \\ 0 & 1 \end{pmatrix} = \frac{1}{2}\begin{pmatrix} 0 & 0 \\ 1 & 0 \\ -1 & 0 \\ 0 & 0 \end{pmatrix} \tag{PA 1.49}$$

And then we complete each term from the left:

$$\left(I_1 \otimes \langle R|_2\right) D \left(I_1 \otimes |R\rangle_2\right) = \begin{pmatrix} 1 & 0 & 0 & 0 \\ 0 & 0 & 1 & 0 \end{pmatrix} \frac{1}{2} \begin{pmatrix} 0 & 0 \\ 0 & -1 \\ 0 & 1 \\ 0 & 0 \end{pmatrix} = \frac{1}{2} \begin{pmatrix} 0 & 0 \\ 0 & 1 \end{pmatrix}$$

$$\left(I_1 \otimes \langle L|_2\right) D \left(I_1 \otimes |L\rangle_2\right) = \begin{pmatrix} 0 & 1 & 0 & 0 \\ 0 & 0 & 0 & 1 \end{pmatrix} \frac{1}{2} \begin{pmatrix} 0 & 0 \\ 1 & 0 \\ -1 & 0 \\ 0 & 0 \end{pmatrix} = \frac{1}{2} \begin{pmatrix} 1 & 0 \\ 0 & 0 \end{pmatrix} \qquad \text{(PA 1.50)}$$

So, finally the reduced trace density matrix is:

$$\mathrm{Tr}_2\left[D\right] = \frac{1}{2} \begin{pmatrix} 0 & 0 \\ 0 & 1 \end{pmatrix} + \frac{1}{2} \begin{pmatrix} 1 & 0 \\ 0 & 0 \end{pmatrix} = \begin{pmatrix} 1/2 & 0 \\ 0 & 1/2 \end{pmatrix} \qquad \text{(PA 1.51)}$$

which is exactly the form of the mixed state that Bob is expecting.

PA1.7 REDUCING MATRICES BY EYE

$$D = \begin{pmatrix} 0 & 0 & 0 & 0 \\ 0 & 1 & 1 & 0 \\ 0 & 1 & 1 & 0 \\ 0 & 0 & 0 & 0 \end{pmatrix}$$

$$\mathrm{Tr}_B\left[D\right] = \begin{pmatrix} |0\rangle\langle 0|_A & |0\rangle\langle 1|_A \\ |1\rangle\langle 0|_A & |1\rangle\langle 1|_A \end{pmatrix} = \begin{pmatrix} 1 & 0 \\ 0 & 1 \end{pmatrix}$$

$(\|0\rangle\langle 0\|)(\|0\rangle\langle 0\|)$ 0	$(\|0\rangle\langle 0\|)(\|0\rangle\langle 1\|)$ 0	$(\|0\rangle\langle 1\|)(\|0\rangle\langle 0\|)$ 0	$(\|0\rangle\langle 1\|)(\|0\rangle\langle 1\|)$ 0
$(\|0\rangle\langle 0\|)(\|1\rangle\langle 0\|)$ 0	$(\|0\rangle\langle 0\|)(\|1\rangle\langle 1\|)$ 1	$(\|0\rangle\langle 1\|)(\|1\rangle\langle 0\|)$ 1	$(\|0\rangle\langle 1\|)(\|1\rangle\langle 1\|)$ 0
$(\|1\rangle\langle 0\|)(\|0\rangle\langle 0\|)$ 0	$(\|1\rangle\langle 0\|)(\|0\rangle\langle 1\|)$ 1	$(\|1\rangle\langle 1\|)(\|0\rangle\langle 0\|)$ 1	$(\|1\rangle\langle 1\|)(\|0\rangle\langle 1\|)$ 0
$(\|1\rangle\langle 0\|)(\|1\rangle\langle 0\|)$ 0	$(\|1\rangle\langle 0\|)(\|1\rangle\langle 1\|)$ 0	$(\|1\rangle\langle 1\|)(\|1\rangle\langle 0\|)$ 0	$(\|1\rangle\langle 1\|)(\|1\rangle\langle 1\|)$ 0

$$|0\rangle\langle 0|_A = (|0\rangle\langle 0|)_A(|0\rangle\langle 0|)_B + (|0\rangle\langle 0|)_A(|1\rangle\langle 1|)_B = 1$$

$$|0\rangle\langle 1|_A = (|0\rangle\langle 1|)_A(|0\rangle\langle 0|)_B + (|0\rangle\langle 1|)_A(|1\rangle\langle 1|)_B = 0$$

$$|1\rangle\langle 0|_A = (|1\rangle\langle 0|)_A(|0\rangle\langle 0|)_B + (|1\rangle\langle 0|)_A(|0\rangle\langle 0|)_B = 0$$

$$|1\rangle\langle 1|_A = (|1\rangle\langle 1|)_A(|0\rangle\langle 0|)_B + (|1\rangle\langle 1|)_A(|1\rangle\langle 1|)_B = 1$$

$$\mathrm{Tr}_A[D] = \begin{pmatrix} |0\rangle\langle 0|_B & |0\rangle\langle 1|_B \\ |1\rangle\langle 0|_B & |1\rangle\langle 1|_B \end{pmatrix} = \begin{pmatrix} 1 & 0 \\ 0 & 1 \end{pmatrix}$$

$(\lvert 0\rangle\langle 0\rvert)(\lvert 0\rangle\langle 0\rvert)$	$(\lvert 0\rangle\langle 0\rvert)(\lvert 0\rangle\langle 1\rvert)$	$(\lvert 0\rangle\langle 1\rvert)(\lvert 0\rangle\langle 0\rvert)$	$(\lvert 0\rangle\langle 1\rvert)(\lvert 0\rangle\langle 1\rvert)$
0	0	0	0
$(\lvert 0\rangle\langle 0\rvert)(\lvert 1\rangle\langle 0\rvert)$	$(\lvert 0\rangle\langle 0\rvert)(\lvert 1\rangle\langle 1\rvert)$	$(\lvert 0\rangle\langle 1\rvert)(\lvert 1\rangle\langle 0\rvert)$	$(\lvert 0\rangle\langle 1\rvert)(\lvert 1\rangle\langle 1\rvert)$
0	1	1	0
$(\lvert 1\rangle\langle 0\rvert)(\lvert 0\rangle\langle 0\rvert)$	$(\lvert 1\rangle\langle 0\rvert)(\lvert 0\rangle\langle 1\rvert)$	$(\lvert 1\rangle\langle 1\rvert)(\lvert 0\rangle\langle 0\rvert)$	$(\lvert 1\rangle\langle 1\rvert)(\lvert 0\rangle\langle 1\rvert)$
0	1	1	0
$(\lvert 1\rangle\langle 0\rvert)(\lvert 1\rangle\langle 0\rvert)$	$(\lvert 1\rangle\langle 0\rvert)(\lvert 1\rangle\langle 1\rvert)$	$(\lvert 1\rangle\langle 1\rvert)(\lvert 1\rangle\langle 0\rvert)$	$(\lvert 1\rangle\langle 1\rvert)(\lvert 1\rangle\langle 1\rvert)$
0	0	0	0

$$|0\rangle\langle 0|_B = (|0\rangle\langle 0|)_A(|0\rangle\langle 0|)_B + (|1\rangle\langle 1|)_A(|0\rangle\langle 0|)_B = 1$$

$$|0\rangle\langle 1|_B = (|0\rangle\langle 0|)_A(|0\rangle\langle 1|)_B + (|1\rangle\langle 1|)_A(|0\rangle\langle 1|)_B = 0$$

$$|1\rangle\langle 0|_B = (|0\rangle\langle 0|)_A(|1\rangle\langle 0|)_B + (|1\rangle\langle 1|)_A(|1\rangle\langle 0|)_B = 0$$

$$|1\rangle\langle 1|_B = (|0\rangle\langle 0|)_A(|1\rangle\langle 1|)_B + (|1\rangle\langle 1|)_A(|1\rangle\langle 1|)_B = 1$$

Bibliography

Entropy is a key concept in many fields of engineering and natural science and accordingly, is introduced, described, and mulled over from many perspectives in countless texts and online content. Here, we list the references we have found particularly helpful when writing this book.

CHAPTERS 1–5

This scope of the opening five chapters covering thermodynamics from both classical and statistical mechanics viewpoints is pretty much what one might expect to encounter in an undergraduate physics course at a UK university, and many, many suitable texts are available.

The Elements of Classical Thermodynamics (Cambridge UP) by **A.B. Pippard** is an influential book first published in 1957; tersely written, it wilfully eschews explanations in statistical and quantum terms, in so doing exposing the power and austere beauty of classical reasoning.

Statistical Physics (Wiley) by **F. Mandl** and *Fundamentals of Statistical and Thermal Physics* (McGraw-Hill) by **Frederick Reif** cover both classical and statistical facets of the field and have served both authors well as undergraduates and beyond.

Statistical Physics: an Entropic Approach (Wiley) by **Ian Ford** is a more modern text: we particularly like the title!

Sustainable Energy – Without the Hot Air (UIT Cambridge) by **David MacKay** is a breezily written but purposefully quantitative survey of our current patterns of power generation and consumption, with an eye to a carbon-neutral future.

The Two Cultures and the Scientific Revolution (Cambridge UP) by **C.P. Snow** offers insight into the prevailing cultural and societal attitudes in 1950s Britain.

CHAPTERS 6–12

These chapters outline the basics of statistical thermodynamics, with a particular slant on quantum theory and its applications. The references here reflect that approach.

Understanding Quantum Physics, A Users Manual (Prentice-Hall International Editions) by **Michael A. Morrison**: a favourite introductory text on quantum theory, with a side slant on the philosophy and meaning of the theory. This is a detailed but highly readable text written in a light style that is very effective.

Quantum Theory (Prentice Hall) by **David Bohm**: another classic text, which takes a slightly unconventional approach but has a reassuringly high words: equations ratio, reflecting the author's determination to understand the meaning of the theory.

Statistical Physics: Berkeley Physics Course Vol 5 (McGraw-Hill) by **Frederick Reif**: a great introduction to statistical physics, with some reference to quantum theory.

Theoretical Concepts in Physics: An Alternative View of Theoretical Reasoning in Physics (Cambridge UP) by **Malcolm S. Longair**: an enthusiastic and engaging broad brush over the span of key ideas in theoretical physics. The book focusses on theoretical physics as a practised and ongoing professional activity.

The Theory of Thermodynamics (Cambridge UP) by **John R. Waldram**: a solid (no criticism) and extensive introduction to thermodynamics spanning classical, statistical and quantum approaches. This has been a go-to reference for a range of topics.

Gibbs vs Boltzmann Entropies by **Edwin T. Jaynes** American Journal of Physics 33, 391 (1965); doi: 10.1119/1.1971557: http://dx.doi.Org/10.1119/1.1971557 (last referenced March 2024) – an interesting and important paper that highly influenced our views.

Harvard University Statistical Mechanics Lecture 6 (2021) by **Matthew Schwartz**, https://scholar.harvard.edu/files/schwartz/files/6-entropy_0.pdf, last referenced March 2024. This is a beautifully constructed lecture covering a great deal of ground in a few pages. The lecture discusses "many ways to think about entropy", and with due respect to our teachers at university, we wish we had benefitted from a course like this when we were undergraduates.

CHAPTER 13

Principles of Biochemistry (Worth) by **Albert Lehninger**: one of the first texts to present a thermodynamical perspective on the chemistry underlying living processes. Sumptuously illustrated, we rather fear it has become a one-stop-shop for tourists to the field like ourselves.

What is Life? (Cambridge UP) by **Erwin Schrödinger** is a hugely influential book based on 1943 lectures by Schrödinger delivered at Trinity College, Dublin, pointing out the association between life and entropy production and famously anticipating the need for the "aperiodic crystal" now referred to as DNA for the transmission of hereditary information.

Gaia: a New Look at Life on Earth (Oxford UP) by **James Lovelock**, first published in 1979, develops the idea that all life on Earth can be viewed as a single tightly-coupled thermodynamic system.

CHAPTER 14

An Introduction to Information Theory: Symbols, Signals and Noise (Dover) by **John Pierce** is a very readable survey of the main concepts, first published in 1961. The tone and style are somewhat "of its time", but we recommend swallowing your scruples to see what you can learn.

Information Theory: a Tutorial Introduction (Sebtel) by **James Stone** is a more modern account with, as the title suggests, many helpful worked examples.

Maxwell's Demon: Entropy, Information, Computing (Princeton UP) eds. **Harvey Leff** and **Andrew Rex** is a collection of articles chronicling the Demon's adventures from his first appearance.

CHAPTER 15

Introductory Nuclear Physics (Wiley) by **Kenneth Krane** and *Nuclear Physics: Principles and Applications* (Wiley) by **John Lilley** each cover thermonuclear fusion and primordial and stellar nucleosynthesis.

Cosmology: the Origin and Evolution of Cosmic Structure (Wiley) by **Peter Coles** and **Francesco Lucchin** surveys the theoretical methods and observational data underpinning modern cosmology in a very accessible way: be aware, though, that this is a field where ideas continue to evolve rapidly, so keep a studied eye on YouTube!

The Emperor's New Mind: Concerning Computers, Minds and the Laws of Physics (Oxford UP) by **Roger Penrose** has as its main thesis the idea that human consciousness is non-algorithmic, requiring new laws of physics for its elucidation. Along the way is science writing of the highest quality, including a superb essay on the origin of the Second Law.

CHAPTER 16

Gravity: an Introduction to Einstein's General Relativity (Pearson) by **James Hartle** and *A First Course in General Relativity* (Cambridge UP) by **Bernard Schutz** each builds the general relativity scaffolding required to study black holes and fill in some of the gaps in the material presented here; both cover Hawking radiation from a relativist's perspective.

Quantum Mechanics (Addison-Wesley) by **Sara McMurry** is an introductory textbook covering the theory of atomic spectra as part of a thorough survey of the applications of quantum theory.

An Introduction to Black Holes, Information and the String Theory Revolution (World Scientific) by **Leonard Susskind** and **James Lindesay** present lectures notes focussing on recent developments in tackling the black hole information paradox. The treatment of Hawking radiation uses full-blown quantum field theory.

A First Course in String Theory (Cambridge UP) by **Barton Zwiebach** is an entry-level text to this fascinating area of theoretical physics.

Index

D

E

I

J

K

L

M

N

O

P

Q

R

S

T

U

V

W

Z

Printed in the United States
by Baker & Taylor Publisher Services